PLATE TECTONICS

A Paradigm Shift in Thinking

Dr James Maxlow (PhD), 2026

PLATE TECTONICS

A Paradigm Shift in Thinking

Dr James Maxlow (PhD), 2026

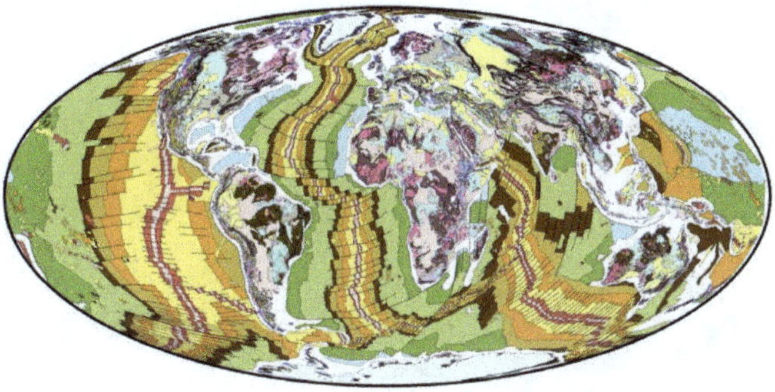

Academica Press
Washington

Library of Congress Cataloging-in-Publication Data
Names: Maxlow, James (author)
Title: Plate tectonics : a paradigm shift in thinking | Maxlow, James
Description: Washington : Academica Press, 2026.
Identifiers: LCCN 2026936398 | ISBN 9781680536140 (hardcover) | 9781680533774 (e-book)

Permission was granted by the Commission de la Carte Geologique du Monda, Paris to use and digitize the Geological Map of the World at 1:25,000,000 M scale (1990), published by the Commission for the Geological Map of the World and UNESCO © (1990).

Digitizing the Geological Map of the World (1990) © was originally carried out by Simon Brown of Geoviz International.

About the Author

James Maxlow was born in Middlesbrough, England in May 1949. James' passion for geology was no doubt inherited from a family history of "*ironstone workers*" supplying iron ore mined from the Eston Ironstone Mine to the foundries and steel rolling mills of Middlesbrough, England, during the early 1800s to mid-1900s.

James immigrated to Australia with his parents in 1953 where he grew up in Melbourne. He studied Civil Engineering at the then Swinburne College but soon became disillusioned with engineering and redirected himself to geology at the then Royal Melbourne Institute of Technology, graduating in 1971. It was in Melbourne where he later met and married his lovely wife Anita and, during their work and travels throughout Australia, they had three wonderful children, Jason, Karena, and Jarred.

During his research years, James met and communicated with many wonderful scientists from around the world. Most notable of which was the late Professor Samuel Warren Carey from Tasmania, the father of modern tectonics, Jan Koziar from Poland who was the first scientist to measure and calculate an ancient Earth radius using modern seafloor mapping, and the late Klaus Vogel from Germany, the father of modern small Earth modelling studies. It was during James' studies that Professor Carey recognized the potential of his research into tectonics. Carey then kindly "*passed on his baton*" to James in order to further his ongoing research into tectonics.

James gained his Master of Science in geology in 1995, followed by a Doctor of Philosophy in 2002 at Curtin University of Technology, Perth, Western Australia, including a letter of commendation from the university Chancellor for original thought-provoking research.

Preface

"*The main cause of intellectual stagnation [is] the herd instinct which causes many scientists to stay with mainstream ideas and concepts rather than risk alienating their peers by investigating controversial new ideas. The peer review system thus tends to maintain the intellectual status quo by rejecting new ideas.*" Gold, 2009.

In modern science, Plate Tectonics is a well-established unifying theory that embraces and integrates much of what we observe and measure in the Earth sciences. In particular, it embraces all data, concepts, theories, and hypotheses relating to the origin and subsequent geological history of each of the continents and oceans. Plate tectonics is also widely considered in science to be synonymous with the general theory of Global Tectonics.

However, as noted by Trümpy in 2000, and rarely appreciated in science today, "*The theory of plate tectonics was developed primarily by geophysicists at sea, who took little account of the Alpine* [geological] *evidence.*" While the fundamentals of plate tectonics are not disputed in this book, since first established in the mid-1960s, plate theory has indeed continued to be driven by geophysics, at the expense of geology, geography, and biogeography.

Because of this, it is maintained that non-geological scientists at the time may have made a poorly informed decision to use the long-since rejected Continental Drift theory to explain the mechanics of newly observed crustal plates during the early 1960s. These scientists further substantiated this decision by adopting palaeomagnetics as the basis for plate assemblage studies, as well as rejecting and continuing to discredit all alternative proposals.

Today, the theory of plate tectonics continues to remain the predominate paradigm in geology and is used to explain a wide range of global observations, such as the movement of continents, formation of mountains, distribution of volcanoes, and magnetic apparent-polar-wander, to name but a few.

During the 1960s, even though advanced for the time, global data gathering was a relatively new science, and the flow-on from early oceanographic and bathymetric observations was the introduction of magnetic seafloor mapping of the oceans, and ultimately to the recognition of seafloor spreading.

In order to explain seafloor spreading, in particular the Atlantic Ocean, a decision was made by scientists to reconsider, and subsequently adopt, Continental Drift theory as the mechanism for the newly conceived concept of plate tectonics. From this decision, continents were then considered to randomly fragment, collide, accumulate and dissipate at the Earth's surface over time, driven by large convection currents within the mantle.

Unfortunately, this decision to use continental drift as the basis of plate tectonics, based primarily on oceanography and geophysics of the Atlantic Ocean, may have been premature and hasty. In particular, when faced with an incomplete oceanic data set and limited input from crustal geology.

Historically, it would seem that the human species have difficulty in recognizing when speculation, based on limited observational evidence, such as the previously rejected continental drift theory, may in fact be entirely wrong. In hindsight, what science continues to fall back on is that it is far better to force "*new knowledge*" into the mold created from the old so as not to upset the present "*status quo*".

History also shows us that, quite often, we humans do get it wrong by unknowingly believing too much in our own speculation beyond observation. Speculation, as we should all be aware, breeds on earlier speculations and we should be acutely conscious of the need to continually modify our speculative ideas as soon as new evidence comes to light.

The research presented here does not directly challenge, or discredit plate tectonics, or the merits of modern global data gathering. It does, however, offer new ways of interpreting and understanding the vast amount of additional global observational data now available, and offers a refinement to the plate tectonic paradigm way of thinking.

With this preamble in mind, the prime purpose and scope of the research presented in this book is to utilize, investigate, and interrogate all modern global observational data; data collected since the introduction of plate tectonics, to see if this data is still commensurate with the fundamental principles and premises that continue to govern plate tectonic theory today.

To ignore this global observational data simply because it doesn't fit conventional plate theory is testament to scientific complacency.

Limitations of Plate Theory

Ongoing, generally unrecognized, or even acknowledged limitations to continental drift-based plate tectonic theory include, but are not limited to:

- Over the past 60 years, plate tectonics have continued to be presented as a conceptual framework, using idealized cartoon sketches to portray plate assemblages and basic mechanisms. To date, this framework has worked well, and scientists and non-scientists alike readily relate to these sketches to visualize plate motion, subduction, and mountain formation over time. But an argument will be made that plate tectonists haven't gone beyond this conceptual phase and, as a consequence, these conceptual sketches have very little to no relevance to the sciences or industries today, in particular the mining and exploration industries.
- Plate tectonics is portrayed as a random process of fragmentation, collision, accumulation and dissipation of crustal plates across the Earth's surface over time, currently considered to be opening within the Atlantic Ocean and closing within the Pacific Ocean. Yet, modern geological mapping of the continents and oceans, in particular the Geological Map of the World (1990), demonstrates conclusively that a random plate motion is not conducive to the seafloor geology shown on this map.
- It is now known that all oceans contain a mid-ocean-ridge, coincident with plate boundaries, and each plate is shown to be increasing their surface areas over time. This increase in plate surface area is symmetrical about the mid-ocean-ridges identified within each ocean and, this cumulative surface area observation is not supported by continental drift theory.
- The maximum age of exposed seafloor volcanic crust, generally located along the continental margins, is early-Jurassic in age. There are no observable instances where the age of the modern seafloor crust is older than 170 million years, whereas, on a continental drift model, there should be remnants of older seafloor crusts.
- While vast amounts of research monies go into assembling crustal plates back in time, assemblages are poorly constrained by geophysics and remain conceptual, in particular for Precambrian times. For each assemblage, there are often multiple plate-fit options

to contend with, for example, the selection of eleven published Rodinian assemblages shown below. As a result, these meaningless assemblages cannot be used confidently by either science or industry to display global, or even local, observational data.

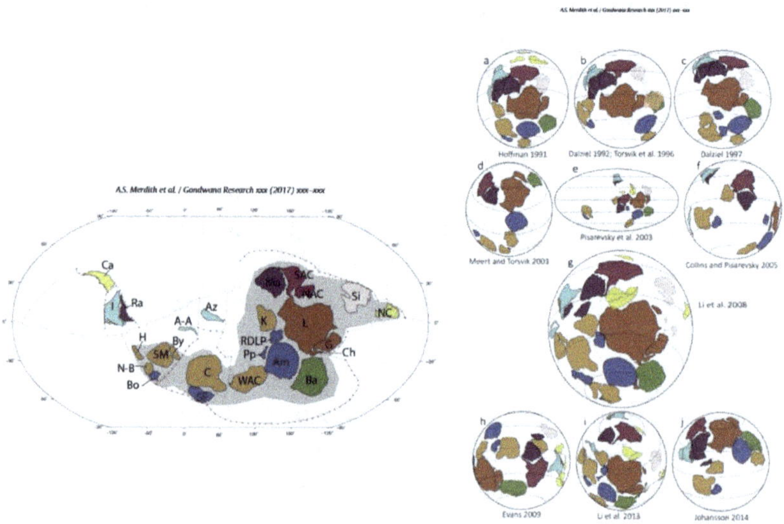

Models of Rodinia at approximately 1000Ma, published by Gplates and Merdith *et. al.,* 2017, using published Euler Pole rotations, or reverse-engineered if poles were not available.

- Since the mid-1960s, there has been a vast array of global observational data collected, in particular from each of the geosciences. Because plate tectonics only have a conceptual framework to work with, these data cannot be modelled globally, in particular for past geological times. As such, both science and industry are missing out on a vital platform to display historical data about our past geology, geography, biogeography, climate, and so on.

These limitations to plate theory prompt an urgent need for consideration of a paradigm shift in plate tectonic thinking. It is further emphasized that the research presented here is based exclusively on modern global observational data. This research is a data modelling exercise, focused on modelling global observational data independently of any present or pre-existing theory. In this context, it is important to appreciate that this research is not a theory modelling exercise.

The critical analysis of modern global observational data presented here will also allow the data to tell its own story which, as will be shown, reveals a new tectonic picture of the Earth that more closely aligns with empirical global observations and fundamental plate tectonic principles. In this analysis, it will be shown that the resulting picture overcomes a great number of known limitations and problems still facing plate tectonics today and, in particular, discredits the use of continental drift as the basis of plate theory.

The work presented in this book represents the results of extensive research by the author, a retired professional geologist and researcher, over a period of twenty-five years. Of particular importance is that accurate, spherical geological models of ancient Earth, extending back to the early Archaean, will be presented. These models are considered to be unique and will be used to display a wide selection of modern global observational data in order to quantify the plate assemblage modelling studies presented here.

Because no one can know for certain what has happened to our Earth over the four to five billion years of its existence, it is further considered that, in order to understand more about what modern global observational data has to offer, we will need to start from the present-day Earth and, step by step, reconstruct observed global data for increasingly older geological periods, thus working our way back in time. This *actualist-principle* approach, which emphasizes reliance on observed geology, is already the basis for modern understanding of the geological evolution of our Earth: *the present is the key to the past.*

The strict approach taken here is considered necessary in order to promote increased objectivity in the modelling of plate assemblages and interpretation of all global observational data. If conventional continental drift-based plate theory is truly consistent with the modern data, then there is no problem, and the data modelling will highlight this. If it is not, then there exists a problem, which demands to be taken seriously and, within the Earth sciences in particular, must be fully addressed by the scientific community as a whole.

The data used throughout this book is sourced from well-renowned international datasets including the *Geological Map of the World* at 1:25,000,000 M scale (1990) published by the Commission for the Geological Map of the World and UNESCO, the International Global Palaeomagnetic Databases of McElhinny & Lock (1996) and Pisarevsky (2004), the distributions of ancient shorelines based on the published data of Scotese (1994) and Smith *et al.* (1994), and palaeobiogeographic data sourced from the Paleobiology Database (PaleoBioDB) (2015). The distribution of metals is sourced from the USGS Mineral Resource Data Set (MRDS) (2015) and oil and gas resources sourced from various publications.

Acknowledgements and special thanks go to Philippe Bouysse of the Commission de la Carte Geologique du Monde, Paris, for granting permission to use and digitize the *Geological Map of the World* at 1:25,000,000 M scale (1990). Simon Brown of Geoviz International, Perth, Western Australia, who originally digitized the published Geological Map of the World, which was then used as a base map in my research. Thanks, are also due to many others, in particular John B. Eichler, who has provided invaluable assistance with editing of early versions of this text and providing constructive assistance on content and structure, stimulating discussion on mountains, causal mechanisms, crustal processes, dinosaurs, as well as ongoing encouragement towards completion of this book.

Dr James Maxlow

Contents

Part One

Introduction, Historical Aspects, and Geological Modelling

"*The theory of plate tectonics was developed primarily by geophysicists at sea, who took little account of the Alpine [geological] evidence.*" Trümpy, 2000.

Introduction

"*Science, does not - in the proper sense - discover new facts or regularities in nature, but rather offers some new ways of seeing and understanding the physical world.*" Zarebski, 2009.

Since its inception in 1965, plate tectonics has earned a reputation for being a paradigm revelation in the Earth sciences. So, why do we need another paradigm shift in plate tectonic thinking? The answer to this, of course, is that science never knows when it needs a paradigm shift in thinking until additional scientific observation and data collection provides new concepts, ideas, or data to convince scientists otherwise.

A paradigm shift in thinking is defined as *a fundamental change in the basic concepts and practices of a discipline or field, leading to a new way of understanding or doing things.* This concept, popularized by Thomas Kuhn (1962), occurs when "*a new worldview replaces an old one, often driven by new evidence, innovations, or discoveries that challenge the status quo.*" According to Kuhn, "*a scientific revolution occurs when: (i) the new paradigm better explains the observations and offers a model that is closer to the objective, external reality; and (ii) the new paradigm is incommensurate with the old.*"

All paradigm shifts come about through what Kuhn then refers to as a "*crisis*" where the current paradigm is unable to explain what is empirically observed. As a result, it is justifiable to consider that a new and different paradigm may then replace the dominant paradigm.

Classic Example of a Paradigm Shift

When we think of paradigm shifts in scientific thinking, a classic historical example is the shift in thinking from Ptolemy's geocentric model of the universe to the Copernican heliocentrism model.

The *Almagest*, written by Claudius Ptolemy (c. AD 100 – c. 170), is reputed to be one of the most influential second-century mathematical and astronomical scientific texts in history, dealing with the apparent motions of the stars and planetary paths. It is said to have *canonized a geocentric model of the Universe that was accepted for more than 1,200 years from its origin in Hellenistic Alexandria, in the medieval Byzantine and Islamic worlds, and in Western Europe through the Middle Ages and early Renaissance until Copernicus.*

Geocentrism is now a superseded astronomical model of the Universe, where Earth was depicted as being located at the center of the cosmos and all else revolves around it. It is also known as the geocentric model, and often simply as the Ptolemaic system. Under most geocentric models, the Sun, the Moon, stars, and planets all orbit Earth. This geocentric model was the predominant description of the cosmos in many ancient European civilizations, such as those of Aristotle in Classical Greece and Ptolemy in Roman Egypt, as well as during the Islamic Golden Age.

Acceptance and preservation of the geocentric model for such a long period of time is understandable, given the early development of astronomy and the general lack of interest, or understanding, by the general populace of the time. Preferring instead to rely on religious instruction.

A number of observations were proffered in support of the idea that Earth was the center of the Universe, and, without having modern astronomy to fall back on, these observations would still be relevant today.

- First, from anywhere on Earth, it was observed that the Sun appeared to revolve around Earth once per day.
- Second, while it was observed that the Moon and the planets have their own motions, they were also observed to revolve around Earth about once per day.
- Third, the stars appeared to be fixed on a celestial sphere rotating once each day about an axis through the geographical poles of Earth.
- And, more importantly, fourth, Earth seemed to be unmoving from the perspective of an earthbound observer; it felt solid, stable, and stationary.

These astronomical observations and predictions of Ptolemy's geocentric model were seen as fundamental facts and served as the basis for preparing astrological and astronomical charts for over 1,500 years. This geocentric model remained the dominant model well into the early modern age but, from the late 16th century onward, it was gradually superseded by the heliocentric model of Copernicus, and supported by Galileo, and Kepler.

As would be expected, there was much resistance to the transition between these two theories, since the general observations mentioned above were highly relevant and, for a long time, the geocentric model was said to produce more accurate astrological results. Additionally, many felt that a new, unknown theory could not change the accepted consensus for geocentrism.

Problems with Ptolemy's system were well recognized in medieval astronomy, and an increasing effort to criticize and improve it in the late medieval period eventually led to the Copernican heliocentrism, developed in Renaissance astronomy. Nicolaus Copernicus (1473 – 1543) was a Renaissance polymath, and Catholic cleric, who formulated a model of the universe that placed the Sun, rather than Earth, at its center. The publication of Copernicus's model in his book *De revolutionibus orbium coelestium* [On the Revolutions of the Celestial Spheres], just before his death in 1543, was a major event in the history of science, triggering the Copernican Revolution and making a pioneering contribution to a scientific revolution in astronomy.

In his book, Copernicus: "*discussed the philosophical implications of his proposed system, elaborated it in geometrical detail, used selected astronomical observations to derive the parameters of his model, and wrote astronomical tables which enabled one to compute the past and future positions of the stars and planets. In doing so, Copernicus moved heliocentrism from philosophical speculation to predictive geometrical astronomy. In reality, Copernicus' system did not predict the planets' positions any better than the Ptolemaic system. This theory resolved the issue of planetary retrograde motion by arguing that such motion was only perceived and apparent, rather than real: it was a parallax effect, as an object that one is passing seems to move backwards against the horizon.*"

The model of the universe that Nicolaus Copernicus proposed represented a major shift in understanding the cycles of heavenly bodies. These changes became known in the 19th century as the Copernican Revolution. In the 20th century, these changes also became recognized as a paradigm shift from the Ptolemaic model of the heavens, which described the cosmos as having a stationary Earth located at the center of the universe, to the heliocentric model with the Sun at the center of the Solar System. Today, the heliocentric model is a fundamental principle of modern astronomy, with later scientific advancements, like Isaac Newton's laws of motion and gravity, providing a more complete understanding of why planets orbit the Sun.

Prior to Copernicus, the accumulated astronomical observations of the time had led scientists to believe that all of the heavenly bodies had increasingly complex epi-cyclic motions, generated as they travelled around a stationary Earth. The scientists of the time went to great lengths to measure and quantify these cycles and, partly because of the influence of religious belief, did not consider the far simpler Sun-centric proposal of Copernicus.

Put simply, while the vast number of observations made by astronomers throughout the 1500-year time period were not questioned or rejected by Copernicus, all he did was propose a very simple modification to the geocentric model. He simply replaced the Earth as being the centre of the known cosmos, with the Sun, thus simplifying the planetary motions, in particular, to what we know and readily accept now (Figure 1).

Figure 1. A geocentric model of the cosmos showing each of the planets and the Sun revolving around Earth, plus the modification proposed by Copernicus to show that, by switching the positions of the Sun, Earth and Moon, each of the planets actually revolve around the Sun.

Modern Plate Tectonics

I have chosen to dwell on this largely historical paradigm in context with modern plate tectonics because plate tectonic theory is currently considered a paradigm in its own right, so how dare anybody challenge it. Plate tectonics is now held in such high esteem throughout all sciences that it would, and still is, extremely difficult to change, or influence, this dogmatic mindset. Often quoted reasons for continuing to reject alternatives to plate tectonics, such as the expanding Earth theory, are still predictably touted, yet these reasons are now so outdated that they are beyond comprehension.

The need for a paradigm shift in tectonic thinking was first recognized, well before plate tectonics came into being, during the 1950s by Professor Samuel Warren Carey, who later became Emeritus Professor of geology at the

University of Tasmania. In order to assemble the continents and quantify continental drift, Carey made a 60-centimetre diameter scale model of the present-day Earth to investigate the potential fit of the continents during closure of each of the oceans.

In addition to closing the Atlantic Ocean, his investigation was extended to also consider fitting the various continents together within the Indian and Pacific Oceans. It is important to mention that Carey had made an earlier observation that the trans-Atlantic fit was not as good as Alfred Wegener and others had previously claimed, hence the need to accurately model the Earth. His comments and conclusions from his research are reproduced in full as follows:

> "*At an early stage in my investigations I went to some pains to ensure that I compared and transferred shapes and sizes of the continental blocks accurately. I have spent tedious years plotting large oblique stereographic projections about diverse centres not only for Africa and South America but for every piece of the Earth's surface. I combined this with spherical tracings from the globe, working on a spherical table. The reward for this zeal for accuracy was frustration. Again and again over the years I have assembled Pangaea but could never attain a whole Pangaea. I could make satisfactory sketches like Wegener's classic assembly, but I could never put it all together on the globe, or a rigorous projection. I could reconstruct satisfactorily any sector I might choose but never the whole. If I started from the assembly of South America...by the time I reached Indonesia there was a yawning gulf to Australia, although I felt sure from the oroclines that Indonesia and Australia belonged together...If I started from Australia and Indonesia I had no hope of closing the Arctic Sphenochasm [where the split occurred]..., which I was convinced was basically correct...I was painfully aware that there was a crucial link missing from the global synthesis. I was tempted to abandon the quantitative assembly and resort to sketches which would show every block related as I inferred they should be, even though I knew I could not bring them together that way with the rigour I sought.*"

> "*But in the end the rigorous approach has paid off. For it has revealed a discrepancy which had not been apparent. It was not my method that was at fault, but my assumption that the earth of Pangaea*

was the same size as the earth of today. The assembly of Pangaea is not possible on the earth of the present radius, but on a smaller globe, a globe such as is demanded by the orocline analysis, these difficulties vanish."

Even with this research of Carey, as late as 1953, the theory of continental drift still continued to be rejected by physicists on the following grounds.

- First, it had been considered that floating masses [continents] on a rotating geoid would collect at the equator and stay there. This may have explained one, but only one, mountain building episode between any pair of continents, but it failed to account for earlier orogenic episodes.
- Second, masses floating freely in a fluid substratum, like icebergs in the ocean, should be in isostatic equilibrium (in which the forces of gravity and buoyancy are in balance). But gravitational measurements showed that many areas are not in isostatic equilibrium.
- Third, there was the problem of why some parts of the Earth's crust should have solidified, while other parts were still fluid. Various attempts to explain this have been foundered on other difficulties.

Unfortunately, with the subsequent universal acceptance and promotion of plate tectonic theory, the very important physical observations and conclusions of Carey, along with objections by physicists, continue to be neglected, ignored, or totally rejected to this day.

By 1967, barely two decades after discovery of the Atlantic mid-oceanic rift, and a decade after discovery of magnetic seafloor striping, plate tectonics was largely seen to be true, and therefore not needing to be further proven by modern geophysics. Even with these inherent limitations, the theory of continental drift was adopted by plate theory in order to explain the formation and movement of the continental crusts, without further consideration of alternative proposals.

In addition, just as Carey had predicted, in order to promote plate tectonics, researchers then had to resort to cartoon sketches of plate assemblages, as well as physical conceptual ideas, to portray quantitative plate assemblages and physical mechanisms. This acceptance is still blatantly obvious in all plate assemblages and conceptual frameworks portrayed in literature today.

Just like the scientists of pre-Copernicus' times, it would seem that Earth scientists of today may in fact be guilty of unknowingly forcing available physical evidence into a pre-conceived singular, continental drift-based plate tectonic mold. As long as the existing paradigm can be stretched out of shape to encompass any new facts, even to the extent of ignoring critically important observations, it would seem that the existing paradigm will remain. This is indeed what is happening today with continental drift-based plate tectonics.

This conceptual approach was fine for the mid-1960s, when plate tectonics was first proposed, but, to an observer, it would seem that science has now adopted continental drift, a previously rejected theory, as a unique and comprehensive mechanism for our tectonic understanding of the Earth, without consideration for the vast amount of new global observational data now available.

In reality, at no stage has the scientific community been encouraged, or has it seriously deviated from conventional continental drift-based plate theory in order to see what else this modern global observational data may reveal, beyond our conventional plate tectonic way of thinking.

Chapter 1

Conventional Wisdom

"*Scientists still do not appear to understand sufficiently that all earth sciences must contribute evidence toward unveiling the state of our planet in earlier times, and that the truth of the matter can only be reached by combing all this evidence. . . It is only by combing the information furnished by all the earth sciences that we can hope to determine 'truth' here, that is to say, to find the picture that sets out all the known facts in the best arrangement and that therefore has the highest degree of probability. Further, we have to be prepared always for the possibility that each new discovery, no matter what science furnishes it, may modify the conclusions we draw.*" Alfred Wegener. The Origin of Continents and Oceans (1915).

In 1915, Alfred Wegener, a German polar researcher, physicist, and meteorologist, was making serious arguments for the idea of Continental Drift in the first edition of his book, *Die entstehung der kontinente und ozeane* [The Origin of Continents and Oceans]. In his book, as did mapmakers before him, he noted how the shape of the east coast of South America and the west coast of Africa looked as if they were once joined.

When Wegener initially presented his arguments for the idea of continental drift, he became the first to gather significant fossil and geological evidence to support his simple observations for the breakup and subsequent movement of the continents through time. From the beginning, Wegener went further to suggest that the present, Atlantic Ocean continents once formed a single land mass, later called Pangaea. This land mass was inferred by Wegener to have subsequently broken up and drifted apart, "*...thus releasing the continents from the Earth's mantle.*" Wegener likened this drifting to "*...icebergs of low-density granite floating on a sea of denser basalt.*"

At that time, Wegener's ideas were not taken seriously by most geologists or scientists alike and were rejected and ridiculed for many decades. They rightly pointed out that there was no apparent mechanism for continental drift. Without detailed evidence, or a force sufficient to drive the movement, the theory was discounted by scientists because: *...the Earth might have a solid crust and mantle and a liquid core, but there seemed to be no way that portions of the crust could move around the surface of the Earth.*

Unfortunately, Wegener could not explain the forces that drove continental drift, and vindication for his efforts did not come until well after his untimely death. Other responses were less than sympathetic, including: "*Utter damned rot*" (the then President of the American Philosophical Society). "*If we are to believe this hypothesis we must forget everything we learned in the last seventy years and start all over again*" (Thomas Chamberlin). "*Anyone who valued his reputation or scientific sanity would never dare support such a theory*" (a British geologist).

Although continental drift of the Atlantic Ocean continents was initially rejected for many decades, when Wegener introduced his theory, he did, in fact, set in motion a completely new train of thinking and speculation about the origin of our continents and oceans. As Wegener correctly promoted, "*the fit of the Americas against Africa and Europe was real and had to be explained.*" Time has, of course, shown that it was only the mechanism behind continental drift that was difficult to explain, not the actual fit of the continents.

Since then, with changing ideas about the Earth, Wegener's theory of continental drift has been adopted and credited with having given rise to the modern theory of plate tectonics. Most people have now come to accept plate tectonic theory, without question, and without prior concern for the considerable amount of initial and still relevant rejection of continental drift.

1.1. Plate Tectonics

The theory of Plate Tectonics has since been extensively promoted in science to explain a diverse range of observed global tectonic observations, and this theory is widely accepted by both scientists and the general public alike. The theory is considered by most scientists to adequately link all geological features, from the age and composition of seafloors to the rise of mountains, as well as the past distributions of plant and animal species.

In plate tectonic theory, the Earth's crust is shown to be broken into a series of seven or eight major and many minor plates, made up of both continental and seafloor crusts (Figure 1.1). These crustal plates are said to move in relation to one another at one of three types of plate boundaries called; convergent or collisional boundaries, where plates are said to collide, resulting in the formation of mountains; divergent boundaries, where ocean crusts break apart and new volcanic crust is erupted along seafloor spreading centres; and conservative or transform-fault boundaries, where plates are faulted relative to each other. Earthquakes, volcanic activity, mountain-building, and seafloor trench formation are said to occur along each of these plate boundaries.

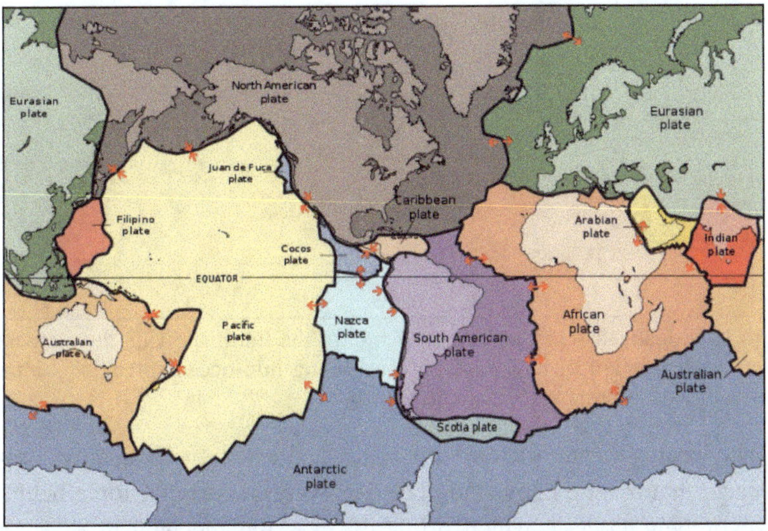

Figure 1.1. Map of the Earth showing the distribution of tectonic plates comprising continental and seafloor crustal rocks.

During the 1960s, it was also recognized that the Earth's crust is increasing in surface area along a centrally located seafloor ridge within the Atlantic Ocean. Early researchers, such as Hess and Dietz, reasoned, like Holmes, Carey, and others, that, because of this increase, a similar area of crust must be shrinking somewhere else. Hess suggested that new seafloor crust continuously spread away from the seafloor ridge in a conveyor belt-like motion, which was referred to as the *conveyor belt principle* (Figure 1.2).

Hess reasoned that, many millions of years later, the seafloor crust eventually descends below the continental margins where seafloor trenches—very deep, narrow canyons, located along the margins of some continents—are formed, for example around the margins of the Pacific Ocean.

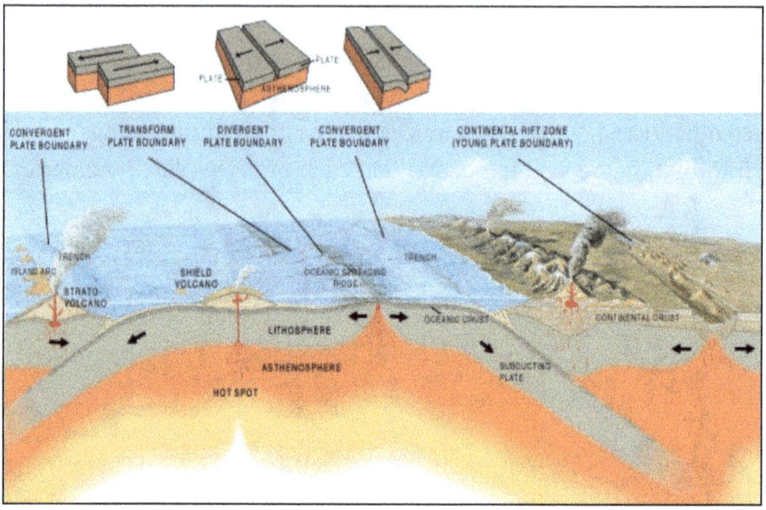

Figure 1.2. Idealized schematic cross section of the Earth showing the plate tectonic model for generation of new seafloor crust along mid-ocean spreading centres and disposal of excess crust within subduction zones.

The proposal put forward by Hess suggested, without any empirical evidence, that mantle convection currents were the driving force behind this conveyor belt process, which primarily uses the mechanism of spreading along the ridges to drive the currents. Even though largely hypothetical, Hess concluded that the Atlantic Ocean was increasing in surface area while the Pacific Ocean was shrinking.

He further suggested that, as old seafloor crust is consumed in trenches around the Pacific Ocean, new magma rises and erupts along the spreading ridges to form new crust. In effect, the ocean basins were said to be perpetually being recycled, with the creation of new crust and the destruction of old seafloor crust occurring simultaneously.

It is now considered by plate tectonists that tectonic plates are able to move—drift—because of a relative density difference between the outer, rigid seafloor crust and the underlying lower, low strength, low rigidity part of the Earth's crust. Dissipation of heat from the underlying mantle was also suggested to be the original source of energy driving plate movement, which resulted in convection or large-scale upwelling and doming. It should be noted that this is a theoretical construction with no empirical under pinning.

When new crust forms at the mid-ocean spreading ridges, this crust was considered to be initially less dense than the underlying lower crust but becomes denser with age as it progressively cools and thickens. The greater density of old crust relative to the underlying crust was said to allow it to sink into the deep mantle at subduction zones. The weakness of the lower crust was then considered to move the tectonic plates towards a subduction zone.

Although subduction is now believed by plate tectonists to be the strongest force driving plate motion, it was also acknowledged by many researchers that it cannot be the only force since there are a number of plates, such as the North American Plate, that are moving yet are nowhere being subducted. The same is true for the enormous European and Asian Plate, and especially the Antarctician Plate. Even though plate tectonics is currently considered the main tectonic theory in the Earth Sciences, the mechanisms for plate motion, and conservation of surface areas, are still a matter of intensive research and debate amongst many Earth scientists.

As initially concluded by Carey during the 1950s, in order to make continental drift-based plate tectonics work, you must, in his own words, "*abandon the quantitative assembly* [of crustal plates] *and resort to sketches which would show every block related as I inferred they should be, even though I knew I could not bring them together that way with the rigour I sought.*"

This conclusion of Carey is still relevant today, but plate tectonists continue to rely heavily on an outdated, idealized, conceptual framework of cartoon sketches to make plate theory work. The majority of scientists, and the general public, accept these sketches because they are easy to visualize and easy to understand.

This is all very well, but there is now a vast amount of new global observational data available that is not being utilized, or addressed, by these cartoon sketches of plate assemblages, or by physical conceptual ideas. In particular, geological, geographical, and bathymetric mapping of the oceans and continents.

This new global data is physical, time-dependent evidence that needs to be utilized to further constrain, enhance, or replace these historical cartoon sketches with hard, physical observations and evidence. Hence, there does appear to be what Kuhn refers to as a "*crisis*" in the Earth sciences, where the current paradigm is unable to explain what is empirically observed.

Chapter 2

New Discoveries

"*Geology—from the Greek gê, earth and logos, study—is the science comprising the study of solid Earth, the rocks of which it is composed, and the processes by which they evolve. Geology gives insight into the history of the Earth, as it provides the primary evidence for tectonics, the evolutionary history of life, and past climates.*" Wikipedia.

Since the mid-1900s there has been a revolution in all technologies, which has enabled vast amounts of global observational data about the Earth to be gathered, stored, and interpreted. Advances in space technologies, remote sensing, computers, and mass media communications have all vastly increased availability of this knowledge. Only a very small percentage of this data was available during the mid-twentieth century, when plate tectonic and the historical continental drift and long-since abandoned Expanding Earth theories were being developed and their merits were being debated.

As is now well known, planet Earth is broadly made up of an inner and outer core, a layered mantle, an outer continental crust, seafloor crust, the ocean waters, and an atmosphere (Figure 2.1). The continental crust typically varies from 30 to 50 kilometres thick, which is less than 1 percent of the present Earth radius. Similarly, the seafloor crust, in turn, is about 5 to 10 kilometres thick, which is about 0.2 percent of the present Earth radius.

In geology, the crust is traditionally seen as the outermost solid shell of the Earth and is also considered to be chemically distinct from the underlying mantle. The continental and seafloor crustal rocks are referred to as the lithosphere. The boundary between the base of the continental and seafloor crusts and the mantle is conventionally placed at the Mohorovičić discontinuity—shown as the Moho in Figure 2.1—a boundary defined by a contrast in seismic velocity. The underlying, relatively rigid, uppermost part of the mantle is, in turn, referred to as the asthenosphere.

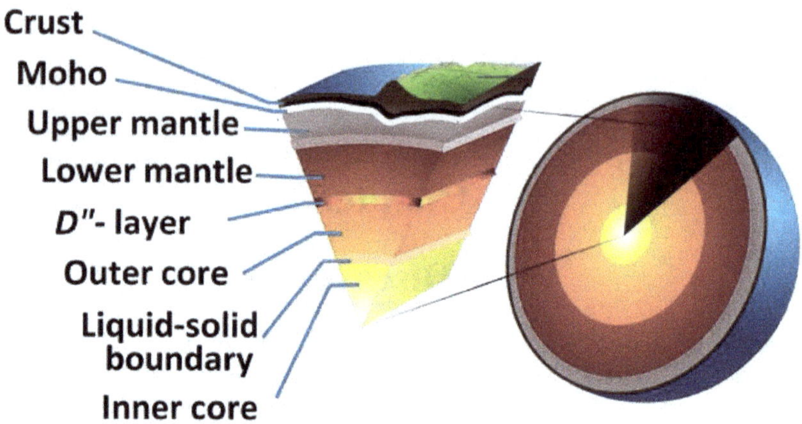

Figure 2.1. Schematic model of the Earth's layered structure showing the crust, upper mantle, lower mantle, and outer and inner core.

With the advent of modern geological, geophysical, and oceanographic studies, it is now known that the seafloor crusts are completely separate and distinct from the continental crusts. These seafloor crusts are shown to comprise almost entirely of intruded iron-rich basaltic lava, which represents intruded, cooled, and solidified upper mantle rocks. The seafloor crust should then be considered as exposed mantle rocks.

In contrast, the continental crusts broadly consist of granite, as well as vast areas of sedimentary and metamorphic rocks originally derived from more ancient granite, metamorphic, and volcanic rocks. In this context, the silica-rich continental crustal rocks are indeed chemically distinct from the underlying iron-rich upper mantle rocks and similarly, chemically distinct from the iron-rich seafloor crusts.

This knowledge about the seafloor crust is in stark contrast to past, pre-plate tectonic thinking, where seafloor crusts were originally thought to represent drowned and oceanised continental rocks covered by seawater. The outlines of the continents were then considered to be the result of periodic uplift of the crusts to form continents, followed by periodic down-warping and drowning of the crusts to form seas or oceans.

In this context, the positions of all continents were considered to be fixed on the Earth surface. Inter-continental land connections enabling species migration were then considered to be the result of periodic land-bridges joining the various continents—which have now long since mysteriously disappeared.

2.1 Geologic Time Scale

The basis of the science of geology is the geologic time scale (Figure 2.2).

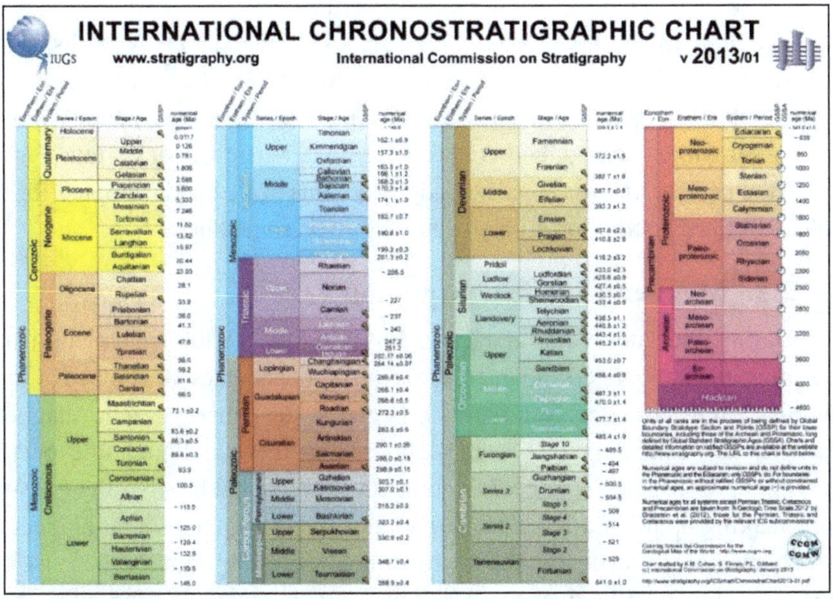

Figure 2.2. The International Chronostratigraphic Chart showing the subdivision of geological time into the various eons, eras, periods, epochs and ages, along with their measured absolute ages.

The geologic time scale is defined as a system of time measurements that relates stratigraphy—the succession of rock strata—to time. This time scale is used by all Earth scientists from around the world to define the timing and relationships between various geologic events that have occurred at different times throughout Earth's history.

Geologic time relates to intervals of time lasting many millions, to tens or even hundreds of millions of years. With the advent of radiometric age dating—the use of radiometric decay of various elements within minerals to physically measure time—science is now able to measure the actual ages of various rocks or events. These ages are then used to provide further absolute numerical ages to the more commonly used intervals of geological time.

Radiometric dating of rocks from all over the world show that the oldest rocks found on Earth are around 4 billion years old. Dating of extra-terrestrial rocks, such as meteors and Moon rocks, potentially extends the age of the Earth to around 4.54 billion years old. Throughout this book, reference to the age of the Earth will be limited to the age of the oldest known terrestrial rocks, and 4 billion years, rather than 4.5 billion years, will be adopted here as the oldest known geological age of rocks on Earth.

The modern chart of geologic time scales (Figure 2.2) was published by the International Commission on Stratigraphy, 2013. There are many more charts published by various organizations and each country generally has its own, often more detailed, charts to represent their specific rock strata. Figure 2.2 is the official, internationally recognized time scale and is adopted for the spherical Earth modelling studies that will be presented here.

2.2. Magnetic Seafloor Mapping

During the late-1940s and extending into the 1950s, scientists, using sensitive magnetometers adapted from airborne devices developed during World War II to detect submarines, began to recognize strange magnetic patterns across the Atlantic Ocean seafloor. This finding, though unexpected, was not entirely surprising because it is now known that basalt—the volcanic rock making up the seafloor crust—contains the iron-rich mineral magnetite, which can locally distort compass readings. More importantly, because the presence of magnetite gives the basalt measurable magnetic properties, these newly discovered magnetic seafloor patterns provided an important means to study the distribution of intruded volcanic rocks throughout each of the seafloors.

As more and more of the seafloors were mapped during the 1950s to 1980s, these magnetic patterns turned out to be not random or isolated occurrences but instead revealed a predictable zebra-stripe like pattern. These stripes were found to be symmetrical about centrally located mid-ocean-ridges within each of the oceans, as shown schematically in Figure 2.3. From this mapping, alternating stripes of magnetized basalt rock were shown to be laid out in parallel rows on either side of the mid-ocean-ridge, where one stripe showed a normal magnetic polarity and the adjoining stripe showed a reversed polarity. While the magnetic stripes were initially not linked to inversions of the Earth's magnetic field, this connection was later explained by Vine and Mathews in 1963, who first made the correlation of polarity change and the changing magnetic field of the Earth. The overall magnetic pattern, as defined by these alternating bands of normally and reversely polarized rock, then became known as magnetic striping.

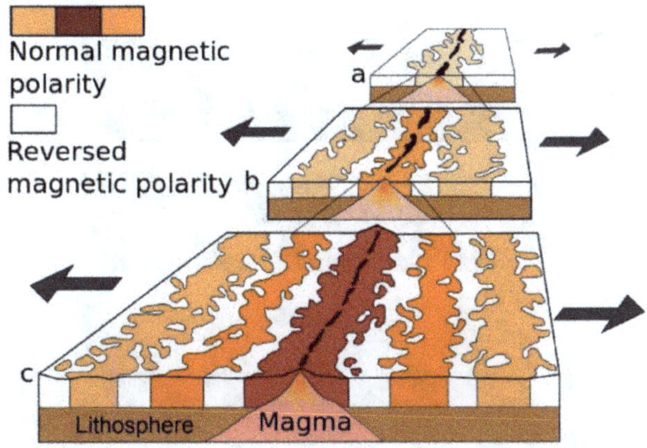

Figure 2.3. Symmetrical magnetic striping centred over a mid-ocean-ridge showing a progressive opening and increase in surface area of the seafloor crusts.

Seafloor spreading was first recognized from early seafloor magnetic mapping by Carey in 1958, and again by Heezen in 1960. In 1961, scientists—most notably the American geologist Harry Hess—began to theorize that mid-ocean-ridges mark structurally weak zones, where the seafloor was considered as being *ripped apart lengthwise* along the crest of the mid-ocean-ridges (Figure 2.4). From this, it was suggested that new volcanic lava from deep within the Earth must rise through these structurally weak zones to eventually erupt along the crest of the ridges and be quenched by the sea water to form new basaltic seafloor crust.

This *seafloor spreading hypothesis* was based primarily on the magnetic mapping evidence of Vine and Mathews. It was also supported by several additional lines of evidence available at the time, including evidence from age dating, seismic activity, and bathymetric surveys.

At or near the crest of the mid-ocean-ridges, the seafloor crustal rocks were shown to be very young, and these rocks become progressively older when moving away from the ridge crests. The youngest rocks at the ridge crests always have present-day normal magnetic polarity. Moving away from the ridge crests, rocks parallel to the ridges were shown to have alternated in magnetic polarity from normal to reverse to normal and so on. This observation quantified the earlier suggestion that the Earth's magnetic field has reversed many times throughout its history.

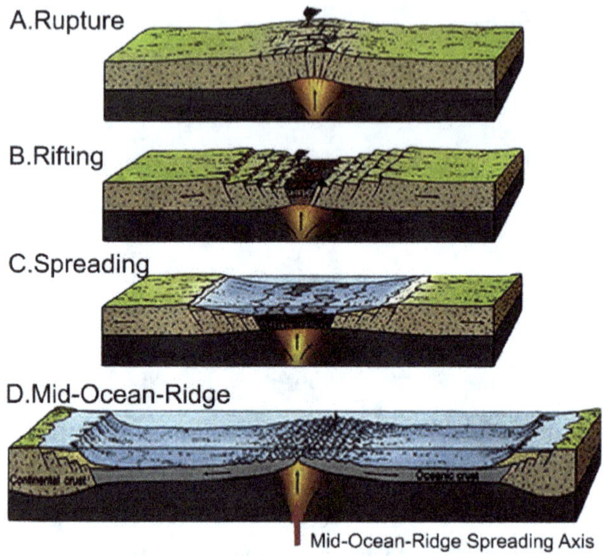

Figure 2.4. Schematic cross-sections showing structurally weak seafloor crustal zones opening along the crest of a mid-ocean-ridge to form a new ocean.

It was further appreciated, from dating the ages of the various seafloor crustal rocks, that this process has operated over many millions of years. Subsequent mapping has shown that this process is continuing to form new seafloor crusts along the entire 65,000-kilometre-long system of centrally located mid-ocean-ridges, now known to be present throughout all of the oceans.

By explaining both the zebra-stripe like seafloor magnetic patterns and the configuration of the mid-ocean-ridge system, the seafloor spreading hypothesis quickly gained converts. Furthermore, this seafloor crustal mapping is now universally appreciated to represent a natural tape recording of both the history of reversals in the Earth's magnetic field and a history of opening of each of the oceans.

A consequence of this observation of seafloor spreading is that new volcanic lava is shown to be continually intruded along the full length of all the mid-ocean spreading ridges. It is interesting to note that this observation was initially considered to support the theory of Earth Expansion, where new crust was considered to have formed at the mid-ocean-ridges as a consequence of an increase in Earth radius.

Subsequent tectonic work has now favoured continental drift-based plate tectonic theory, where excess crust generated at the mid-ocean-ridge spreading centres was considered to eventually disappear along seafloor trenches, located along the margins of some continents, by subduction of the seafloor crustal rocks.

2.3. Commission for the Geological Map of the World

The Commission for the Geological Map of the World (CGMW) is an international association in the field of geology. Its main role is the production of small-scale geological or geoscience-related maps of the world, or parts of the world. It is an organization affiliated to the International Union of Geological Sciences (IUGS) and International Union of Geodesy and Geophysics (IUGG) and is a UNESCO-related scientific, non-government organization.

In January 1983, the wall map concept of the Geological Map of the World was revived at the UNESCO general assembly and the 1st edition of this map at 1:25M scale map was published in 1990. For the first time, continental geology was shown, alongside ocean geology, on a single map (Figure 2.5). The 2nd edition (2000), benefited from new imaging techniques based on satellite altimetry combined with on-board multibeam echo sounder bathymetry, representing a technological breakthrough. The ocean floor then took on a new aspect, clarifying the structure of the oceans at a much smaller scale than before.

In the 3rd edition (2010), the morphology of the ocean floor was even finer, thanks to data freely available from the world's major geophysical databases. In addition, the considerable growth and greater reliability of geochronological techniques provided better resolution of the main subdivisions of the Precambrian, leading to a significant upgrade of the 2nd edition.

Completion of oceanographic work by the Commission for the Geological Map of the World and UNESCO during the 1980s led to publication of the original Geological Map of the World in 1990 (digitized with permission in Figure 2.5, Maxlow 2002). A legend for each of the colours depicted in this map is shown in Figure 2.6 which coincides with the intervals of geological time shown in Figure 2.2.

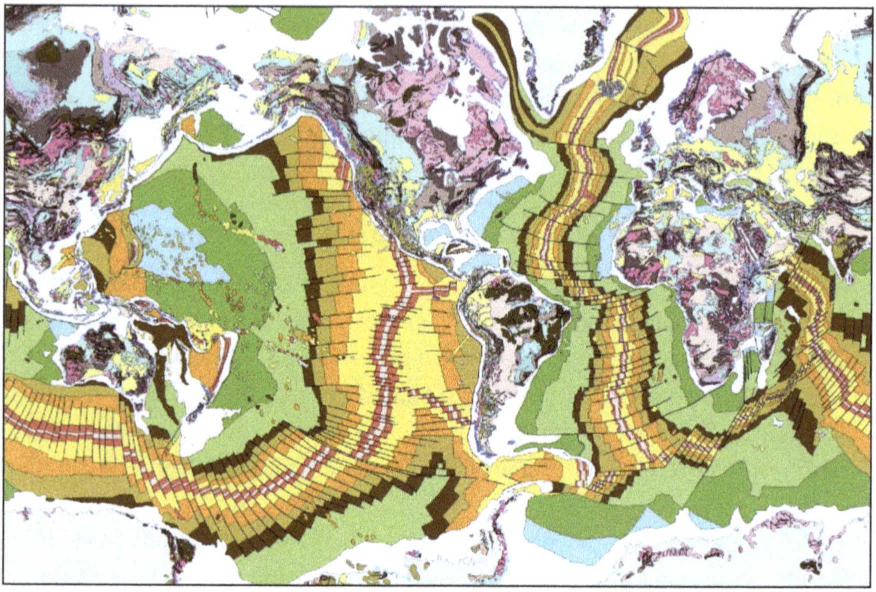

Figure 2.5. Geological Map of the World. Digitized with permission from the Commission for the Geological Map of the World and UNESCO, 1990. Maxlow 2002.

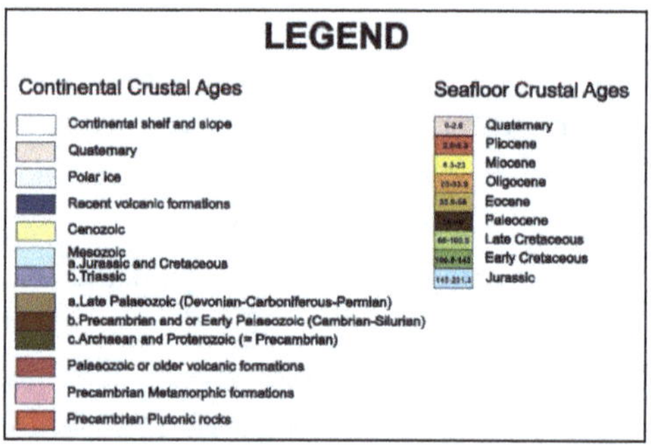

Figure 2.6. Geological timescale legend showing the various colours of the continental and seafloor crustal ages as shown in Figure 2.5. Seafloor crustal ages are in millions of years before the present-day. (Maxlow, 2002)

Similarly, an Age of Oceanic Lithosphere map was published in collaboration with the National Geophysical Data Centre (NGDC) and the National Oceanic and Atmospheric Administration (NOAA) in 2008 (Figure 2.7).

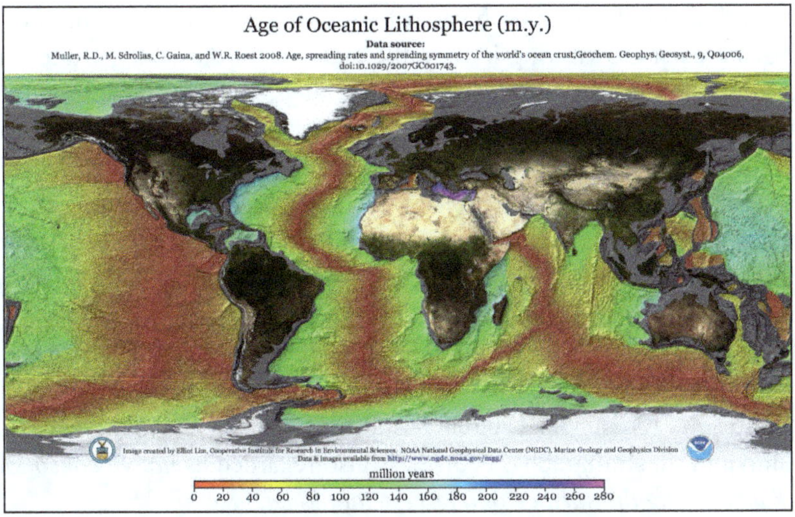

Figure 2.7. Age of Oceanic Lithosphere map. (NOAA & NGDC, 2008)

In both maps, the same continental and seafloor magnetic evidence was taken a step further. By age dating the samples of rock collected from the seafloor crust at regular intervals across the bottom of each of the oceans, and by comparing these ages with the magnetic striping, the seafloor crustal mapping was then summarized as intervals of geological time.

The coloured seafloor striping shown in Figure 2.5 represents age data displayed as variable intervals of geological time, coinciding with the known geological periods and epochs, while in Figure 2.7 the same age data is displayed as a continuous rainbow coloured time-spectrum.

What this mapping means is that the yellow seafloor stripes in Figure 2.5, for example, located between the younger red stripes and the older orange stripes, represent basaltic lava that was progressively intruded along the entire length of the ancient mid-ocean-ridge spreading centres during the Miocene Epoch. The Miocene Epoch is the interval of time extending from 5.33 to 23.03 million years ago.

During that time, the younger red and pink rocks respectively did not exist. By inference, the two adjoining yellow Miocene stripes were then joined together throughout all of the oceans and remained assembled along their common mid-ocean-ridges throughout this interval of time—progressively widening over time.

A good analogy for the seafloor stripes in Figure 2.5 is growth rings on a tree. For most trees, a new growth ring is added around the outside perimeter of a trunk and branch for each year of growth. By moving back in time, and removing each growth ring in turn, the diameter of the younger tree can be measured, and by counting the remaining rings, the age of the tree can be determined.

Similarly, the volcanic seafloor stripes in Figure 2.5 can be visualized as a record of the growth patterns for each crustal plate, where each stripe represents millions to tens of millions of year's plate growth. It is important for the reader to keep this concept clearly in mind because it is critical to understanding the formation of the seafloor crust over time.

In contrast to seafloor crusts, the continental geology in Figure 2.5 represents bedrock geology. In traditional geological usage the term bedrock refers to solid rock underlying unconsolidated surface materials, such as soil or alluvium. In stratigraphy, bedrock is also regarded as *native consolidated rock underlying the surface of a terrestrial planet.*

In Figure 2.5, these basic definitions are further extended to consider that the bedrock geologic map of the world shows the distribution of differing rock types, that is, rock that would be exposed at surface if all soil and superficial deposits were removed.

None, or very little of this magnetic striping, age dating, or geological mapping evidence was available when plate tectonics were first proposed during the mid-1960s. The global distribution of this seafloor mapping, age dating, and geology was completed later in order to assist with, and substantiate, the presence and assemblage of the various crustal plates and continents on a continental drift-based plate tectonic Earth model.

While this map was originally conceived to provide a key role in defining and quantifying plate movement in our oceans, it is unfortunate that, from a geoscientific perspective, it remains underutilized and continues to be ignored in all modern plate tectonic studies.

A close look at this map demonstrates conclusively the indisputable presence, configuration, and distribution of all identified plates, which quantify the underlying basis of plate tectonics. Yet, when you have a second look, there is no evidence for the randomness of plate motion insisted on by continental drift theory.

It is again unfortunate that the geological content of this map is rarely, if ever, used to constrain plate motion histories in any plate assemblage study. This geological mapping is factual data and is considered to be critical in the understanding of the growth of all plates, and plate movement over time.

Chapter 3

A Paradigm Shift in Thinking

"*The oceanic crust of any epoch, originating as a result of spreading, is fully fixed in the geological record, and continues to increase the surface area.*" Blinov, 1983.

As previously mentioned, a paradigm shift in thinking is defined as a fundamental change in the basic concepts and practices of a discipline or field, leading to a new way of understanding or doing things. And a paradigm shift—or paradigm change—happens when scientific activity and experimentation begin to contradict premises that experts previously considered unshakable. Kuhn referred to this contradiction as a "crisis situation" and that it is present in all paradigm shifts. Without recognition of a crisis, no paradigm shift will occur.

In plate theory, the Earth's crust is arranged into a series of seven or eight major and many minor plates, made up of both continental and seafloor crusts (Figure 1.1). Earthquakes, volcanic activity, mountain-building, and seafloor trench formation are said to occur along each of the plate boundaries.

This theory is considered by most scientists to adequately link all geologic features, from the age and composition of seafloors to the rise of mountains, as well as the past distributions of plant and animal species. Continents are then considered to randomly fragment, collide, accumulate and dissipate across the Earth's surface over time, driven by large convection currents within the mantle. But keep in mind that such convection currents have never been directly observed and are still only a theory.

As more and more of the seafloors were mapped during the 1950s to 1980s, the mapped magnetic patterns turned out to be not random or isolated occurrences but instead revealed a predictable zebra-stripe like pattern. These stripes were found to be symmetrical about centrally located mid-ocean-ridges within each of the oceans. The completion and publication of the Geological Map of the World (1990) (Figure 2.5), based on these magnetic patterns, has since confirmed the presence and distribution of all identified plates, however, has also shown that a random plate motion is not conducive with the seafloor geology shown on this map.

This limitation presents a major problem for plate tectonic plate assemblages and can be rightly seen as a "crisis situation," which will require a paradigm shift in our thinking to resolve it.

Publication of the Geological Map of the World was first completed by the Commission for the Geological Map of the World and UNESCO in 1990 (Figure 2.5) and is reproduced in spherical format in Figure 3.1. The various views in Figure 3.1 show the present-day continental and seafloor bedrock crustal geology of the Earth.

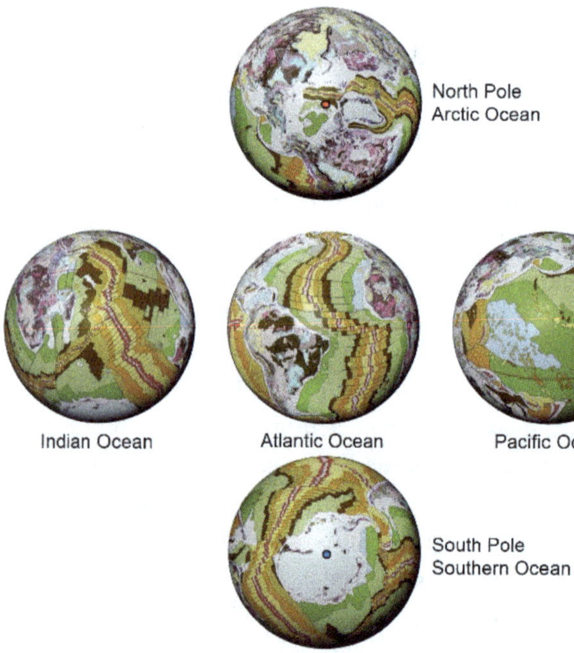

Figure 3.1. Present-day Geological Map of the World shown in spherical format, based on the Geological Map of the World, 1990. Maxlow, 2001.

The colours depicted on this spherical geological model represent rocks that have been deposited, intruded, or extruded during set intervals of geological time. The colours do not represent specific rock types, although, at the map scale used, there is a broad correlation between the most ancient continental rocks shown as red and pink colours, belts of ancient deformed and folded rocks shown as khaki colour, younger sedimentary rocks shown as browns, blues and yellow colours, and the seafloor crusts shown as various coloured stripes.

The colours depicting the continental crustal geology in Figure 3.1 represent rocks that were formed during the five major intervals of geological time, including the most ancient Archaean Eon—beginning around 4,000 million years ago—followed by the Proterozoic Eon and then the Palaeozoic, Mesozoic, and Cenozoic Eras.

Similarly, the colours depicting the seafloor crustal geology represent rocks that were formed during the geological time periods and epochs, ranging from the Jurassic—beginning around 200 million years ago—through to present-day times. These periods and epochs also represent subdivisions of the Mesozoic and Cenozoic Eras. For the actual age relationships of these geological times, refer to the colour legend (Figure 2.7) as well as the International Chronostratigraphic Chart (Figure 2.2).

In Figure 3.1, each ocean is shown to contain a mid-ocean-ridge, and these ridges are centrally located within each ocean, coinciding with the pink Pleistocene stripes. These observations are further substantiated by seafloor bathymetric surveys, which show that the mid-ocean-ridges coincide with the distribution of an extensive, centrally located, network of submarine mountain ranges.

As is known from plate tectonic studies, these mid-ocean-ridges subdivide the entire Earth's crust into very large, convex, plate-like crustal fragments. Each plate generally includes both continental and surrounding seafloor crust, and the plates are mainly centred on and around each of the continents. These observations are unequivocally quantified by this geological mapping.

3.1 Preliminary Observations

There are a number of preliminary observations that can be made from an examination of this Geological Map of the World map. These observations include:

- The pattern of colours representing the ages of the crustal rocks confirms that seafloor crusts are vastly different from the continental crustal rocks.
- Plates, in general, comprise a centrally located continental crust surrounded by seafloor crust.

- The continental crusts are shown to be more ancient and complex than the seafloor crusts.
- The intrusion of volcanic lava along each of the mid-ocean-ridge spreading centres represents quenched and cooled mantle rock, not oceanised continental rocks.
- The coloured patterns of seafloor crusts in Figure 3.1 are shown to be symmetrical about the mid-ocean-ridges, currently centred beneath the present-day pink-coloured stripes, and their ages become progressively older when moving away from these ridges towards the continents.
- This age distribution occurs because new volcanic lava is continually being added along the full length of the mid-ocean-ridge spreading centres.
- The oldest seafloor crust in each ocean is early-Jurassic in age—pale blue-coloured areas—and these rocks are mainly located along the present continental margins.
- The coloured seafloor striping confirms that all oceans contain a mid-ocean-ridge, and each ocean, and hence each plate, is increasing its surface area with time (Figure 3.2).
- This increase in surface area is symmetrical about the mid-ocean-ridges within each ocean.
- It is important to keep in mind that there are no observable instances where the age of the modern seafloor crust is older than 170 million years.
- From these observations, the symmetrical pattern of seafloor stripes, in each of the oceans, demonstrates that all continents are moving away from each other as the surface area of each crustal plate increases over time.

To reiterate, the Geological Map of the World quantifies the observation that the Earth's crust comprises a series of seven or eight major and many minor plates, made up of both continental and seafloor crusts. However, the proposal that continents randomly fragment, collide, accumulate and dissipate again at the Earth's surface over time, driven by large convection currents within the mantle, is not supported by this factual geological mapping.

Each of the oceans is shown by the red arrows in Figure 3.2 to be opening and increasing in surface area away from the mid-ocean-ridges. A random plate motion history would imply that seafloor crustal generation would not necessarily be symmetrical in each ocean, seafloor crustal patterns would not be parallel to continental outlines, seafloor crustal ages would not necessarily decrease in age away from the continents, and there should be a random patchwork of pre-Jurassic seafloor crusts remaining in many oceans.

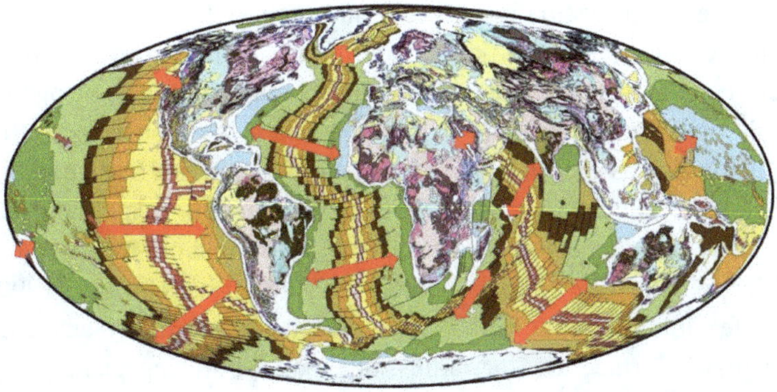

Figure 3.2. Commission for the "Geological Map of the World" & UNESCO (1990) map, shown in Mollweide projection, where all oceans are shown to be opening and increasing in surface area.

The significance of this Geological Map of the World is that a means to independently constrain plate motion and plate assemblage histories has been available since 1990; constrained to each of the coloured seafloor time stripes, back to the early-Jurassic Period. Yet, to date, geosciences have failed to utilize this map or recognize the significance of this mapping, relying instead on palaeomagnetic apparent-polar-wander patterns to constrain plate motion and plate assemblage on a static radius Earth model.

From this geological mapping, it is considered technically feasible to construct spherical geological models of the ancient Earth, and to assemble each of the identified plates back in time to the early-Jurassic Period, independently of palaeomagnetics.

3.2 A Shift in Thinking

Traditional plate tectonic thinking insists that, in order to maintain a constant Earth surface area, an equivalent amount of old crust must be disposed of in sympathy with new crust being generated along the mid-ocean-ridge spreading centres. This has formed the basis for traditional plate theory, where excess crust is disposed of by subduction within trenches, primarily located around the margins of the Pacific Ocean.

What the Geological Map of the World provides is the opportunity to broaden our approach to plate understanding and assemblage by utilizing the history of plate growth, shown by the coloured seafloor stripping, to accurately constrain plate assembling back in time.

When moving back in time, the pattern of seafloor crusts, as highlighted by the magenta arrows in Figure 3.3, would suggest that, rather than constrain assemblage to closing the Atlantic Ocean only, geology shows that the area of every plate must reduce, and all continents must move closer together.

This observation prompts the need to seriously consider shifting our thinking towards assembling all plates on a variable radius Earth model, rather than a static radius continental drift-based model. Something that has traditionally been vehemently rejected in the past, without basis.

Prior to completion of the Geological Map of the World in 1990, there was no feasible method available to researchers to quantify any change in ancient Earth surface area or radius, other than by using palaeomagnetics, hence these early researchers had no option other than to constrain Earth surface area and radius to constant values.

Palaeomagneticians, such as McElhinny and Brock in 1975, had long since concluded from palaeomagnetic determinations of ancient Earth radius that "*...there has been no significant change in the ancient radius of the Earth with time.*" Publication of the Geological Map of the World, however, disputes this mandate and provides a means to independently test this conclusion by measuring surface areas of coloured seafloor stripes and, independently determine potential ancient Earth radii back to the early-Jurassic.

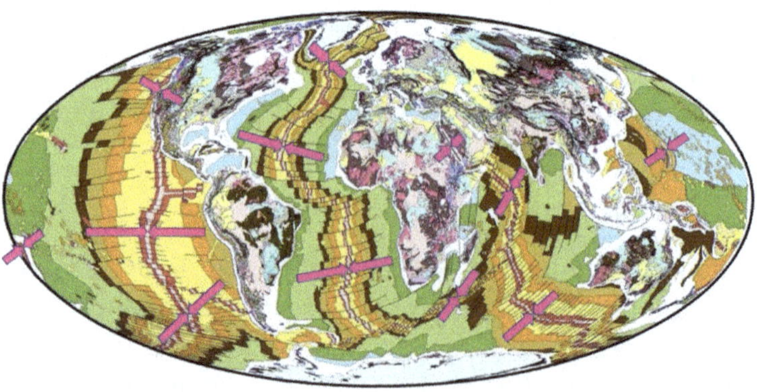

Figure 3.3. Commission for the "Geological Map of the World" & UNESCO (1990) map, shown in Mollweide projection, where, moving back in time, all oceans must close and surface areas decrease.

Rather than confining our thinking to traditional continental drift-based plate tectonic thinking, there is a critical need to test this empirical observation before continuing to arbitrarily reject the proposal that Earth radius must remain constant over time.

Considering the vast amount of global observational data now available to science, rejection of an increasing radius Earth plate tectonic proposal, based on unsubstantiated historical objections has, in reality, no basis in science.

3.3. Measuring Seafloor Surface Areas

Measuring surface areas of intruded seafloor basaltic lava to determine a potential rate of increase in surface area of new basaltic seafloor crust was carried out by Garfunkel in 1975, Steiner in 1977, and Parsons in 1982. In each case, early versions of geological mapping of the oceans were used, and it was assumed that an equal amount of seafloor crust must be disposed of elsewhere in order to maintain a constant radius and surface area Earth model. By taking their results a step further, measuring seafloor surface areas to determine an ancient Earth radius was pioneered by Koziar during the early 1980s and later replicated by Blinov in 1983. This research was again in direct response to preliminary completion of geological mapping of seafloor crusts throughout all of the oceans.

While conventional plate tectonic crustal modelling traditionally considers that plate assemblage on a constant radius Earth model is feasible, at no point has the alternative consideration for a changing Earth surface area and radius been investigated. This alternative consideration maintains that, for an Earth undergoing an increase in surface area and radius over time, volcanic lava intruded along the mid-ocean-ridge spreading centres is preserved and added to the outer margins of each of the plates. The increase in surface area of each of these plates is then a direct result of an increase in Earth surface area and radius, with no requirement for any net disposal of excess crust by substantial subduction processes.

To account for this increase in surface area of seafloor crust, past researchers have speculated that a change in Earth radius may have ranged from a partial increase in radius, where some subduction occurred, to full increase in radius over time. For full increase in radius, no pre-existing crust or substantial form of subduction is required to account for the observed geological and geophysical features present on the present-day seafloor, whereas a partial increase in radius was originally proffered to account for some subduction-related observations.

Other suggestions have also included a possible pulsating Earth process, where an increase in radius was said to have caused crustal stretching and later contractions were said to have given rise to crustal uplift to form mountains. In this process it was argued that, in order to have mountains there must have been compression and in order to have mid-ocean-rifting there must have been an increase in Earth radius. A pulsating increase-contraction process was then said to have given rise to the various mountain-building episodes which were in turn accompanied by opening of each of the modern oceans.

While conceptually feasible, this concept is now contrary to modern global mapping and observation. There is no evidence for this cyclical process shown in the seafloor spreading patterns in Figure 3.1. This pulsating Earth process also conflicts with current understanding of mechanisms for mountain building and known laws of physics.

For the alternative suggestion of an increasing surface area Earth, when moving back in time, the seafloor crusts must be restored and assembled back to their original configurations along each of the mapped mid-ocean-ridge axes on a smaller radius Earth. There is no other physically valid or proper method available to explain this unique seafloor mapping phenomenon, as shown throughout all of the oceans. Similarly, each of the continents must be moved closer together and, at the same time, the surface area of each of the plates must be reduced.

The completed Geological Map of the World mapping is therefore unique in that it shows precisely where the edges of each plate, and hence mid-ocean-ridges, were at any moment back in time. Something that conventional plate tectonic modelling studies cannot do.

The intent of the rest of this chapter is to test this alternative suggestion, whereby new seafloor crust forms in sympathy with an increase in Earth surface area and hence radius over time. It is anticipated that this information will provide an important opportunity to reassemble and constrain the positions of adjoining crustal plates with a high degree of precision, as well as to measure changes in seafloor surface areas over time, for any time period ranging from the present-day back to around 200 million years in the past.

3.4. Measuring Potential Ancient Earth Radii

Geological mapping of the Earth shows there are basically two types of crusts covering the Earth's surface: relatively thick continental crust and relatively thin seafloor crust. Each of these crusts have different physical characteristics and each has completely different age characteristics—continental crusts are predominantly older, and seafloor crusts are predominantly younger. Modern geological mapping and direct observation also shows that new seafloor crust is being added continuously along the entire length of the mid-ocean-ridge spreading centres and in doing so progressively increases the surface areas of each plate and ocean over time.

With completion of the Geological Map of the World, it is a relatively simple process to determine ancient Earth radii by measuring the cumulative surface areas of the seafloor plates. The added advantage of this geological map is that the seafloor crusts also have known ages, which are in turn used to accurately constrain the moment in time when these rocks were first formed. Once the areas of each of the coloured seafloor crustal stripes are known, the cumulative areas of seafloor crust are then taken away from the present-day Earth surface area to mathematically derive ancient Earth radii.

To determine potential ancient Earth radii for each of the time intervals shown by the coloured seafloor spreading stripes in Figure 3.1, focus will initially be made on the distribution of seafloor crusts, extending in time from the present-day back to about 200 million years in the past. This 200-million-year time-frame covers the age of the oldest known seafloor crusts—to around 170 million years old—plus continental shelf sediments deposited around the margins of the continents.

The term geological crustal budget is a term used to account for the surface areas of each of the coloured crustal rocks in Figure 3.1, measured against time. What is meant by this term is that if something is created, in this case an area of new volcanic seafloor crust added along the mid-ocean-ridge spreading axes, it must be accounted for. The overall global geological crustal budget for each of the time-periods involved must also be able to be balanced—in this case the total ancient surface area of the Earth at any moment in time.

To measure the surface areas of each of the coloured seafloor stripes, all projection distortion must be eliminated from the bedrock geological map and, strictly speaking, this information must be displayed in spherical format. The method adopted here requires the existing Geological Map of the World, shown in Mercator projection format in Figure 2.6, to be displayed as a 24-gore sinusoidal projection map (Figure 3.4).

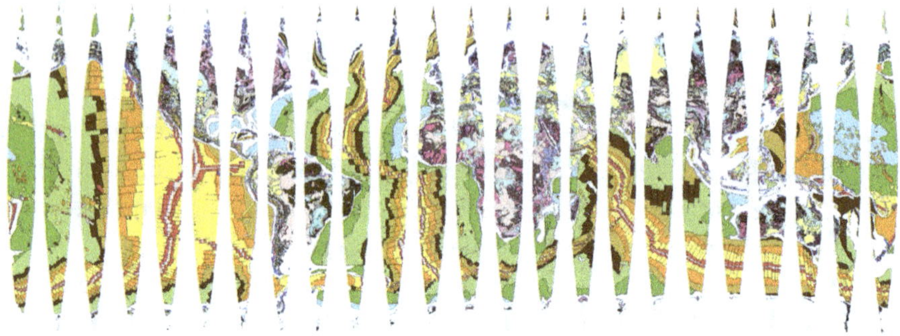

Figure 3.4. 24-gore sinusoidal map projection of the Geological Map of the World. This projection enables the geological map to be displayed in distortion-free spherical format and forms the primary base-map for both surface area measurement and small Earth model constructions. Maxlow, 2001.

This sinusoidal projection format is unique in that it represents an unfolded sphere. The sinusoidal map gives undistorted, true-to-scale, geological information from anywhere within the map area enabling the information to be both measured and confidently modelled on spherical models.

The term gore simply means that each curved stripe, which in this figure represents fifteen degrees longitude at the equator, tapers to a width of zero degrees longitude at each pole. In Figure 3.4, twenty-four gores are used to represent the original Mercator map area. The unique quality of this type of map projection is that it can then be cut and pasted directly onto a globe during model construction, as seen in Figure 3.1.

By using a sinusoidal map, the outline of each coloured seafloor stripe can be digitized in turn. The surface areas of successive intervals can then be measured, and a potential ancient Earth radius derived for each time period shown. The measured raw data and derived radii from this exercise are shown in Table 3.1, and a graphical plot of the cumulative surface areas against time is shown in Figure 3.5.

The graph shown in Figure 3.5 displays the distribution and accountability of measured surface areas of all Earth's continental and seafloor crusts. This graph covers approximately 200 million years of Earth history, extending from the present-day back to the beginning of the Jurassic Period.

In this figure, each of the measured seafloor stripes is progressively added together, when moving forward in time, extending from the early-Jurassic to the present-day. In other words, the area of new crust intruded along the mid-ocean-spreading-ridges is considered cumulative over time and is progressively added to the area of previous seafloor crust for each successive time interval shown.

Table 3.1. Empirical surface area data derived from mapping by Larson et al. (1985) and CMGW & UNESCO (1990). Areas were originally digitized using a CAD based Graphical Design System software package and ancient Earth radius calculated from this data. Arbitrarily assigned errors are ±5%. Maxlow, 1995.

Age		Surface Area			Ancient Radius
Chron	Years ($x10^6$)	dS ($x10^7$ km^2)	SdS ($x10^7$ km^2)	S_a=S_0-SdS ($x10^7$ km^2)	R_a (km)
C0	0	0	0	51	6370.8
C2	-1.9	0.5342	0.5342	50.4658	6337.15
C3a	-5.9	1.3328	1.867	49.33	6265.43
C6b	-23	4.9213	6.7883	44.2117	5931.49
C15	-37.7	4.1624	10.9507	40.0493	5645.37
C25	-59.2	4.1649	15.1156	35.8844	5343.77
C29	-66.2	1.0462	16.1618	34.8382	5265.3
C34	-84	4.7956	20.9574	30.0426	4889.49
M0	-118.7	5.6758	26.6332	24.3668	4403.46
M17	-143.8	1.9348	28.568	22.432	4225.02
M38	-205	1.9386	30.5066	20.4934	4038.31

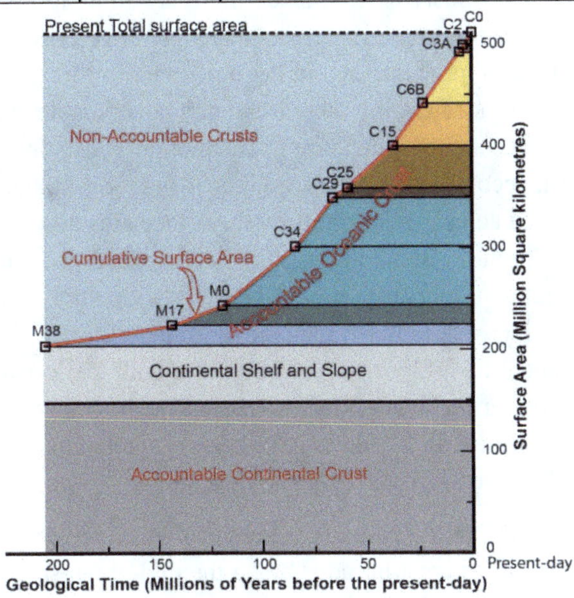

Figure 3.5. Cumulative surface areas of both continental and seafloor crustal rocks as measured from the Geological Map of the World sinusoidal base-map (Figure 3.2). The time scale represents millions of years before the present. Maxlow, 1995.

On this graph, the cumulative surface areas of each of the crusts can be compared to the present-day total surface area, shown as a dashed line at the top of the chart. Here, areas of both accountable oceanic and accountable continental crusts are shown below the red surface area curve. In this context, accountable crust represents all continental and oceanic crusts currently existing on the Earth's surface today. The cumulative surface area of each of these accountable-crusts must then equal the present surface area of the Earth at the present-day time zero.

Also shown on this graph is a pale blue area, located to the upper left, referred to as non-accountable crusts. This area is located above the measured red cumulative surface area curve and below the dashed total surface area line at the top of the graph. This non-accountable crust represents the additional surface area of inferred crust that must have been present in order to maintain a constant surface area Earth. No record of this non-accountable crust exists on Earth today and there is very limited evidence to suggest that it ever did.

If a constant surface area Earth is to be maintained, this non-accountable crust represents the additional surface area of continental and seafloor crusts that must have previously existed since the early-Jurassic Period. This area also represents the total area of crusts that must have been removed by an inferred subduction process. A subtle outcome of this graph is that an explanation for the accelerating increase in non-accountable crust shown, and by inference an accelerating increase in subduction, must also be explained.

In contrast, accountable crust on an increasing radius plate tectonic Earth model reflects the actual variation in Earth surface area and radius over time. On this model, non-accountable crust is not required and may have never existed. The rate of increase in surface area with time is then simply a reflection of an increase in Earth radius over time.

The accountability of Earth surface areas for both seafloor and continental crusts can be further understood by comparing the basic premises of the four main tectonic theories, all of which have been promoted in science at various times during the past 100 years of geological study

1. **Constant Earth radius** (Conventional plate tectonic Earth): In this theory, excess crust intruded along the mid-ocean spreading centres is continuously being disposed of along subduction zones, displacing and recycling pre-existing crust into the upper mantle. To maintain this view of a constant surface area it must be presumed that non-accountable crust existed prior to subduction. Limited, if any, record of this crust exists and an explanation must also be provided for the accelerating rate of increase in the actual, observed surface areas shown.

2. **Partial increase in Earth radius**: In this theory, some excess crust is disposed of along subduction zones during limited increase in Earth radius. To maintain a partial increase in surface area, it must be presumed that some non-accountable crust existed prior to partial subduction. The accelerating increase in observed surface area since the early-Jurassic Period must again be explained.
3. **Pulsating Earth radius**: Similarly, in this theory crust is generated at spreading centres during a period of increasing Earth radius and is either compressed to form mountains or disposed of along subduction zones during a reduction in Earth radius. No evidence for a pulsating Earth radius is shown from the available surface area mapping data and new ideas on mountain building negate the requirement for crustal compression to form mountains.
4. **Increasing Earth radius**: For an increasing radius Earth, new volcanic lava intruded at mid-ocean spreading centres is cumulative with time. The increase in total surface area is a direct result of an increase in Earth radius. On an increasing radius Earth, non-accountable crust never existed, subduction is not required, and the cumulative surface areas are a physical measure of the actual change in Earth radius. The measured change in Earth surface area in turn suggests there has been an accelerating rate of increase in Earth radius over time.

From this brief comparison of the various Earth radius options available, it can be seen that both the historical expanding Earth and continental drift-based plate tectonic theories represent the extreme endpoints of a spectrum of global tectonic possibilities. Continental drift-based plate tectonic theory is currently the accepted theory in science and has been extensively presented in publications elsewhere.

Owen, in 1976, adequately presented partial expansion and Milanovsky in 1980 adequately presented the pulsating Earth theory. Each of these conventional plate tectonic, partial expansion, pulsating, and expanding Earth theories were introduced well before completion of the Geological Map of the World and, to remain viable, these theories must be able of explain the new observations shown by this mapping.

The measured surface areas in Table 3.1 can be readily converted to an ancient Earth radius for each time interval measured. The derived Earth radius data are shown plotted as a dashed black line in Figure 3.6, along with a curved red line representing an idealized exponential curve which has been statistically fitted to the smoothed measured data. The discrepancy between the measured dashed line and the red curve at the lower left is suggested to represent masking of seafloor volcanic crusts by sediments deposited around the margins of the continents.

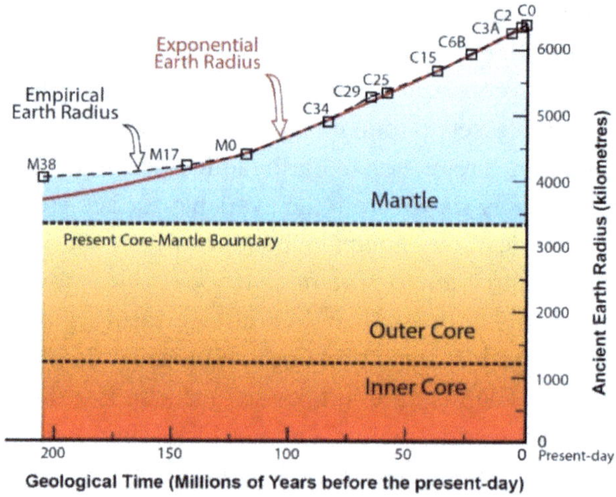

Figure 3.6. Ancient radius of the Earth, extending from the early-Jurassic to the present-day, derived from digitizing the areas of each coloured seafloor strip in turn, as seen on the published post-Triassic seafloor mapping. The red line represents an idealized exponential curve fitted to the data. Maxlow, 1995.

While this exercise, from the prospective of conventional plate tectonics, may be viewed with skepticism and considered largely hypothetical, the implications stemming from measuring plate surface areas, as well as deriving ancient Earth radii measurements, are that we now have a means to accurately measure potential ancient Earth radii for each of the time periods depicted in the coloured seafloor crustal mapping data. By removing each coloured seafloor stripe from each of the modern oceans and reducing Earth radius back in time to the early-Jurassic Period, we are also in a unique position to test the proposal that all remaining plates will unite together on a smaller radius Earth model.

If spherical modelling of the continental and seafloor plate assemblages is successful, these spherical models will then confirm the need for a paradigm shift in plate tectonic thinking, plus form a unique platform for displaying additional global observational data from a wide range of geological and related disciplines.

Chapter 4

A Paradigm Shift in Plate Modelling

"*But in the end the rigorous approach has paid off. For it has revealed a discrepancy which had not been apparent. It was not my method that was at fault, but my assumption that the earth of Pangaea was the same size as the earth of today. The assembly of Pangaea is not possible on the earth of the present radius, but on a smaller globe...these difficulties vanish.*" Carey, 1958.

In order to model the plate motion history of the Earth, Weijermars (1986, 1989), and Scotese *et al*, (1988), used early versions of published seafloor mapping to produce empirical and computer-based plate reconstructions respectively on a static radius Earth model. Scotese has since gone on to become a leading authority on conventional plate tectonic crustal modelling.

During plate reconstruction, both authors were confronted with problems of plate boundary mismatch and constraints to the Earth's overall crustal budget. Balancing the Earth's crustal budget was overcome by introduction of presumed, pre-existing crust in order to maintain a static radius Earth premise.

In order to justify modelling on a static radius Earth Weijermars, in particular, provided "*six conclusive arguments against any fast Earth expansion hypothesis,*" prior to modelling studies. These arguments, understandably, did not scientifically resolve the proposal that seafloor mapping data may in fact be better suited to modelling on smaller radius spherical models.

Conventional plate tectonic studies do not use the Geological Map of the World (Figure 2.5) or Age of Oceanic Lithosphere (Figure 2.7) maps to constrain plate assemblages. Heavy reliance, in particular for pre-Triassic assemblages, is instead placed on palaeomagnetic apparent-polar-wander studies to locate ancient pole positions for adjacent continental crustal fragments. While at an advanced stage of sophistication, these apparent-polar-wander studies result in multiple plate-fit options for the various continents and are severely constrained by a lack of longitudinal control during plate assemblage. Crustal fragmentation, dispersal, and reassemblage to form supercontinents is then portrayed as a random, non-predictable process of continental drift across the surface of the Earth.

In contrast, in addition to providing a means to measure ancient Earth surface areas and radii, the coloured Geological Map of the World map provides a unique means to constrain the precise location of all crustal plates at any moment back in time to the early-Jurassic Period, with only one unique plate-fit.

By returning the seafloor volcanic rocks back to the mantle, from where they originally came from, the surface areas of each of the oceans must, by established protocol, be progressively reduced and each of the continents moved closer together. The uniqueness of adopting an increasing radius model of the Earth is that there is no need to consider where, or when, pre-existing crusts occurred or, similarly, where they must go back in time. All that is required is to let the configuration of the coloured seafloor crustal mapping dictate the precise crustal plate assemblages on small Earth models.

Contrary to objections still proffered in the sciences to discredit the largely historical Earth Expansion theory, a proposed causal mechanism for new mass distribution will be given in Chapter 11.

The plate modelling technique outlined in this chapter is empirical research, adopted to simply test the proposal that geological mapping can be used to constrain plate assemblage back in time, and adoption of this proposal does not need further justification at this stage.

To construct small Earth plate tectonic models, spherical polystyrene foam spheres were found to be the best means of assembling the various crustal plates. Cutting and pasting information from anywhere on a spherical model is true-to-scale, and distortion free, and crustal plates can be accurately constrained anywhere on the globe during model construction by using standard cartographic techniques.

To test the application and viability of increasing radius Earth modelling and, in particular, the ancient Earth radii, as determined from empirical seafloor mapping data, a series of spherical small Earth models (Figure 4.1) were initially made for the present-day and the beginning of the Pliocene, Miocene, Oligocene, Eocene, Paleocene, Late-Cretaceous, Mid-Cretaceous, Early-Cretaceous, Late-Jurassic, and early-Jurassic Epochs and Periods. These small Earth models were constructed using the seafloor crustal information displayed on the Geological Map of the World sinusoidal base map (Figure 3.4), originally digitized with permission from the Commission for the Geological Map of the World and UNESCO (1990).

Figure 4.1. Spherical small Earth models of the Earth extending from early-Jurassic to the present-day. These models demonstrate that the seafloor crustal plates of the Atlantic Ocean, assembled on small Earth models, coincide fully with seafloor spreading and geological data and accord with derived ancient Earth radii. Maxlow, 1995.

To minimize geological complexity shown by continental geology, the small Earth modelling presented in this chapter will initially focus on seafloor crustal mapping only. All continental geology is coloured grey and is left blank. In later chapters, these models will be reproduced in more detail in conjunction with modelling continental crustal geology. The models in Figure 4.1 will now form the basis for discussions on constructing and assembling seafloor crustal plates on plate tectonic small Earth models.

4.1 Assumptions

In order to accurately quantify any variation in the Earth's ancient surface area and radius, and similarly to constrain crustal plate assemblages back in time, it is argued that it is necessary to consider the area and pattern of seafloor and continental crusts.

The main assumptions used to construct and assemble crustal plates on each of the plate tectonic small Earth models presented here are summarized below. These will be further investigated for bias and realism in later chapters when additional geological, geophysical, and geographical information is plotted on each of the completed models.

For the coloured seafloor crusts shown on each of the constructed early-Jurassic to present-day small Earth models it is assumed that, for an Earth undergoing an increase in surface area and radius with time:

1. The Earth's overall seafloor and continental crustal budget increases with time and all increases in surface area are cumulative with time.
2. Once formed, the surface areas of seafloor and continental crust, as represented by the bedrock geological mapping in Figure 3.2, are fully fixed in the geological rock-record.

3. Seafloor and continental crusts are not removed by subduction processes.
4. There are no pre-existing seafloor or continental crusts to account for.

The only stipulation required during cutting and pasting of crustal information between each successive small Earth model is that:

5. Continental and seafloor crustal plates must undergo small vertical and surface area adjustments to allow for a progressive change in surface curvature of the Earth with time.

What each of these assumptions, as well as the cutting and pasting stipulation, mean is that, for an Earth that is progressively increasing in radius over time, seafloor crust accumulates as a direct result of an increase in surface area of each of the plates. Once formed, the seafloor crusts remain fully fixed in the geological rock-record, the crusts are not removed or returned to the mantle by subduction processes, and similarly there is no need to invoke the existence of non-accountable or pre-existing crusts.

To be consistent with the geological evidence, it is acknowledged that both continental and seafloor crusts need to undergo some, to variable crustal distortion, in particular during on-going changes to surface curvature. It is also acknowledged that localized crustal interaction between plates during changing surface curvature, as distinct from extensive plate subduction, may also occur along plate or crustal boundaries, but only at a relatively local scale and not to the extent that is required in conventional plate tectonic theory.

What the crustal distortion and crustal interaction means is, when moving forward in time from the older, smaller radius models to the younger, larger radius models, the surface curvature of the Earth would need to progressively flatten during changing Earth radius. On an increasing radius plate tectonic Earth, changing surface curvature may then represent an important mechanism for most, if not all geological processes observed on the Earth's surface today. These processes include folding and faulting currently observed in all continental crusts, as well as uplift of the crust along continental margins to form mountain belts and escarpments. The processes also apply to the seafloor crusts where distortion of seafloor volcanic lava is accommodated by intrusion and preservation of new lava at ever changing Earth radii and surface curvatures.

In reality, from an increasing radius plate tectonic Earth perspective, radius increases at an extremely small rate per year—shown later to be currently increasing at 22 millimetres in radius per year. During this process, the crusts creep and distort at an imperceptibly small rate in order to keep pace with the ever-changing surface curvature. The joints, fractures, faults, and folds in existence on Earth today simply absorb this minute curvature adjustment. To model back in time, all of these geological processes must then be reversed and any changes in surface curvature must be accommodated for on each of the small Earth models.

Potential change in surface curvature will be discussed in more detail later, sufficient to say that it is anticipated that, when moving forward in time, progressive flattening of the Earth's surface curvature will result in a gradual deformation of the crust over a vastly extended period of geological time. During small Earth model construction, when moving back in time, these small crustal changes must be accommodated for and removed.

Each of these assumptions, as well as the stipulations used to construct the small Earth models in turn, are consistent with the main assumption that, for an Earth undergoing an increase in radius, the actual surface area of each of the plates increases as a direct result of an increase in Earth radius with time.

4.2. Model Construction

Model construction is carried out and further discussed in two main phases. The first construction phase coincides with the past 200-million-year time interval, extending from the present-day back to the beginning of the Jurassic Period. This phase uses coloured, relatively well age-dated, seafloor geological mapping to constrain the assemblage of all crustal plates and covers the closure of each of the modern oceans. The second construction phase will extend small Earth model coverage further back in time, beyond the early-Jurassic Period, to the beginning of the Archaean Era—around 4,000 million years ago.

The second construction phase is more difficult to achieve because, while age dating of the continental crust exists in detail, its use in model construction is complicated by many overprinting phases of sedimentary deposition, erosion, metamorphism, and tectonism. During this extended interval of ancient times there were also no seafloor crusts to guide crustal assemblage. During the second construction phase, reducing surface areas instead focused on an identified network of continental sedimentary basins and assembling the remaining, more ancient, continental crusts on pre-determined smaller radius Earth models.

The construction method used to assemble the first phase of crustal plates on spherical scale models is very simple and very basic (Figure 4.2).

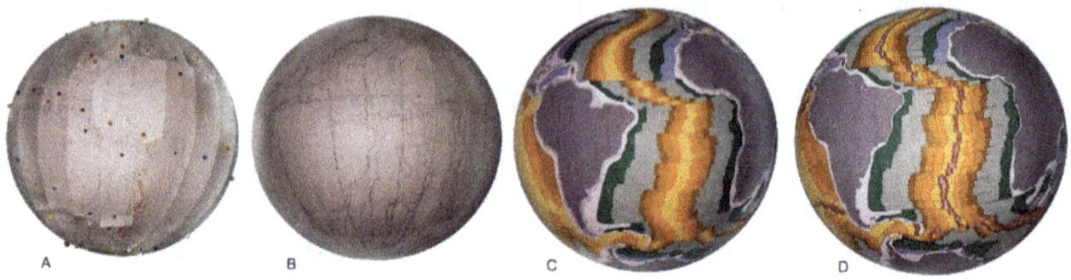

Figure 4.2. Small Earth scale model construction details of various ages and size models: Figure A, preliminary pin-up of plate fragments; Figure B, final paste-up; Figure C, painting of seafloor striping in progress; Figure D, completed present-day small Earth model. Maxlow, 2001.

Each coloured seafloor stripe is simply removed from the sinusoidal base map in turn and the remaining pieces of map are reassembled by pinning and pasting the crustal plates onto a previously determined smaller radius Earth model. This process, in effect, can be likened to a spherical jigsaw puzzle.

In Figure 4.2, tracings of the sinusoidal base map were originally used during my first cutting and pasting phase and these models were later hand painted. The pieces of base map cut from each of the paper gores were then simply reassembled together on a scale model sphere.

For each small Earth model, as the youngest coloured seafloor crustal stripes are progressively removed, the remaining plate boundaries are reassembled along their common mid-ocean-ridge spreading axes on a reduced radius model. Of interest, on an increasing radius Earth, the mid-ocean-ridge spreading axes do not move. They instead retain their global spatial integrity throughout time. It is the configuration of the mid-ocean-ridge spreading zones that dictate precisely the assemblage of each adjoining crustal plate boundary.

4.3. Observations

For the completed plate tectonic small Earth models (Figure 4.3), it is observed that, as the coloured seafloor stripes are progressively removed in turn, the remaining-coloured stripes neatly close together on each successive model, preserving the integrity of each of the remaining plates.

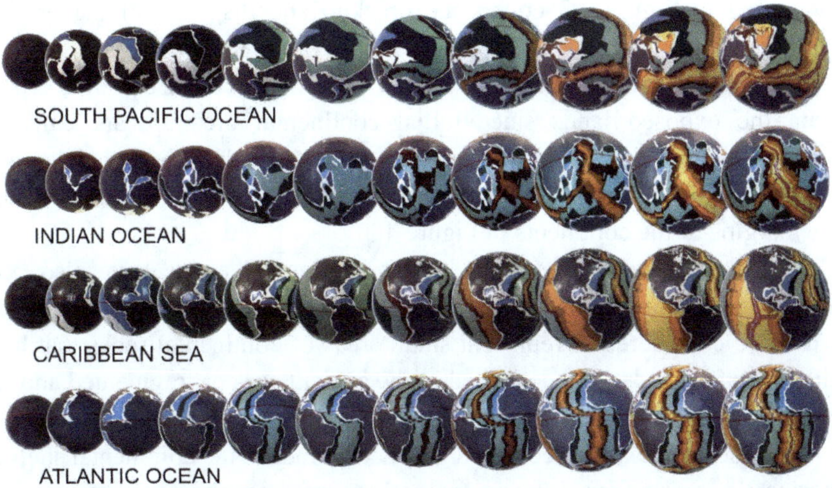

Figure 4.3. Spherical small Earth models of the Jurassic to present-day increasing radius Earth. Each small Earth model demonstrates that the seafloor crustal plate assemblage coincides fully with seafloor spreading and geological data and accords with the derived ancient Earth radii. Maxlow, 1995.

Each crustal plate assembles together in a unique, orderly, and predictable manner during systematic closure of each of the oceans. By removing the coloured seafloor stripes in succession and refitting the plates together on smaller radius Earth models, each plate is then shown to reunite precisely along their respective mid-ocean-ridge spreading axes, estimated to be at better than 99 percent fit-together for each model. The remaining one percent misfit is simply a combination of map error, cartographic discrepancies, as well as the influence of unaccounted crustal distortion or graphical limitations inherited during model making.

The distinguishing feature of this modelling technique is that, prior to the Jurassic Period—model at far left in Figure 4.3—all continental crusts are united as a single Pangaean supercontinent covering the entire ancient Earth. On an increasing radius plate tectonic Earth, this Pangaean supercontinent is shown to simply wrap itself around to enclose the entire ancient Earth with continental crust.

From this observation, it is considered that rupture and breakup of Pangaea during increase in both Earth radius and surface area of each plate—shown later to occur around 250 million years ago during the late-Permian—then led to the formation of the modern continents and opening of the modern oceans. The 200-million-year interval of time shown by the small Earth modelling in Figure 4.3 covers the breakup and dispersal of the original Pangaean supercontinent and opening of each of the modern oceans.

As well as the presence of seafloor volcanic rocks, sediments, eroded from the exposed lands since initial continental breakup, are currently preserved along the margins of each continent and in submerged remnant plateau within the oceans. These sediments are shown as white areas around the margins of the continents in Figure 4.3.

Small, remnant fragments of older continental crusts have also been identified and mapped by others in several areas on the seafloors. These older crusts are considered to represent small areas of continental rocks that have either been eroded and redeposited in the ancient seas or fragmented and left behind as the Pangaean supercontinent first ruptured, broke-up, and dispersed.

On each of the plate tectonic small Earth models, the continental shelves, marine plateau, and remnant ancient seafloor crusts are shown to merge during the Jurassic Period to form a global network of intracontinental marine sedimentary basins on an ancient small Earth at approximately 55 percent of the present Earth radius. This mergence suggests that pre-existing continental sedimentary basins, plus marine basins, represent a network of ancient continental seaways.

By progressively returning the eroded sediments from these continental and marine sedimentary basins back to the lands, where they came from, additional small Earth modelling will later show that each continent further re-unites with adjoining continents along their mutual margins, at approximately 50 percent of the present Earth radius, during the late-Permian to early-Triassic Periods—250 million years ago.

Similarly, during this construction phase, as the seafloor crusts are progressively removed and each plate and continent re-united with its adjoining continental neighbour, a network of continent-to-continent *docking-points* are recognized. These docking-points represent sections of coastline where two or more previously separated continents precisely touch and lock together. These docking points then represent important cartographic and geographic reference points identified during model construction and are used to further constrain the positions of the continents on each successive small Earth model.

It is observed during model construction that the location and significance of these docking-points continued throughout the Mesozoic Era. It will be shown that their on-going usage will become increasingly important during construction of pre-Jurassic small Earth models using continental crusts.

From the spherical model studies briefly introduced in this chapter, it is important to reiterate an important observation. By removing the seafloor volcanic lava from each of the coloured stripes in succession and refitting the remaining crustal plates together on small Earth models, all plates are shown to reunite precisely, with one unique global assemblage. Each crustal plate retains its unique integrity and is shown to assemble with a better than ninety-nine percent fit-together, for each model constructed.

4.4. Fact or Mere Coincidence

It would seem that we have now reached a critical stage in justifying a paradigm shift in plate tectonic thinking. Irrespective of the continued rejection by the peer review system, it would seem that modern geological mapping of the continents and oceans aptly supports an increasing radius plate tectonic Earth mechanism which, in turn, gives further justification in continuing to test this proposal.

A close look at the Geological Map of the World has demonstrated the indisputable presence, configuration, and distribution of all identified plates. This observation has quantified conclusively the underlying basis of plate tectonics. Yet there is no evidence for any randomness of plate motion, as insisted on by continental drift theory.

If the Earth were not increasing in size, or was undergoing a partial or pulsating increase, then this unique fit-together of all plates and continents would not occur. Similarly, if the Earth were not increasing in size, then large gaps or overlaps in the reconstructed plates would occur and the need to fragment continental and seafloor crusts to accommodate for the seafloor crustal evidence would also become increasingly apparent and necessary. Each of these gaps or overlaps, in turn, would then outline areas where presumed, pre-existing subducted crusts would have once been.

Instead, the early-Jurassic to present-day plate tectonic small Earth models demonstrate conclusively that large gaps or overlaps in the crustal plates do not occur. The fact that large gaps or overlaps do not occur, on any of the models, demonstrates the significance of the coloured seafloor geological mapping as a valuable tool for reconstructing plate assemblages and constraining the unique fit-together of all crustal plates back to at least the early-Jurassic Period.

This unique assemblage of seafloor crustal plates also identifies the need to further consider investigation of an increasing radius plate tectonic Earth model, extending beyond the Triassic Period, to evaluate continental crusts.

At this point I am going to disregard all peer review criticism of an increasing radius Earth proposal, in particular the need to justify any variation in Earth mass and volume. I am going to simply apply empirical science to model the modern oceans and seas in detail on increasing radius plate tectonic small Earth models, prior to modelling the continents and supercontinents.

Chapter 5

Modern Oceans and Seas

"*It is difficult to believe that chance alone can explain this fitting together of the continental margins.*" Barnett, 1962.

In all continental drift-based plate tectonic reconstructions large, ancient, essentially theoretical Tethys, Iapetus, and Panthalassa Oceans are required to maintain a constant radius Earth premise. The Panthalassa Ocean is considered by others to have been considerably larger than the present Pacific Ocean, having to accommodate for the surface areas of the closed Atlantic, Southern, and Indian Oceans, and this ocean merged imperceptibly with a Tethys Ocean. With time these, and other lesser oceans, are then depicted as periodically opening and closing as ancient continental fragments randomly collide, amalgamate, re-fragment, and disperse across each of the ancient oceans throughout Earth history.

In this chapter, a distinction is made between the term's modern oceans and ancient continental seas, where continental seas refer to bodies of seawater that covered low-lying parts of the ancient continental lands—often referred to as epeiric seas.

On an increasing radius plate tectonic Earth, the conventional Panthalassa, Tethys, Iapetus and lesser oceans cannot be reconciled in their entirety. Instead, these same ancient oceans initially formed a network of much smaller continental seas, covering low-lying areas of the ancient Pangaean supercontinental lands. These seas are represented by a primitive Panthalassa Sea, located between Australia, Asia, and North America; a Tethys Sea, located within the present Eurasian continent; and an Iapetus Sea, located between the West African and North American cratons.

On an increasing radius plate tectonic Earth, prior to the late-Permian—around 250 million years ago—small Earth modelling shows that there were no modern oceans, only ancient continental seas. The transition from ancient seas to modern oceans only came about when the Pangaean supercontinent first started to rupture and break up to form the modern continents and intervening modern oceans. It is envisaged that breakup then initiated draining of waters from the ancient continental seas into the newly opening modern oceans, plus expulsion of new waters from along the newly formed mid-ocean-ridge spreading zones.

Modern oceans on an increasing radius plate tectonic Earth initially opened within ruptured areas of continental crust, where the crust had failed to keep pace with increases in Earth surface area and surface curvature. On each of the small Earth models it is shown that these rupture zones were initially located within the present northwest Pacific and North Atlantic Ocean regions respectively. These rupture zones then progressively opened and rapidly extended in surface area throughout the Mesozoic and Cenozoic Eras, forming what are now the modern Pacific and Atlantic Oceans.

By late-Triassic to early-Jurassic times, rupture of the supercontinental crusts had also occurred in the north-polar Arctic Ocean and south-polar South Pacific and South Atlantic Ocean regions. These rupture zones commenced as passive margin extensional basins accompanied by extensive deposition of shallow marine sediments. By the late-Jurassic, opening of rupture zones then shifted from a polar to a more elongate meridional position centred on the North Atlantic and Pacific Ocean regions.

On-going rupture and subsequent rifting saw the preservation of early-Jurassic seafloor crust within the newly emerging North Pacific Ocean, followed by late-Jurassic seafloor crust in the North Atlantic and Indian Oceans. Similarly, the Southern Ocean commenced opening during the Paleocene Epoch.

Throughout the Cretaceous Period, the plate tectonic small Earth models show that each of these modern oceans continued to increase their surface areas and were accompanied by further breakup and rifting of the modern continents. Within these newly formed oceans, mid-ocean-ridge spreading axes were then initiated and continued to spread as either:

1. Asymmetric spreading patterns along the perimeters of the North and South Pacific Oceans.
2. Symmetric spreading patterns in the North and South Atlantic and Southern Ocean.
3. A combination of both asymmetric and symmetric-type spreading patterns in the Indian Ocean (Figure 5.1).

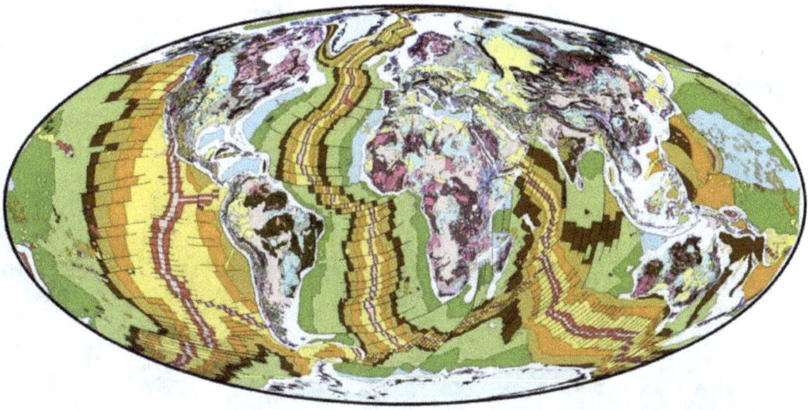

Figure 5.1. Geological Map of the World in Mollweide projection (digitized with permission from the Commission for The Geological Map of the World and UNESCO, 1990).

Not until the Mesozoic to Cenozoic transition did modern symmetrical, mid-oceanic-ridge spreading patterns become firmly established throughout all of the modern ocean basins—commencing with the brown-coloured stripe in Figure 5.1. This consistency of timing and sequence of ocean basin development agrees fully with the constraints imposed by the coloured seafloor spreading evidence.

From this brief overview of observations from the constructed plate tectonic small Earth models, the observed history of each of the modern oceans will now be looked at in more detail. In these descriptions it is emphasized that, when talking about movement, or migration of continents on an increasing radius plate tectonic Earth, this movement refers to an apparent migration of the continents as a direct result of opening and widening of the modern oceans. Movement of the continents simply accommodates for opening of the oceans and has nothing to do with continental drift or mantle convection theory.

5.1. Arctic Ocean

The Arctic Ocean originated as a very ancient marine sedimentary basin—a localized low-lying depression in the surface crusts—which first formed and commenced opening as the ancient Pangaea supercontinental crust started to rupture around 250 million years ago. This early Arctic Ocean basin was originally located in mid- to high-northern latitudes—shown as areas of white in Figure 5.2. Successive plate tectonic small Earth models show that over time, as the ocean continued to open, the basin and surrounding continents remained within the north polar region.

Figure 5.2. Arctic Ocean plate tectonic small Earth spreading history, extending from the present-day back to the early-Jurassic. Maxlow, 1995.

In Figure 5.2 the older Arctic Ocean, shown as white in the lower right models, is made up of a large expanse of marine sediments deposited around the margins of the ancient continents, plus two seafloor basins shown as green colours. These green coloured seafloor basins are known as the Amerasia and Eurasia Basins. The basins are Cretaceous in age and are shown to crosscut and displace the extensive white areas of marine sediments, which were first deposited along the newly emerging continental shelf margins prior to exposure and preservation of the volcanic seafloor crusts.

Conventional plate tectonic reconstructions of the Arctic Ocean show the region to be a northern extension of a theoretical ancient Panthalassa Ocean. Convergence of Siberia against Alaska is then considered to have closed off this basin, prior to re-opening again in conjunction with opening of the North Atlantic Ocean. This convergence of continents amounted to about 790 kilometres, across whole regions where no evidence of gross crustal shortening exists. Rowley and Lottes (1988) conclude that the area is, in reality, characterized by extension and strike-slip basins, with no evidence whatsoever of Mid-Mesozoic or younger structures associated with a theoretical continental collision.

On an increasing radius plate tectonic Earth, opening of the Arctic Ocean occurred as a result of crustal rupture and breakup of the ancient Pangaea supercontinent and was located between the newly formed North American and northern European continents. This breakup was then followed by on-going seafloor crustal stretching and opening of the Arctic Ocean within this region. The initially small Arctic Ocean basin progressively increased in surface area over time and its boundaries continued to extend further to the southeast during the Mesozoic Era.

The presence of exposed coloured seafloor crustal rocks within the Amerasia and Eurasia Basins shows that there was an initial period of seafloor spreading in each of these areas during the Cretaceous Period. The lack of any further coloured seafloor crusts suggests that spreading then effectively ceased during late-Cretaceous times. Today, there are no active spreading centres in the older Arctic Ocean regions. Spreading is now located within the adjoining Nansen-Gakkel mid-ocean-ridge, which is a northern extension of the North Atlantic Ocean mid-ocean spreading ridge.

From the late Cretaceous to the present-day, the North Atlantic mid-ocean-ridge progressively extended into the Arctic Ocean region, crosscutting pre-existing marine sedimentary basins in this area. This progression into the Arctic Ocean gave rise to further fragmentation of the ancient Canadian continental crust and opening of new seas located between Greenland, Canada, and Russia.

During this opening phase, sediments eroded from the lands were initially deposited within the early Arctic Ocean basin. The deposition of sediments then progressively changed from deposition within a shallow marine basin to deposition within marine continental shelf settings which then bordered a true deep-ocean-spreading centre.

Throughout much of this time, the present-day Alaskan and Siberian Peninsulas remained joined, forming an important land-bridge between the two continents and effectively isolating the Arctic Ocean from the opening Pacific Ocean. During that same time, continental crustal stretching between the two land masses was accompanied by translational faulting along each of the peninsulas during opening of the Arctic Ocean basins. The continental crust that now makes up Greenland and the Canadian Arctic Islands also fragmented, and these fragments began to gradually rift apart as the newly formed seas continued to open.

From 65 million years ago to the present-day, the Arctic Ocean basin was then characterized by an on-going phase of symmetric-style seafloor and continental crustal extension and opening. This phase resulted in further separation of the Canadian Arctic Islands, opening of Hudson Bay, and further rifting between Greenland and Canada. Separation of the Alaskan and Siberian Peninsulas across the Bering Strait occurred within the past 2 million years and crustal separation, rifting, and extension is continuing to the present-day.

5.2. Atlantic Ocean

The ability to match the east coast of South America with the west coast of Africa has long been recognized by mapmakers for many centuries. This remarkable fit-together now forms the conceptual basis for both early continental drift and conventional plate tectonic studies.

The closing of the Atlantic Ocean and reconstruction of these continents also forms the basis for assemblage of the Pangaea and Gondwana supercontinents. This fitting together of the South American and African continents is further substantiated by an extensive array of geological evidence dating back to the investigations of Wegener.

On conventional reconstructions of the Atlantic Ocean, the corresponding margins of northern Brazil in South America and Guinea in Africa are traditionally fitted together according to their geological matches. But fitting these coastlines together produces a narrow triangular gap which widens south between the continental margins of South America and Africa south of the Niger Delta region.

To minimize the misfit, the margin of South America may also be fitted against southern Africa south of the Niger Delta region. This misfit then produces a narrow triangular gap between the Guinea and north Brazil coastlines, widening northwards. Unfortunately, this misfit also produces a significantly greater area of misfit in the Florida and Central American regions.

It is significant to note that in 1958, when Carey first reassembled these continents on a spherical model representing the Earth's modern dimensions, Carey noted these very same misfits. He commented that "*...if all the continents were reassembled into a Pangaean configuration on a model representing the Earths modern dimensions, the fit was reasonably precise at the centre of the reassembly and along the common margins of north-west Africa and the United States east coast embayment but became progressively imperfect away from these areas.*" *Carey concluded from this research that the fit of these ancient continents* "*...could be made much more precise in these areas if the diameter of the Earth was smaller at the time of Pangaea.*"

Opening of the North and South Atlantic Oceans on an increasing radius plate tectonic Earth is shown sequentially in Figure 5.3. Opening is shown to have commenced in the lower right models during the early- to late-Jurassic in the North Atlantic region, located between North Africa and North America. This opening North Atlantic region later extended south to merge with the opening South Atlantic Ocean, as well as west into the Gulf of Mexico and Caribbean Sea regions.

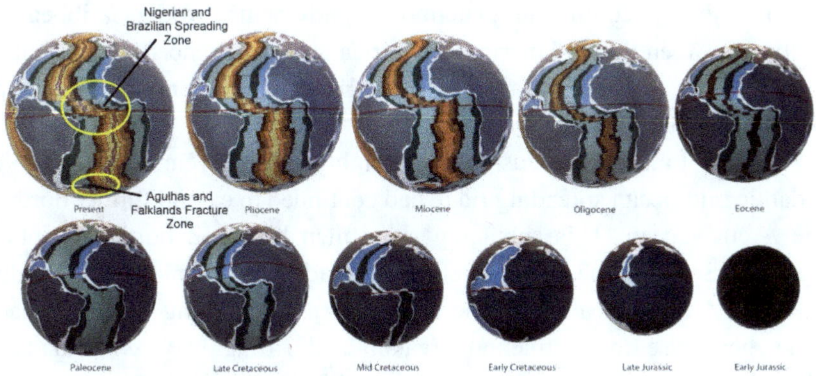

Figure 5.3. Atlantic Ocean plate tectonic small Earth sequential spreading history, extending from the present-day back to the early-Jurassic. Maxlow, 1995.

On small Earth models, it is significant to also note that, just as Carey concluded, when the North and South Atlantic Oceans are closed, misfitting between the continental margins of North and South America, Europe, and Africa are entirely eliminated. The opening of these oceans during the Mesozoic and Cenozoic Eras is then shown to be progressive and symmetrical. Subsequent opening of the Atlantic Ocean is also shown to have occurred in conjunction with opening of both the Arctic and Indian Oceans.

5.2.1. North Atlantic Ocean

The opening of the North Atlantic Ocean on an increasing radius plate tectonic Earth (Figure 5.3) commenced as a narrow rift basin—an elongate crack in the continental crusts filled with water—located between the east coast of North America and west coast of Africa. Over time, this rift basin continued to open and progressively extend north into the Arctic Ocean and south into the South Atlantic Ocean. From the early-Jurassic a small counterclockwise rotation of the combined South American and African supercontinent, relative to North America, accelerated opening of the North

Atlantic region and extended the ocean west into what is now the Caribbean Sea. This rotation occurred in sympathy with opening of the Pacific Ocean as each of the adjoining continents continued to adjust for changing Earth radius and surface curvature.

During the early-Cretaceous—around 130 million years ago—the newly formed North Atlantic spreading ridge extended north into the Grand Banks continental shelf, then located between Canada and Iberia. This occurred in response to both continental plate motion and opening of the Caribbean Sea. Rifting between South America and Africa also commenced, which initiated a progressive extension of the spreading ridge into the previously separated North and South Atlantic Oceans.

By the Late Cretaceous—around eighty million years ago—the North Atlantic mid-ocean spreading ridge had continued to extend further north into the Arctic Ocean. This spreading ridge then branched northwest into the Labrador Sea rift zone, located between Canada and Greenland, and northeast into the Mediterranean Sea. The extension and branching of this spreading ridge gave rise to continental breakup and rifting between Canada and Greenland, as well as rifting and rotation of Spain relative to both France and England.

From the late Cretaceous to the present-day, the North Atlantic mid-ocean-ridge continued to open in conjunction with both the Nansen-Gakkel spreading ridge within the Arctic Ocean and the South Atlantic spreading ridge. Spreading within each of these areas has resulted in a progressive enlargement of both the North and South Atlantic Oceans and was accompanied by symmetrical seafloor spreading.

5.2.2. South Atlantic Ocean

The South Atlantic Ocean (Figure 5.3) commenced opening between the southern South American peninsular and the South African region as a long narrow rift zone and has continued too steadily open to the present-day. The pattern of coloured seafloor stripes shows that this opening occurred in sympathy with opening of the Indian Ocean. Because of this opening, these two oceans can be considered as extensions of the same continental crustal rupture and opening event.

Separation between the west coast of Africa and east coast of South America commenced during late-Jurassic—around 155 million years ago. Opening occurred along the now separated Agulhas and Falklands fracture zones, presently located in the southern South Atlantic Ocean. During late-Jurassic times continental rifting and separation of the two continents progressively extended north. The South Atlantic spreading ridge then merged with the North Atlantic spreading ridge during the early-Cretaceous along the Nigerian and Brazilian rift zone. Since that time, the North and South Atlantic Oceans have remained united as a single Atlantic Ocean.

From the late-Cretaceous—around eighty million years ago—to the present-day spreading in both the North and South Atlantic Oceans has continued along a common mid-Atlantic spreading ridge. Subsequent spreading in both regions continued to be symmetrical, with a slow clockwise rotation of South America, relative to Africa, giving a slightly greater spreading rate in the southern South Atlantic Ocean.

5.3. Caribbean Sea

Conventional plate tectonic reconstructions of the Atlantic Ocean traditionally fit the Brazil and Guinean coastlines of South America and Africa together. This fit helps to minimize any misfit in the Caribbean Sea region. The Caribbean region is then seen as a buffer zone between the North American plate, the South American plate, and subducting oceanic plates of the Pacific Ocean. On these reconstructions the Caribbean region is a preserved piece of the ancient Pacific Ocean, referred to as the Farallon plate, which is inferred to have originated from outside of the Caribbean region.

The Caribbean Sea is made up of Mexico, Colombian, and Venezuelan Basins separated by the Antilles Arc (Figure 5.4). On an increasing radius plate tectonic Earth, the development and subsequent opening of the Caribbean Sea is intimately related to the continental plate motion histories of both South America and Africa, relative to North America. Bedrock geological mapping of the seafloor shows that opening of each of the Caribbean basins was most active during the Jurassic Period. This opening was then later reactivated along the Antilles Arc during the Paleocene—around sixty-five million years ago—and has continued to open as a relatively restricted basin to the present-day.

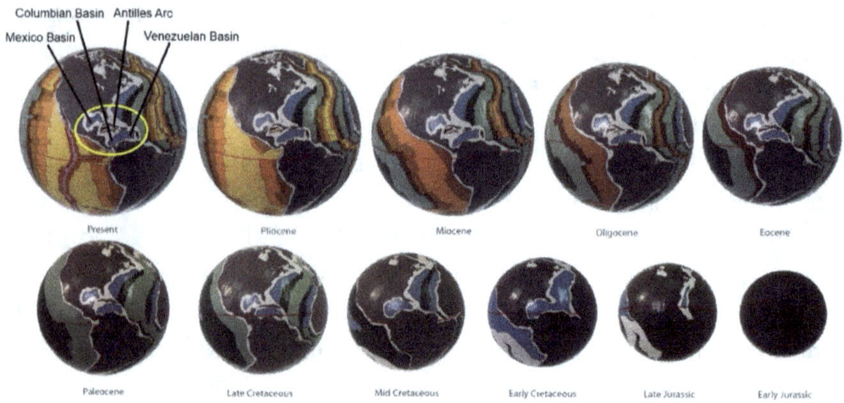

Figure 5.4. Caribbean Sea plate tectonic small Earth spreading history, extending from the present-day back to the early-Jurassic. Maxlow, 1995.

On an increasing radius plate tectonic Earth, early development of the Caribbean Sea was intimately associated with opening of the North Atlantic Ocean along with subsequent rifting of Africa away from the Americas. Opening of the Caribbean Sea commenced during the Triassic to early Jurassic Periods because of north-south crustal stretching and extension between the North and South American continents. Subsequent development of the Caribbean Sea was then closely related to a northwest migration of North America, relative to the still joined South American and African supercontinent. The Colombian and Venezuelan basins opened during the Jurassic to Cretaceous Periods and were later separated by the Antilles island-arc.

This early opening phase lasted until the early-Cretaceous—around 130 million years ago—when rifting and eventual breakup between South America and Africa first began. During this phase, the Nicaraguan and Panama Peninsulas remained joined to South America. After the separation of South America from Africa, South America then began to slowly rotate clockwise, relative to North America, in response to the opening of the South Atlantic Ocean. This rotation gave rise to further crustal extension and opening within the Caribbean basins, as well as isolation of the Antilles island-arc.

5.4. Indian Ocean

Reconstructions of the Indian Ocean and assemblage of the surrounding continents is crucial in understanding the assemblage, breakup, and dispersal history of the conventional plate tectonic Gondwana and Pangaea supercontinents. These conventional reconstructions place the precursor to the Indian continent within an ancient southern Gondwana supercontinent, located adjacent to Africa, Madagascar, and Antarctica. The Indian Ocean is then inferred to have formed during the subsequent breakup and dispersal of the Gondwana continental fragments during a later Pangaean supercontinental assemblage cycle.

On this conventional reconstruction, an inferred ancient Tethys Ocean was located between a southern Pangaea supercontinent, made up of fragments of the Gondwana supercontinent, and a northern Laurasian supercontinent. During breakup of these supercontinents, India is then inferred to have broken away from Gondwana to drift north as an island continent during subsequent opening of the Indian Ocean.

This northward migration of India during the Mesozoic to early Cenozoic Eras requires subduction of some 5,000 lineal kilometres of inferred pre-existing seafloor crust to close the ancient Tethys Ocean. India is then said to have collided with the Asian continent to form the Himalaya Mountains during the Cenozoic Era, leaving behind no trace of this pre-existing ocean crust.

In contrast, on an increasing radius plate tectonic Earth, small Earth modelling shows that the Indian continent remained firmly attached to the Asian continent throughout the Mesozoic and Cenozoic Earth history (Figure 5.5). The Indian Ocean is shown to have opened during a post-Triassic phase of crustal rupture of the Pangaea supercontinent, in sympathy with opening of the adjoining Atlantic Ocean basin.

Opening of the Indian Ocean on an increasing radius plate tectonic Earth (Figure 5.5) commenced during the early-Jurassic Period. This opening occurred because of initial crustal rupture within two separate areas, located adjacent to both India and Madagascar, in what are now the Somali Basin and the Bay of Bengal. Opening, and a southward extension of the Somali Basin continued up to the mid-Cretaceous Period which later became the Mozambique Channel, located between Madagascar and Africa. During that time the Indian, Madagascan, and Sri Lankan continents and islands continued to remain firmly attached to the ancient Asian continent.

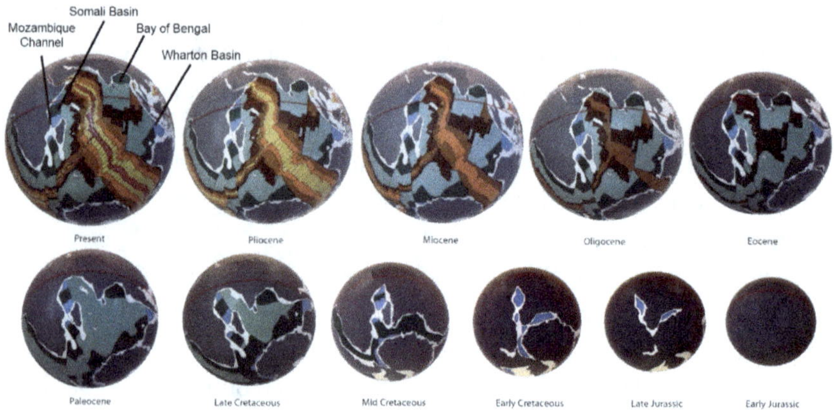

Figure 5.5. Indian Ocean plate tectonic small Earth spreading history, extending from the present-day back to the early-Jurassic. Maxlow, 1995.

Crustal breakup and opening between Africa, Madagascar, India, and Antarctica then progressively extended south during the early- to late-Jurassic. During that time the Somali Basin and Bay of Bengal areas merged south of Madagascar. The merging of these basins formed a triple-junction—the junction of three separate spreading ridges and crustal plates—south of Madagascar and this region now represents the birthplace of the modern Indian Ocean basin. It should be noted that triple-junctions present a mathematically impossible problem for plate movement which is unexplainable in continental drift-based plate tectonic theory but easily understood on an increasing radius plate tectonic Earth.

From the mid-Cretaceous to Paleocene times, seafloor spreading continued between Madagascar, Africa, and India forming the Arabian Sea. During that time the Indian Ocean underwent a rapid north-south and east-west opening between Australia, Antarctica, Africa, and Southeast Asia, with a rapid northward opening and displacement of the Indian and Asian continents relative to Antarctica.

During the Paleocene—65 million years ago—to the present-day, symmetrical seafloor spreading radiated from a new triple junction located within the central Indian Ocean. This new spreading crosscuts and displaces the older, previously established Mesozoic seafloor crust. Spreading then extended into the southern Indian Ocean, east between Australia and Antarctica and northeast into the Gulf of Aden.

In the southern Indian Ocean region, the southwest limb of this spreading ridge split across older spreading patterns and now forms an extension of the South Atlantic Ocean mid-ocean-ridge system. The north south Ninety East spreading ridge, located within the eastern Indian Ocean, was abandoned during the Eocene. A new spreading ridge then extended southeast from the central Indian Ocean triple junction into a newly opening rift zone located between the Australian and Antarctician continental plates. This rifting-initiated separation and northward migration of Australia relative to Antarctica, as well as initiation and relatively rapid opening of the Southern Ocean.

As spreading continued, the Gulf of Aden opened during the Miocene along a northwest extension of the central Indian Ocean and Carlsberg spreading ridges. A second triple junction has since formed at the western end of the Gulf of Aden during the Pliocene. This triple junction is now represented by the actively spreading Red Sea ridge and the East African Rift Valley system.

5.5. Pacific Ocean

At present the Pacific Ocean occupies nearly half the surface area of the Earth and can be arbitrarily subdivided into North Pacific, Central Pacific, and South Pacific Ocean regions. In all conventional Plate Tectonic reconstructions, the precursor to the Pacific Ocean was a much larger ancient Panthalassa Ocean. This largely hypothetical early-Mesozoic Panthalassa Ocean is inferred to have possessed an old seafloor crust, which was formed by spreading along ancient mid-ocean-ridge zones during the Palaeozoic Era extending into the Triassic Period.

The conventional Mesozoic and Cenozoic plate tectonic history is then depicted as an east-west and north-south contraction and reduction in surface area of the ancient Panthalassa Ocean in sympathy with opening of the Arctic, Indian, and Atlantic Oceans. This ancient ocean is thought to have eventually contracted in area to the size of the modern Pacific Ocean and is inferred to be still shrinking.

During that time, all pre-Mesozoic crust is inferred to have been subducted beneath the continents along subduction zones located around the margins of the Pacific Ocean basins. This subduction must have also included a substantial amount of new Mesozoic and Cenozoic seafloor crust, which is currently being generated within the East Pacific Ocean mid-ocean-ridge zone.

On all conventional plate tectonic reconstructions, the North American continent progressively overrides the eastern North Pacific Ocean region. This overriding occurs during a westward displacement of the North American continent during opening of the North Atlantic Ocean. The North Pacific Ocean spreading ridge is then inferred to have been entirely overridden and subsequently dislocated beneath the Pacific margin of the North American continent.

In contrast, on an increasing radius plate tectonic Earth an expansive pre-Mesozoic Panthalassa Ocean, and similarly an ancient Tethys Ocean, did not exist. Subduction of between 5,000 to 15,000 lineal kilometres of pre-existing East Pacific seafloor crust beneath the American continent is also not required.

In Figure 5.6 the Pacific Ocean is instead shown to originate during early-Jurassic times as two separate marine sedimentary basins. A North Pacific basin was located between northwest Australia, Canada, and China, and a South Pacific basin was located between east Australia, South America, New Zealand and Antarctica.

Both marine basins progressively opened to the south and north, along the west coasts of North and South America respectively. These basins then merged to form a single Pacific Ocean basin during the mid- to late-Jurassic Period. Remnants of this early basin history are now preserved as continental margin and marine plateau sediments within the Southeast Asian and Coral Sea regions—shown as white areas in Figures 5.6 and 5.7.

On an increasing radius plate tectonic Earth, the subsequent evolution of the North and South Pacific Oceans involved a period of rapid northeast to southwest crustal extension and opening between North America, South America, and Australia. By the late Jurassic, a deep ocean had extended southeast and south along the west coasts of North and South America. This opening coincided with the initiation, exposure, and preservation of new volcanic seafloor crust in the South Pacific Ocean.

Figure 5.6. North Pacific Ocean plate tectonic small Earth spreading history, extending from the present-day back to the early-Jurassic. Maxlow, 1995.

Figure 5.7. South Pacific Ocean plate tectonic small Earth spreading history, extending from the present-day back to the early-Jurassic. Maxlow, 1995.

Throughout the Mesozoic Era the North Pacific Ocean underwent a relatively rapid enlargement, with an asymmetric spreading axis extending southeast into the South Pacific region. This spreading and mid-ocean-ridge development curved along the west coasts of North and South America and extended west into the Coral Sea region during the Cretaceous.

During the mid- to late-Jurassic, crustal rifting and opening between New Zealand, New Caledonia, South America, Australia, and Antarctica isolated the Coral Sea plateau, as well as the Lord Howe Rise and New Zealand. This rifting phase then left New Zealand as an island continent, which has remained isolated from surrounding continents to the present-day.

Development of the Pacific Ocean on an increasing radius plate tectonic

Earth during the Cenozoic Era is characterized by the initiation and rapid development of symmetric-style seafloor spreading. This development commenced within the Tasman Sea region, located southeast of Australia, during the Paleocene and it continued to extend east towards South America during the Eocene. From there, symmetrical spreading continued north, forming the present East Pacific spreading ridge, and then extended along the west coast of North America to its present location adjacent to California.

The opening and formation of the Pacific Ocean on an increasing radius Earth differs from each of the other oceans. Because of the longer period involved in opening the Pacific Ocean, the large area of older seafloor crust in the North Pacific region has been subject to considerable crustal stretching and distortion because of changing surface curvature of the Earth.

This changing surface curvature was generally absorbed as extension within the thin seafloor crust, but also gave rise to complex plate interaction, island-arc volcanism, and jostling between and along adjoining plates, between the various seafloor and continental plate margins.

Between the North Pacific Ocean plate and the Australian, Southeast Asian, and Chinese plates, complex crustal interaction and movement formed the Southeast Asian island-arc-trench systems, which are now characteristic of the western Pacific region. On an increasing radius plate tectonic Earth, this region represents a complex interplay of plate motions that were generated during on-going adjustments to surface curvature, especially within the older North Pacific Ocean region.

This development of the Pacific Ocean on an increasing radius plate tectonic Earth cannot be reconciled with conventional plate tectonic reconstructions on a constant sized Earth. On an increasing radius plate tectonic Earth, the restrictions imposed by early Mesozoic Panthalassa and Tethys Oceans, as well as circum-Pacific subduction of pre-existing seafloor crust, simply did not exist and were not required during small Earth plate reconstruction.

Instead, a process of asymmetric to symmetric evolution of the East Pacific Ocean spreading ridge readily explains the complex crustal patterns shown by the Pacific Ocean seafloor mapping. Likewise, the island-arc-trench systems and extensional back-arc-basins are readily explained by crustal interaction along the plate margin boundaries during relief of surface curvature, especially in the older crustal regions.

5.6. Mediterranean Sea

The Mediterranean to Middle East region is a complex and contentious region on conventional plate tectonic reconstructions. To reconstruct the continents on a constant radius Earth, both Europe and Asia must be fragmented to close the Atlantic and Indian Oceans.

On all reconstructions, the Mediterranean region then formed the western apex of a large triangular-shaped ancient Tethys Ocean. From there, this ocean is inferred to have widened eastward towards an even larger ancient Panthalassa Ocean which separated the ancient supercontinents of Gondwana in the south and Laurasia in the north.

The evolution of the Mediterranean to Middle East regions on an increasing radius plate tectonic Earth (Figure 5.8) represents the remnants of a more extensive continental Tethys Sea—as distinct from a conventional Tethys Ocean. The Tethys Sea will be shown later to have had an extensive crustal and sedimentary basin history, extending back to the early Precambrian times.

On an increasing radius plate tectonic Earth, the Mediterranean region initially developed during pre-Triassic times because of crustal extension between the ancient African and European continents. Crustal rupture within the ancient continental Tethys Sea then commenced during the Jurassic, which initiated the opening of the modern Mediterranean Sea and crustal development within the Middle East region. The ancient Tethys Sea, in effect, represents the seabed of what is now the continental region of central Europe, the Middle East, and Asia.

The opening of the Caspian Sea followed during the early Cretaceous Period. The Black Sea then followed during the mid-to late Cretaceous and the Aral Sea during the Paleocene. These small seas all represent regions of continental crustal rupture and extension, which was initiated during complex clockwise crustal rotation occurring between the combined African Arabian plate and central Europe.

Figure 5.8. Mediterranean Sea small Earth spreading history, extending from the present-day back to the early-Jurassic. Maxlow, 1995.

Seafloor geological mapping shows that crustal extension within the Mediterranean to Middle East region was only active during Jurassic and early Cretaceous times. From mid-Cretaceous to Miocene times complex crustal motions, including some localized compression between Africa and Europe, formed the Alpine Mountain belts.

This event was followed by a continuation of continental crustal extension and renewed basin opening forming the young sedimentary basins presently located within much of central Europe.

The Alpine Mountain building event coincided with a northward extension of the North Atlantic Ocean spreading ridge which was marked by the opening of the Bay of Biscay, located between Spain and France, during the late-Cretaceous.

Similarly, opening of the Persian Gulf and Red Sea commenced during the Eocene and these are continuing to open to the present-day. Seafloor spreading was re-activated in the western Mediterranean region during the Miocene and was accompanied by continental rifting between West Africa and Spain.

On an increasing radius Earth, separation of the Gondwana and Laurasia supercontinents by a large Tethys Ocean, and similarly a north-south closure of the Tethys Ocean during the Mesozoic and Cenozoic Eras, is not required. Instead, opening of the Mediterranean to Middle East region occurred because of clockwise rotation and crustal extension between Europe and the Middle East, relative to the adjoining African Arabian continent.

This continental crustal motion, extending from the Iberian Peninsula to the Tibetan Plateau, also resulted in rotation of Italy and fragmentation of the Alpine Mountain belts during the mid-Cretaceous to present-day times.

The increasing radius plate tectonic small Earth reconstructions also present a straightforward development history for the Black Sea region, the mountain belts of the Balkans, Turkey, and the Caucasus, the development of the southern Russian platform north of the Black Sea and the development of the Aegean Sea during the late-Cenozoic.

5.7. Southeast Asian Seas

The Southeast Asian region comprises the Philippine, South China, Celebes, Banda, and Java Seas. In this region, the present-day Southeast Asia to New Guinea and Japanese Island chains represent complex island-arc systems.

Conventional plate tectonic reconstructions on a constant sized Earth are particularly vague when it comes to reconstructing Southeast Asia. On these reconstructions, the region is represented by small remnant crustal fragments, which are inferred to have existed at the eastern end of a large ancient Tethys Ocean, where it merged with an even larger Panthalassa Ocean.

During closure of both the Tethys and Panthalassa Oceans, the Southeast Asian Seas are then interpreted to represent marginal or back-arc basins. These basins were further separated from the various island-arcs by spreading along zones of inferred crustal subduction and deep trench development.

The development of the Southeast Asia basins and seas on an increasing radius plate tectonic Earth (Figure 5.9) is complex and progressive. Small Earth modelling studies show that development within this region was intimately associated with the formation and subsequent plate interaction of the North Pacific oceanic plate.

On small Earth plate tectonic models, the region is shown to represent the fragmented remains of ancient intercontinental marine and continental sedimentary basins, as well as associated island-arc volcanic activity.

The Southeast Asian region initially formed during early-Jurassic times. This region represents an area of marine basin opening, accompanied by deposition of sediments, and was located between the early Australian, North American, and Chinese continents.

These basins were first formed in conjunction with the initial rupturing and opening of the North Pacific Ocean. At that time, sediments eroded from the surrounding lands were redeposited within newly formed marine basins.

These sediments were mixed with volcanic rocks, erupted along the precursors to the modern mid-ocean-ridge spreading zones, to form the early island-arc systems. Remnant sediments from this early opening event are now preserved as both continental shelf and marine plateau sediments, which are exposed as the Southeast Asian islands and submerged plateau throughout the Southeast Asian Sea region.

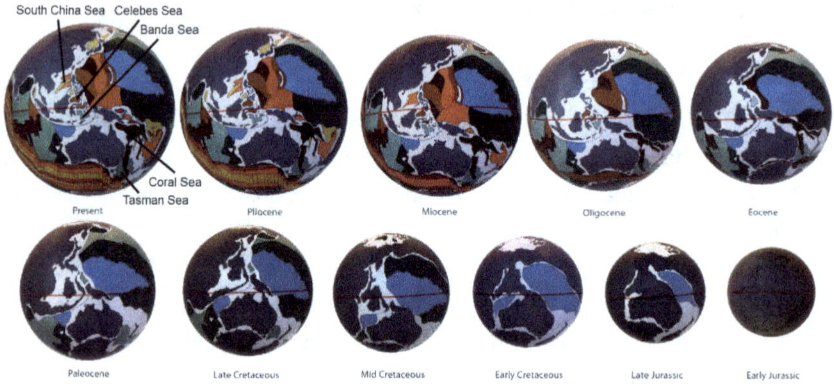

Figure 5.9. Southeast Asian and Southwest Pacific Basin small Earth spreading history, extending from the present-day back to the early-Jurassic. Maxlow, 1995.

Development of the Southeast Asia region was further complicated by the plate motion history of Australia, relative to Asia, during opening of both the Pacific and Indian Oceans. Progressive crustal extension and opening between each of these continents occurred during late-Cretaceous to Pliocene times which then resulted in opening of the South China, Celebes, and Banda Seas, as well as fragmentation of the early Southeast Asian island-arc system.

The crustal fragments making up the Southeast Asian region have since undergone complex clockwise rotation, crustal fragmentation, plate interaction, and on-going island-arc volcanism. These events all occurred during an extended period of southeast to northwest crustal extension between the Asian and Australian continental plates.

This interpretation of the Southeast Asia region contrasts strongly with the conventional plate tectonic requirement for continental collision, the closure of pre-existing Tethys and Panthalassa Oceans, and complex subduction of the Australian and North Pacific plates beneath the Asian and Philippine plates.

5.8. Southwest Pacific Ocean

The Southwest Pacific Ocean region is also structurally complex and comprises the Coral and Tasman Seas. This region is made up of remnants of ancient seafloor crust and marine plateau sediments. These sediments were initially deposited and later fragmented during Cretaceous to Paleocene times and again during Miocene to present-day times. The region has been further complicated by crustal plate motion and plate interaction occurring along the margins of the Indo-Australian and Pacific plates.

The southwest Pacific Ocean region is shown on plate tectonic small Earth models (Figure 5.9) to represent the fragmented remains of earlier marine basins and sediments, deposited within an early pre-Triassic South Pacific Ocean basin. Initially, this area opened as a passive marine basin during pre-Triassic times. The newly formed basin then formed part of the early South Pacific Ocean, prior to merging with the North Pacific Ocean during the Triassic Period.

During pre-Triassic times, West Antarctica and New Zealand were assembled adjacent to Australia and South America. Subsequent rifting and opening between South America, New Zealand, and West Antarctica initiated formation of the South Pacific Ocean. During Triassic to early Jurassic times, New Zealand and New Caledonia were separated from Australia during opening of the Tasman Sea. These remained attached to Ecuador in Central America until final separation from South America during the mid-Jurassic times.

During the mid- to late-Jurassic an early New Zealand and New Caledonian continent comprising the Coral Sea plateau, Lord Howe Rise, and New Zealand island-arc complex were well established. This ancient continent then fragmented further and was partly submerged to form the present Southwest Pacific region.

Asymmetric seafloor spreading first developed in the South Pacific Ocean and this spread extended west during the Cretaceous into the Tasman Sea. This event further isolated the Lord Howe Rise and New Zealand continent from Antarctica and Australia.

On an increasing radius plate tectonic Earth, the Tongan to South Solomon and New Hebrides trench and island-arc systems represent complex zones of plate interaction between the New Zealand and New Caledonia continental plate and the South Pacific Ocean plate margins. This complex interaction developed during on-going changes in Earth surface area and surface curvature over time.

The interaction of these plates resulted in a slow clockwise rotation of the southwest Pacific Ocean region along each of the established trench and arc systems. In addition, the New Zealand and New Caledonian plate was further fragmented and displaced during the Cenozoic Era.

The rapid development of symmetric-style seafloor spreading in the South Pacific Ocean resulted in further opening of the southwest Pacific Ocean region during the Cenozoic. Complex plate interaction and motion along the Kermadec and Tongan trench and arc systems again accompanied this opening during the early-Miocene. This crustal motion was also related to movement along the Alpine Fault system of New Zealand, which continues to the present-day.

On an increasing radius plate tectonic Earth, the southwest Pacific Ocean basin region cannot be reconciled with conventional plate tectonic reconstructions. Instead, this region represents a complex interplay of extensional crustal motion and opening between Australia, North America, South America, and Antarctica in conjunction with opening of the North and South Pacific Oceans during on-going changes to Earth radius and surface curvature.

5.9. Southern Ocean

Opening of the Southern Ocean is a paradox on conventional plate tectonic reconstructions and little mention of this ocean is made in the literature. This paradox arises because there are no subduction zones available to absorb the extensive plate motion required to open this ocean or to explain the northward migration of all the northern continents.

The Southern Ocean is located between the present Australian and East Antarctician continents. The ocean arbitrarily merges with the South Pacific and Indian Oceans to the east and west respectively. On an increasing radius plate tectonic Earth rifting and opening of the Southern Ocean first commenced in conjunction with opening of the Atlantic and Indian Oceans during the late-Jurassic (Figure 5.10). Further opening then initiated final separation between Australia and Antarctica during the Paleocene, sixty-five million years ago.

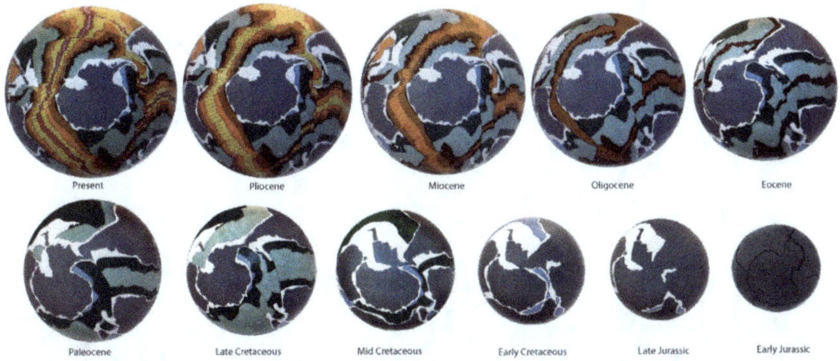

Figure 5.10. Southern Ocean small Earth spreading history, extending from the present-day back to the early-Jurassic. SOURCE: Maxlow, 1995.

Symmetric-style seafloor spreading in the Southern Ocean has since extended west, in conjunction with the eastern arm of the central Indian Ocean spreading ridge, and east to form an extension of the East Pacific spreading ridge.

On plate tectonic small Earth reconstructions, opening of the circum-polar Southern Ocean forms part of the extensive global network of mid-ocean spreading ridges. This opening and northward migration of the southern continents represents a natural consequence of breakup of the Pangaea supercontinent and increase in surface areas of the Southern and adjoining ocean basins during an increase in Earth radius and surface area.

Chapter 6

Pre-Jurassic Earth

"*There is nothing more contentious in global tectonics at this time than the expanding Earth concept.*" Owen, 1992.

When Vogel constructed his small Earth models during the 1980s, he suggested that it was "...theoretically possible for the continents, without shelves, to fit together at approximately 40 percent of the present Earth radius by considering that continental shelves were formed after the continental crust had fragmented." What he meant by this statement is that the continental shelves are where sediments, originally eroded from the lands after breakup of Pangaea, were deposited.

By moving back in time, it would then be necessary to reverse this geological process and return the sediments back to the lands from where they were previously eroded. By returning the sediments to the lands, it is inferred that any exposed volcanic seafloor crust beneath the sediments could, just as was done with the younger small Earth models, also be returned to the mantle. From this process, it is then conceivable to consider that the ancient Earth could be further reduced in surface area and hence radius.

In addition to returning the seafloor volcanic lava and marginal sediments back to their places of origin, it is also conceivable to consider that the bulk of the ocean waters and much of the atmospheric gases must be returned to the mantle. Black smokers discharging hot water and gases from along the seafloor mid-ocean-ridges, as well as volcanic eruptions, are modern-day examples of this new water and gas discharge process in action and by reversing these, and any other discharge processes, back in time, the water and gases must be returned to the mantle from where they came from.

Before modelling continental crust on an increasing radius plate tectonic Earth, two important considerations must be discussed. These include firstly; what the remaining continental crusts are made of, and secondly; an understanding of the mechanics of how these crusts would potentially behave during any subsequent change in Earth radius?

6.1. Continental Crust

In contrast to the simple seafloor crusts, continental crust is made up of a diverse range of present-day to ancient rocks dating to the earliest Archaean times. These rocks include ancient granite and volcanic rocks, deformed and physically altered sediments eroded from the more ancient lands, intrusive and extrusive magmatic rocks, as well as multiple layers of overlying younger sediments deposited in past low-lying regions.

These younger rocks often cover vast areas of older rocks. Studies elsewhere also show that the average composition of the continental crust is that of granite. That is, rocks rich in silica and aluminum—often referred to as SiAl in geology—in the form of quartz and feldspar minerals. This composition contrasts with the seafloor crust which has an average composition of basalt—a lava rich in iron and magnesium—FeMg.

As previously discussed, the continental and seafloor crusts displayed on the Geological Map of the World (reproduced in Mollweide projection in Figure 6.1) represent time-based bedrock geology. This geology is not specific to any one rock type, but displays all rocks according to their ages, which are broadly correlated with the global geological history.

The continental geology shown in this figure is further complicated by subsequent deposition of layers of young sedimentary and volcanic rocks, which may cover and overprint the older crustal rocks lying beneath them. These, in turn, are complicated still further by periods of metamorphism, folding, faulting, weathering, and erosion that may have occurred intermittently throughout Earth history.

The Earth's continental crustal geology is far more complex than depicted in Figure 6.1, and well beyond the needs of this explanation. Like the coloured seafloor crustal patterns, this map simply shows the ages of the various continental crustal rocks, along with the preserved tectonic history. Little of this geological information was available to early researchers during the 1950s and 1960s when continental drift, plate tectonic, and Earth Expansion theories were being developed, in particular the completed global distribution of rocks shown in this map.

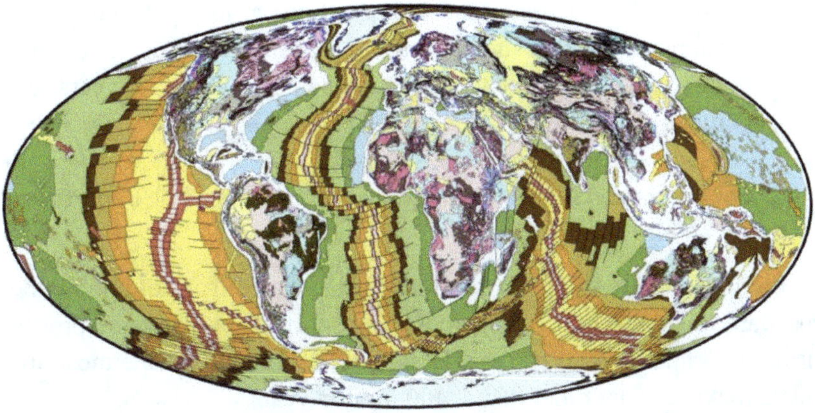

Figure 6.1. Geological Map of the World showing time-based bedrock geology reproduced in Mollweide projection. After CMGW & UNESCO, 1990.

At the global scale in Figure 6.1, the continental rocks can be visualized as a mosaic of continental crustal domains, which are made up of three dominant crustal types. In geology, these crustal types are referred to as cratons, orogens, and basins. These terms vary in their usage and definition throughout the world and within the published geological literature.

To avoid confusion, these terms are defined as:

1. A **craton** is defined as a part of the Earth's crust that has attained relative crustal stability, and the rocks have been little deformed for a prolonged period of time. By definition, cratons must have reached crustal stability by about 2,400 million years ago (the end of the Archaean Eon) and since then have undergone little deformation compared to adjacent parts of the crust.
2. An **orogen** refers to a belt of rocks characterized by regional folding, metamorphism, and intrusion of magmatic rocks. The rocks of an orogen can include deformed, eroded, and reworked parts of older, early formed cratons, as well as volcanic and sedimentary rocks. A distinct tectonic phase of Earth movement, over a relatively short period of time, first establishes an orogen. It is also possible for an orogen to become re-activated during subsequent tectonic events, and the belt normally remains as a permanent zone of relative weakness within the Earth's crust.
3. A **basin** refers to an area that is underlain by a substantial thickness of sedimentary rocks. These rocks possess unifying characteristics of

both sediment type and deformation history. Within a basin, sediments are deposited during a regionally restricted period of time, often extending over tens to hundreds of millions of years, during crustal depression or a related sequence of such events. The term basin is usually synonymous with the term sedimentary basin, and it represents a regional topographical down-warp of the Earth's surface, generally filled with water.

From these definitions, it should be appreciated that not all continental crusts are the same, and not all rocks were formed by the same process or during the same interval of time. Cratons, for instance, are the most ancient crustal rocks, and are older than 2,400 million years.

It is conceivable to imagine that an orogen may have started out as an ancient sedimentary basin but later may have been deformed and metamorphosed and perhaps intruded by molten igneous rocks to form an orogen at any time during the subsequent Earth history. If older than 2,400 million years, these rocks may then be stable enough to be considered a craton.

During plate tectonic small Earth modelling of the continental crusts, the physical characteristics of each of these distinct types of crust, as well as their distribution and tectonic histories, must be recognized and strictly adhered to. During model construction, the general shape, configuration and integrity of each of these crustal domains must also be retained throughout Earth history, in strict accordance with their known distributions and ages.

6.2. Crustal Extension

Crustal extension refers to the ability of crustal rocks to stretch and thin under an applied extensional force. On an increasing radius plate tectonic Earth, it is considered that this extension process is eventually followed by crustal rupture and complete failure once the ability of the crust to stretch and extend is exceeded. This applied extensional force is a direct result of an increase in Earth surface area and change in surface curvature during an increase in Earth mass, volume, and radius.

Researchers elsewhere have shown that the strength of the Earth's thin outer crust is controlled mainly by the ability of the crust to deform, by a process referred to as plastic deformation. Plastic deformation refers to the ability of solids to slowly change shape over extended periods of time when they are under stress. This plastic deformation process is dependent on the prevailing temperature gradient existing within the crust and mantle, the composition of the rocks making up the crust, and on how much geological time is available to perform the necessary crustal deformation.

These observations were made by studying heat-flow data from both active and inactive seismic regions of the Earth. Here, rifts—zones of continental crustal extension—were shown to be a common feature of the continents. Within these rift zones, volcanism and seismic activity were found to commonly precede crustal rifting by several million years. This activity then supported the idea that the crust is heated and weakened prior to rifting and extension.

On an increasing radius plate tectonic Earth, the microscopic amount of crustal extension occurring each year throughout Palaeozoic and Precambrian times progressively accumulates too many tens to hundreds of kilometres of crustal stretching over billions of years of geological time.

As will be shown, throughout Precambrian times, it is estimated that the average increase in circumference of the Earth amounted to much less than one millimetre per year, which was distributed around the entire circumference of the ancient Earth. Under these conditions most rocks making up the Earth's continental crust would have had plenty of time to slowly stretch and deform by atomic scale relaxation, plastic flow, or crystal regrowth. If not, local areas of crust may have gradually accumulated excess stress and subsequently fractured during periods of earthquake activity.

In this context, the ability of the crust to both stretch during increases in Earth surface area, and deform during changes in surface curvature, depends on the amount of geological time available, as well as the softening effect that the warmth from the deeper crust and mantle provides. The various rock types that make up the continental crusts then determine how the crust will extend, and how much time is available determines how successfully they will deform.

In the context of an increasing radius plate tectonic Earth, volumetric growth of the Earths mass occurs deep within the lower-mantle to upper-core regions—to be introduced and discussed in Chapter 11. It will be suggested that volumetric growth is transferred to the outer crust as crustal extension which, on ancient Earth, was localized within a global network of crustal weakness coincidence with a network of sedimentary basins. Within these zones of weakness, crustal extension was also accompanied by high heat flow, which coincided with the accumulation of sediments within each of the sedimentary basins.

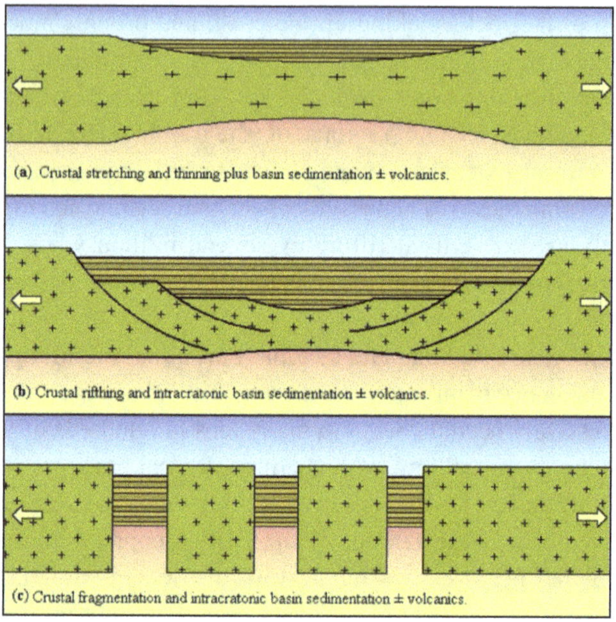

Figure 6.2. Primary mechanisms for crustal extension during increase in Earth radius. Figure a) Simple crustal extension; b) crustal rifting; c) crustal fragmentation. In each figure continental crust is shown as green, the mantle as pink shading and basin sediments are shown as horizontal stripes. Maxlow, 2001.

The main processes and outcomes that affect crustal extension on an increasing radius plate tectonic Earth during progressive increase in Earth surface area are schematically summarized in Figure 6.2. These include:

1. **Simple crustal extension:** This process occurs as extension, thinning and localized down warping of the continental crust to form shallow sedimentary basins (Figure 6.2a). These basins are subsequently filled with sediment eroded from the surrounding elevated land surfaces. Continuation of this process eventually leads to crustal failure and may also be accompanied by localized crustal fragmentation at the point of rupture.
2. **Crustal rifting:** This process occurs mainly within the continental crustal regions as the crust starts to fragment and separate during tectonic activity (Figure 6.2b). The process may form horst and graben-related rift structures, with curved, listric faulting occurring along the margins of the rift zones. On the continents, this process may then form a rift basin, which is in turn filled with sediment

eroded from the surrounding elevated land surfaces plus intrusion of new volcanic lava. The East African rift zone and the Red Sea are good modern examples.

3. **Simple brittle fragmentation:** This process is a simple version of crustal rifting and may be more applicable to the ancient Precambrian granite crusts (Figure 6.2c). Here, the thick, brittle crust simply fractures, fragments, and separates and may again be accompanied by deposition of basin sediments as well as accumulation of intrusive and extrusive volcanic lava within any newly formed basins.

Past objections to crustal extension mechanisms revolve around a long-standing conception in geology that if we were to dig or drill deep enough into the Earth's crust, we would eventually intersect much older bedrock. In Figure 6.2a the ages of basement rocks are as old as the surrounding crustal rocks that are being stretched. By digging or drilling through the overlying younger sediments, the basin would then be founded on a much more extensive area of ancient rocks than what is shown by the original surrounding rocks outcropping at surface.

Similarly, for Figure 6.2b and Figure 6.2c the basement rocks would be as old as either the remnant ancient rocks, or as old as the mantle rocks exposed at the time of crustal breakup. By moving back in time, the exposed surface crusts are simply moved back to where they originated from, by undoing the extensional geological processes that had acted on the crusts during this protracted interval of time until such time as the basin or rift zone closes.

On an increasing radius plate tectonic Earth, each of these examples of crustal thinning and rifting may be accompanied by elevated heat flow originating from the mantle. This heat flow may also be accompanied by intrusion of mantle-derived igneous rocks in the form of granites or intrusive dykes or by volcanic eruptions at surface to form lava flows.

Vast quantities of hot fluids and gases may also be involved, which then accumulate in low-lying areas to form the ancient seas or rise to form the ancient atmosphere over time. This elevated heat flow also represents an important mechanism for the observed geological processes of metasomatism: the alteration of rocks by hot fluids; metamorphism: the alteration of rocks by heat and pressure; and granitization: the alteration of rocks by extremes in heat or pressure to form granites.

6.3. Establishing Potential Pre-Jurassic Earth Radii

Several scenarios for establishing potential pre-Jurassic ancient Earth radii from measured areas of continental crusts are graphically shown in Figure 6.3. These scenarios range across a spectrum of possibilities, ranging from adoption of the present total continental surface area (Curve A), through to an estimate of the total surface area of preserved Archaean crusts (Curve D). Additional constraints also include the approximate relationship between sedimentary rock age and amount of outcrop area on the present land surface (Curve C), plus an intermediary curve determined by removing areas of young sedimentary rocks from around the continental margins (Curve B).

While each of these scenarios has some merit in constraining potential pre-Jurassic Earth radii, none of these can substantiate, with any degree of confidence or certainty, what a primordial Archaean Earth radius may have been. The estimates for a primordial Earth radius shown in this figure vary from between 1,500 kilometres to 3,500 kilometres radius for the early-Precambrian.

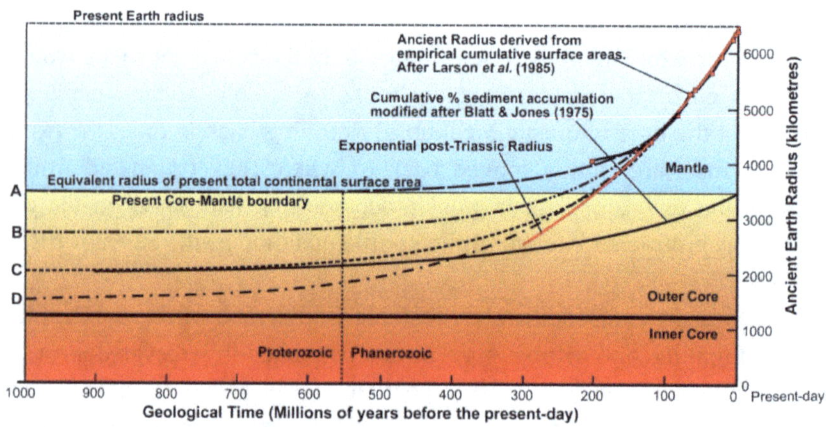

Figure 6.3. Constraints to determining a primordial Precambrian Earth radius. Curves depict Earth radius constrained by Curve A: the present total continental surface area; Curve B: present total continental surface area, less marginal basins; Curve C: published estimates of cumulative sediment areas and Curve D: total estimated area of Archaean crust. Maxlow, 2001.

The method adopted here to determine a potential primordial Earth radius is basically very simple. During pre-Jurassic model construction, by moving back in time, the radius of each preceding model will be progressively reduced in small, incremental stages. An equivalent area of the youngest sedimentary and magmatic rocks remaining will then be removed from the primary sinusoidal base map. The established global network of sedimentary basins will also be progressively reduced in surface area until only the most ancient Precambrian continental cratonic and orogenic rocks remain.

This method is justifiable because younger crustal rocks represent sedimentary and magmatic rocks that were deposited, intruded, or extruded after the older cratonic or orogenic crustal rocks were first formed. During progressive removal of the younger rocks, each of the sedimentary basins and rift zones are then restored to a pre-stretching or pre-rift configuration, and all sediments and magmatic rocks are simply returned to the exposed ancient land surfaces or mantle where they originated from.

By using this subjective construction method, an Archaean plate tectonic small Earth model with a primordial Earth radius of about 1,700 kilometres can be constructed—to be discussed in detail later. This primordial plate tectonic small Earth is made up of an assemblage of the most ancient pink and red coloured Archaean cratonic fragments, plus remnant khaki coloured early Proterozoic orogenic rocks (Figure 6.1). This 1,700-kilometre Archaean Earth radius then represents an approximate limiting radius for the most ancient primordial plate tectonic Earth (Figure 6.4).

The ancient radius of the primordial Earth can then be used to establish a mathematical formula to calculate a potential Earth radius at any moment back or forward in geological time. From this primordial plate tectonic small Earth radius, mathematical modelling studies of Archaean to Triassic continental crusts, plus previously modelled early-Jurassic to present-day Earth seafloor crusts, show that the change in potential Earth radius increases in accordance with an exponential rate of increase in Earth surface area (Figure 6.4).

Derivation of the ancient Earth radius formula is given in Appendix A and is expressed as:

$$\mathbf{R_a = (R_0 - R_p)\, e^{kt} + R_p}$$

Where R_a = the ancient Earth radius at time t, R_0 = the present-day mean Earth radius, R_p = the primordial Earth radius = 1,700 kilometres, e = base of natural logarithm, k = a constant = 4.5366×10^{-9}/year.

In Figure 6.4 the location of constructed Archaean to present-day plate tectonic small Earth models used to constrain this graph are shown as red dots and squares, along with historical estimates of ancient Earth radius by others based on both geophysical methods and visual reconstructions of the continents.

The radius curves of Koziar and Vogel are shown to coincide precisely with the estimates of ancient Earth radius established here to about two hundred million years ago. Establishing these historical curves was limited to early versions of seafloor mapping data available at the time and the authors used estimates of pre-Jurassic continental crustal areas only, without considering the potential for continental crustal extension.

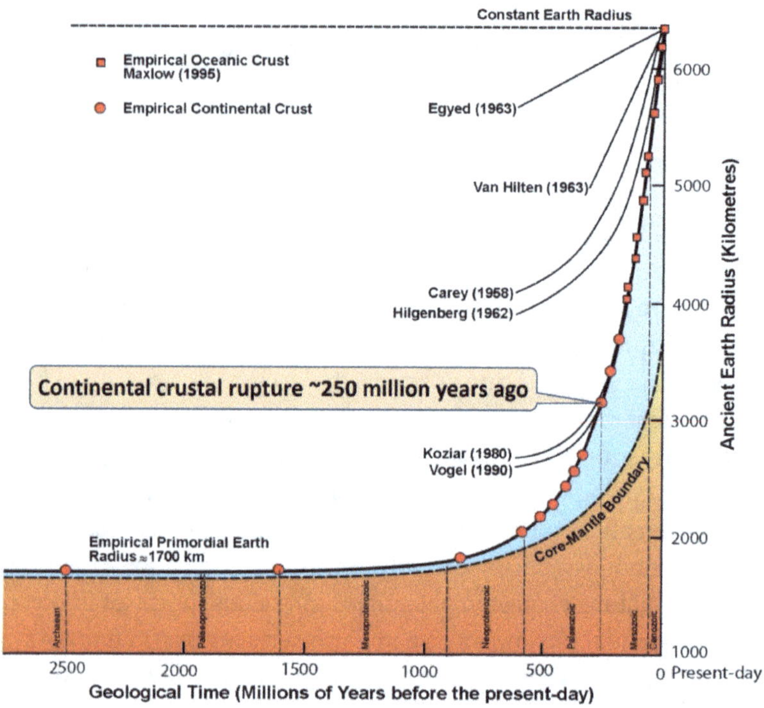

Figure 6.4. Exponential increase in Earth radius extending from the Archaean to present-day. The graph shows post-Triassic increase in radius derived from seafloor mapping and pre-Jurassic change in radius derived from an Archaean primordial Earth radius of 1,700 km. Plate tectonic small Earth models constructed are shown as red coloured circles and squares. An inferred core-mantle boundary remains speculative. Maxlow, 2001.

This graph suggests that during the early-Archaean to mid-Proterozoic times—ranging from about 4,000 to 1,600 million years ago—the Earth's ancient radius remained static, increasing in radius by approximately sixty kilometres during the entire 2,400 million years of ancient Earth history. From the mid-Proterozoic—about 1,600 million years ago—the Earth then underwent a steady to accelerating increase in radius to the present-day. This accelerating increase is now reflected as breakup of the Pangaean supercontinental crust during the late-Permian, the subsequent opening of the modern oceans, and the recent development of modern seafloor crusts.

The graph also suggests that, during Precambrian times, an increase in both radius and surface area of the Earth may have been accommodated for by molecular-scale redistribution and extension of the ancient crusts. This extension is evidenced in the geological rock-record by linear rock fabrics often seen in the most ancient rocks exposed at surface today—such as schist, gneiss and foliated granitoid rocks.

It is suggested that the ability of continental crust to absorb this molecular-scale crustal extension then began to rapidly decline during the Palaeozoic Era, resulting in an overall weakening and subsequent thinning of the crusts, until the ability to extend was finally exceeded during late-Permian to Triassic times.

During these latter times, the continental crust began to fail, rupture, and break apart to form the modern continents and open to form the modern oceans. By removing the inherent strength of the overlying continental crust, the Earth then underwent a rapid to accelerating increase in Earth radius and surface area to the present-day and, by inference, this process will continue well into the future.

6.4. Changing Earth Parameters

Based on this ancient Earth radius formula, the calculated changes in annual rate of change in Earth radius, surface area, circumference, and volume throughout the ancient and future geological times are graphically displayed in Figure 6.5. The calculated values used to generate this graph show that, prior to about 2,200 million years ago, the annual rate of increase in Earth radius was less than one micron per year. By about 1,200 million years ago this annual rate had increased to approximately 100 microns per year.

By the beginning of the Palaeozoic Era—around 540 million years ago, change in radius had then increased to 1.5 millimetres per year and by the late-Permian Period—250 million years ago—the annual rate of increase was about 8.5 millimetres per year. The current rate of increase in Earth radius, as measured from surface areas of seafloor geology, is calculated to be 22 millimetres per year.

These figures can be compared to historical estimates of annual rates of change in Earth radius obtained by various geological methods and published by the authors listed in Table 6.1. These results are all very similar and the mean of each of these published values is 21.9 millimetres per year, essentially the same as the twenty-two millimetres per year as measured here from seafloor mapping.

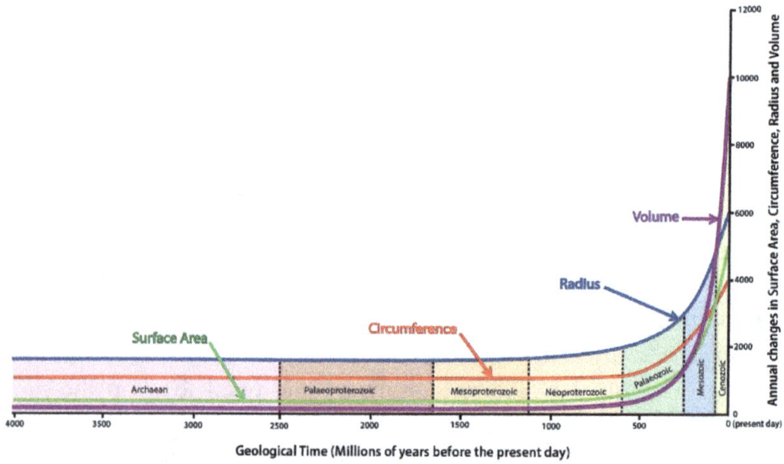

Figure 6.5. The annual changes in Earth radius, circumference, surface area and volume over time derived using the established Earth radius formula. Radius is in kilometres, Circumference is times 10 kilometres, Surface Area is times 100 million square kilometres, and Volume is times 100 million cubic kilometres. Maxlow, 2005.

As well as radius, the present annual rate of increase in circumference, surface area, and volume can also be determined using the established Earth radius formula. The present-day annual rates of increase in these variables are calculated as:

1. Radius annual change = 22 mm/year
2. Circumference annual change = 140 mm/year
3. Surface Area annual change = 3·5 square km/year
4. Volume annual change = 11,000 cubic km/year

Table 6.1. Published historical estimates of the present-day rate of increase in Earth radius. Koziar, pers. comm.

Author	Year	Rate [cm/yr]	Method
Koziar	1980	2.59	Increase in the Earth's surface area (Phanerozoic)
Blinov[1]	1983	1.99	Present annual increase in the surface area of oceanic lithosphere
Blinov[2]	1983	>1.91	Present annual increase in the Earth's circumference
Osipishin & Blinov	1987	1.96	Increase in the Earth's surface area (Meso-Cainozoic)
Koziar[3]	1996	2.7	Present annual increase in the Earth's circumference
Maxlow	2002	2.2	Increase in the Earth's surface area (from the Archean)
Koziar	2011	>2.0	ratio of the lengths of Atlantic Ridge and its African parent margin

[1] correct interpretation of the result obtained by Steiner (1977)
[2] correct interpretation of the result obtained by Kulon (1973)
[3] correct interpretation of the result obtained by Le Pichon (1968)

These present-day annual rates of increase represent the potential actual measured rates that are occurring now, and these rates are continuing to increase exponentially with time. These values compare favourably with estimates derived from published data, which were based on spreading values from early versions of the seafloor mapping program.

While an increase in seafloor spreading and surface area was acknowledged by many of these previous authors, it was presumed that this increase was then balanced by removal of excess crust by subduction to maintain a constant Earth radius.

For example, in 1977 Steiner estimated a global rate of increase in seafloor spreading during the past five million years of 3·19 square kilometres per year. This equates to a calculated rate of increase in radius of 20 millimetres per year. Also, in 1975 Garfunkel calculated an increase in seafloor spreading of 3·15 square kilometres per year and in 1982 Parsons calculated 3·45 square kilometres per year, which equates to rates of increase in Earth radius of 20 and 23 millimetres per year respectively. These figures are further substantiated by early space-based geodetic measurements of intercontinental distances between various observation sites. The rate of increase in Earth radius was calculated by Parkinson (in Carey, 1988) to be 24±8 millimetres per year and a published global mean value of 18 millimetres per year was subsequently calculated from space-based data in 1993 by Robaudo and Harrison. The rate of increase in circumference of the Earth was calculated by Blinov in 1983 to be 120 millimetres per year by considering the relative plate motions in the Pacific, Atlantic, and Southern Oceans. This equates to an annual increase in Earth radius of 19·1 millimetres per year.

The mean value of each of these published surface areas and circumference estimates is then 21.2 millimetres per year, which again compares favourably with the value of 22 millimetres per year measured here from surface areas of modern published seafloor mapping.

6.5. Mass, Density and Surface Gravity

Investigating potential rates of change in Earth mass, density, and surface gravity remains inconclusive because constraints on these variables are unknown. However, it is important to have some understanding of how these variables may have behaved over time. Both mass and density are mathematically bound and are calculated using the standard formula:

1. Mass = Density divided by Volume.

Similarly, surface gravity is calculated using the standard formula:

2. Surface gravity = the product G·M divided by Earth radius squared

Where G is the value for universal gravitation and is assumed to be constant, and M is the mass of the Earth.

In the mass and density equation, volume can be determined using the ancient Earth radius formula introduced above. Because ancient mass and density are both unknown, three scenarios exist for an expression representing potential changes in Earth mass and density, and hence surface gravity over time. These include:

1. Mass remains constant, requiring density to decrease as volume increases exponentially throughout geological time.
2. Density remains constant, requiring mass to increase as volume increases exponentially throughout geological time.
3. Both mass and density are variable as volume increases throughout geological time.

At this stage, the mass and density variables are indeterminate, and can only be investigated by constraining either mass or density to constant values. Potential changes in these mass and density variables range from a constant Earth mass with a decreasing density scenario, to a constant density with an increasing Earth mass scenario. Between these two end points a whole spectrum of possibilities exists, where both mass and density may be variable at the same time.

Constraints can be envisaged by studying other planets within the Solar System. These planets can be divided into inner rocky planets, with densities ranging from 3.3 to 5.5 grams per cubic centimetre, and the giant gaseous planets, with densities ranging from 0.7 to 1.7 grams per cubic centimetre. While each of these planets is different, they do show a marked decline in density values between the rocky planets and the giant planets. These limits to the range of densities may then potentially place constraints on ancient Earth values.

To determine changes to surface gravity it must also be assumed, for the purpose of this exercise, that the value for universal gravitation "G" has been constant or near constant throughout Earth's history. This may not necessarily be so, but any rate of change in this value is potentially small and may also be indeterminate.

6.5.1. Constant Earth Mass Scenario

For a hypothetical plate tectonic Earth increasing its radius under conditions of a constant Earth mass (Figure 6.6), density is shown to decrease exponentially over time. Density would decrease from a peak value of approximately 52.5 times the present value during Archaean and Early Proterozoic times, to a present value of 5·52 grams per cubic centimetre. Density would then approach zero by about 300 million years into the future.

For this same constant plate tectonic Earth mass scenario, surface gravity is also shown to decrease exponentially. Surface gravity would decrease from a peak value of approximately 14 times the present value during Archaean and Early Proterozoic times, to 9.5 times the present value during the Cambrian Period. Surface gravity is then shown to approach a value of zero by about 300 million years into the future.

What this scenario means is that, if we assume that the Earth's mass has remained constant, or near constant, since Earths' formation, it must be accepted that, for the ancient Earth, everything would have been over 50 times heavier during the Precambrian. Surface gravity would have also been over 14 times stronger than it is now. To approach zero density and zero surface gravity by about 300 million years into the future also means that, at that time, the Earth would then evaporate and cease to exist.

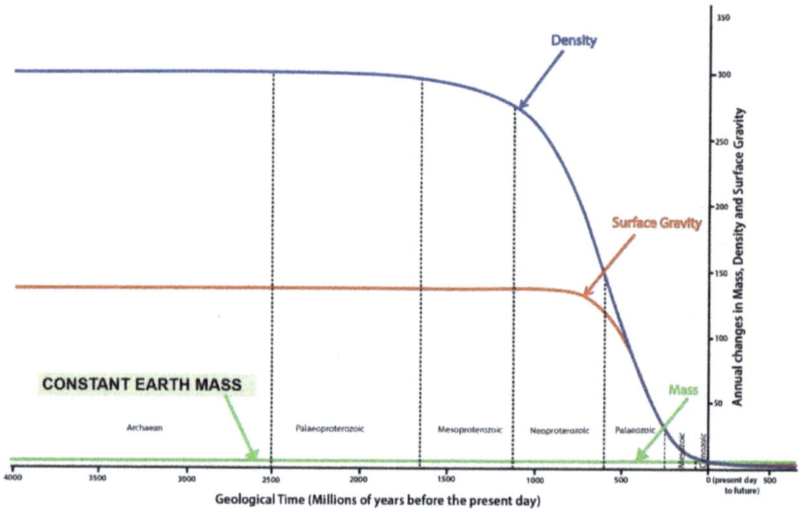

Figure 6.6. Changes in Earth Mass, Density, and Surface Gravity assuming a constant Earth Mass scenario. Annual change in Density is in grams per cubic centimetre, Surface Gravity is in metres per second per second, and Mass is kilograms times 10 raised to the power of 24. Maxlow, 2001.

6.5.2. Increasing Earth Mass Scenario

In contrast, for a hypothetical plate tectonic Earth increasing its radius while retaining a constant density (Figure 6.7), mass is shown to increase exponentially over time. Mass would increase from a primitive Archaean and Mid-Proterozoic value (shown as Mesoproterozoic in Figure 6.7) like the mass of the present Moon, to a present mass of about 54 times the ancient value. This increase in mass would then continue to about 32 times the present Earth mass by about 300 million years into the future.

Similarly, surface gravity is shown in this graph to increase exponentially. Surface gravity would increase from approximately one third of the present value during Archaean and mid-Proterozoic times and reach approximately 3 times the present value by about 300 million years in the future.

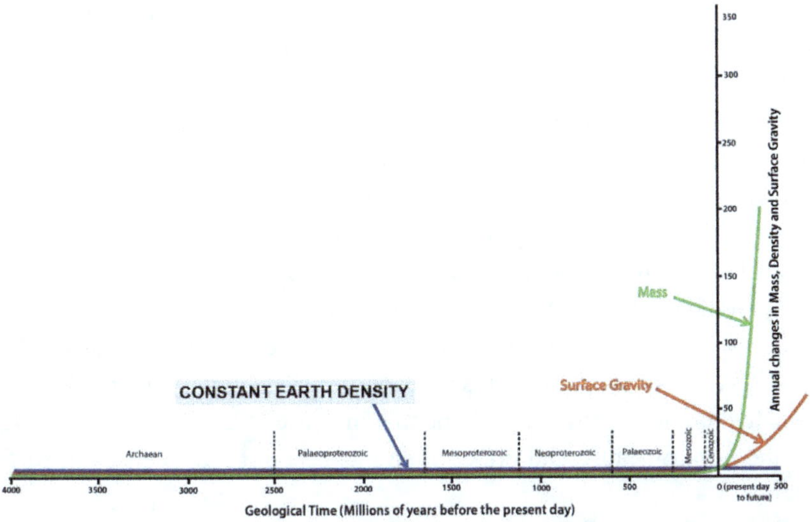

Figure 6.7. Annual changes in Earth Mass, Density and Surface Gravity assuming a constant Earth Density scenario. Annual change in Density is in grams per cubic centimetre, Surface Gravity is in metres per second per second, and Mass is kilograms times 10 raised to the power of 24. Maxlow, 2001.

By assuming that Earth's mass has been steadily increasing since formation of the Earth, and density has remained constant, or nearly constant, then surface gravity on an increasing radius plate tectonic Earth would progressively increase with time. In this plate tectonic increasing Earth mass scenario, surface gravity during Precambrian times would have been about one third of the present value and about half of the present value during the early Mesozoic.

Projecting this plate tectonic increasing Earth mass scenario to the future suggests that the Earth would approach the size of Jupiter or Saturn within about 500 million years' time. During this interval of time, the Earth's oceans, as well as the gaseous elements presently retained in the crust and mantle, may then simply evaporate to form a thick atmosphere or ring structure like those of the present giant planets.

This scenario may be subdued slightly if a density gradient were superimposed on the data, declining in time from higher rocky planet densities to lower giant gaseous planet densities.

An increasing mass scenario may then suggest that the transition from an inner rocky planet to a giant gaseous planet may be a natural evolutionary planetary process in the Solar System. This suggestion may also mean that the

Earth is currently in a transitional phase and may end up as either another giant planet or a failed planet, such as the asteroid belt might resemble.

While the changes in mass, density, and surface gravity shown in these figures represent endpoints of a spectrum of possible causes for an increasing radius plate tectonic Earth, they remain speculative mathematical derivatives. For a hypothetical Earth with a constant mass, however, the densities and surface gravities prevailing during Precambrian times would have been unacceptably high, particularly when compared to the density and surface gravity of the other planets within the present Solar System.

Carey in 1995 concluded from his own research that, because of the limitations shown from estimates of ancient surface gravity, there may be no alternative but to consider an exponential increase in Earth mass with time as both the primary cause and effect of an increase in Earth radius and surface area.

By adopting a constant density and increasing mass scenario for the cause of an increasing radius plate tectonic Earth, as first proposed by Carey, the present approximate annual rates of increase in mass and surface gravity are then calculated to be:

1. Mass annual change = 60×10^{12} t/year
2. Surface gravity = $3{\cdot}4 \times 10^{-8}$ m sec/sec/year
3. Density remains constant.

Chapter 7

Modelling Continental Crusts

"*...it is theoretically possible for the continents, without shelves, to fit together at approximately 40 percent of the present Earth radius by considering that continental shelves were formed after the continental crust had fragmented.*" Vogel, 1983.

Construction of continental plate assemblages in the context of what will be described here is not applicable to conventional plate tectonic studies. Instead, heavy reliance is made in plate tectonics for assemblage of crustal fragments of pre-existing supercontinents, plus accretion of new or pre-existing crusts, during continental drift. Constraining assemblage of these crustal fragments is then reliant on definition of palaeomagnetic apparent-polar-wander paths, with emphasis on assemblage of crustal fragments, rather than maintaining the strict definition of geological continental crustal domains.

Construction of small Earth models on an increasing radius plate tectonic Earth, extending from the early-Jurassic Period back to the early-Archaean, will involve the progressive removal of all younger continental basin sediments, and magmatic rocks, and simply returning these rocks to the ancient lands or back to the mantle where they came from.

Each continental basin and igneous complex will then be restored to a pre-extension or pre-rift configuration on a smaller radius plate tectonic Earth model. By moving back in time, the adjacent margins of each sedimentary basin or igneous complex will then be progressively moved closer together while still preserving the spatial integrity of adjacent, more ancient, cratonic or orogenic crusts.

To avoid confusion, the term subduction, when used, will be restricted to its conventional usage. In order to maintain a constant radius Earth model, conventional plate tectonic usage insists that thousands of square kilometres of excess seafloor crust must be disposed of beneath the continents. Similarly, subsequent closure of oceans during continental drift is inferred to give rise to continental collision to form mountains. In contrast, subduction, in its strictest sense, does not occur on an increasing radius plate tectonic Earth, and observed subduction-related observations are basically related to crustal interaction or obduction processes during changing Earth surface curvature. These will be discussed in detail in later chapters.

7.1. Assumptions

For construction of continental crust on each of the Archaean to Triassic plate tectonic small Earth models presented here, it is assumed that, for an Earth undergoing an exponential increase in surface area and radius through time:

1. Sediments deposited in continental sedimentary basins, as well as magmatic intrusions and volcanic eruptions, represent new rocks or crusts that accumulate primarily within areas of extensional continental crust.
2. Continental basin sediments, magmatic intrusions, and volcanic eruptions are not removed from the Earth's surface by subduction processes and, in conjunction with additional on-going geologic processes, are essentially preserved in the rock-record.
3. Moving back in time, all younger sediments and intruded igneous rocks must be returned to their respective places of origin.
4. Moving back in time, the surface area of sedimentary basins and igneous complexes must be progressively reduced to their pre-rift, pre-extension, or pre-breakup configurations.

What these increasing radius plate tectonic Earth assumptions mean is that, by moving back in time, all sediments must be returned to the ancient lands where they were initially eroded from and similarly, all magmatic and volcanic rocks must be returned to the mantle or lower crustal regions where they originally came from. The surface area of each of the sedimentary basins or rift zones must also be reduced to simulate the return of these crusts to a pre-extension or pre-rifted state.

During model construction, the only stipulation required during cutting and pasting of crustal regimes between each successive small Earth plate tectonic model is that:

1. Continental crustal plates must undergo small vertical and surface

area adjustments to allow for a progressive change in surface curvature of the Earth with time.

7.2. Model Construction

The series of 24 small Earth models (Figure 7.1) presented in this chapter are based on the continental and seafloor crustal geology shown on the Geological Map of the World (1990). These models represent earlier seafloor crustal models—models 13 to 23—recreated to include continental geology, plus additional pre-Triassic models—models 1 to 12—extending back to the early-Archaean. One model—model 24—has also been extended to 5 million years into the future.

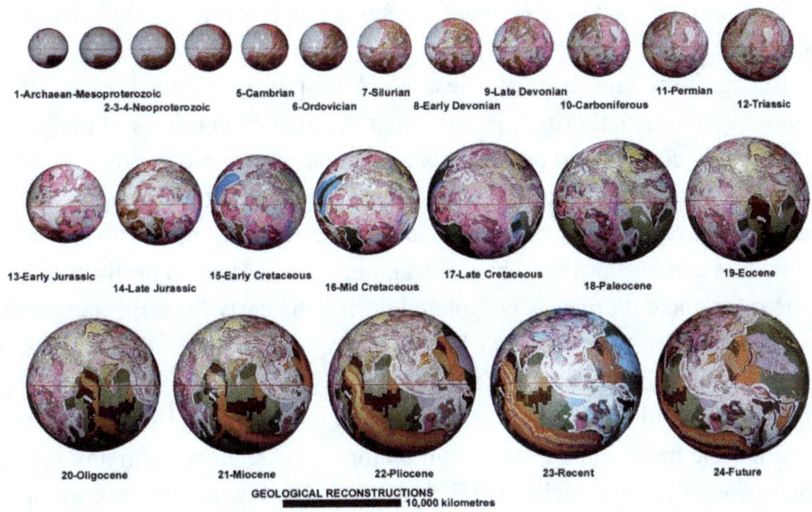

Figure 7.1. Spherical Archaean to future small Earth geological models. Models range in age from the early-Archaean to present-day, plus one model projected 5 million years into the future. Maxlow, 2001.

Shown on these models are both seafloor and continental time-based bedrock geology. The models in this particular figure show the closure of the Indian Ocean after removing all seafloor crusts. These models also show the relative sizes of the ancient Earth through time after progressively removing seafloor volcanic rocks, igneous rocks, and eroded continental sediments, and returning these rocks and sediments back to either the mantle or ancient lands.

During the earlier seafloor modelling studies, as the oceanic plates were progressively re-assembled back in time, a network of continent-to-continent docking-points was identified. These docking-points formed important positional constraining points during assemblage of the continental plates and were further complimented by constraints from seafloor mapping. These docking-points were maintained throughout the post-Triassic seafloor assemble times, and their usage is extended here to include the pre-Jurassic continental crustal constructions.

Examples of continent-to-continent docking-points used on each of the pre- and post-Triassic models (Figure 7.2) include: south-west Australia to Wilkes Land of East Antarctica; Tasmania to both south-east Australia and Victoria Land of East Antarctica; South Africa to New Schwabenland of East Antarctica adjacent to the Weddell Sea; west Africa to South America; India to Queen Maud Land of East Antarctica; Siberia to Alaska; the Canadian Northwest Territories to Greenland; Greenland to Norway; Spain to Morocco; and Arabia to Africa.

Recognition and use of these docking-points allowed each of the remaining continental crusts, in particular the ancient cratons and orogens, to be accurately constrained on each plate tectonic small Earth model, relative to each other. For each successive small Earth model, this recognition then enabled the relationship between each of the docking-points to be strictly maintained across adjoining crustal fragments or continents until assemblage of the most ancient crusts is completed during the early Precambrian times.

The stepwise construction of present-day to Archaean models presented here continues back in time until removal of all seafloor volcanic and continental sedimentary basin sediments and magmatic rocks is complete. At that time, the most ancient Archaean cratonic rocks, along with any remnant Proterozoic orogenic rocks, were then assembled together as a complete, primordial, pan-global Archaean plate tectonic small Earth model.

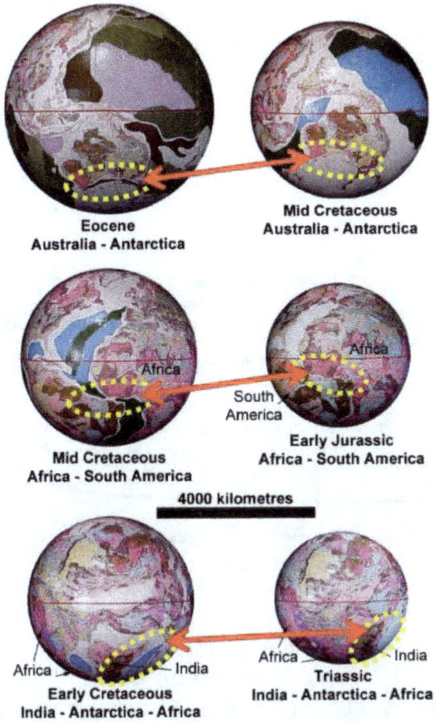

Figure 7.2. Examples of continent-to-continent docking-points used to constrain continental reconstructions. These include Australia and Tasmania to Antarctica; Africa to South America; and India to Antarctica, Madagascar, and Africa. Maxlow, 2001.

Modelling studies suggest that this Archaean plate tectonic small Earth model had a primordial radius of approximately 1,700 kilometres, which is about 27 percent of the present Earth's radius and of a similar size to the present Earth's Moon.

7.3. Preliminary Observations

In the following discussion, a scenario for crustal development on an increasing radius plate tectonic Earth is presented, based on preliminary observations from spherical modelling studies. Crustal development on an increasing radius plate tectonic Earth is shown to be an evolving process, which is intimately related to changing Earth radius, surface area, and surface curvature over time. This crustal development process can be considered as four overlapping phases of Earth history (Figure 7.3).

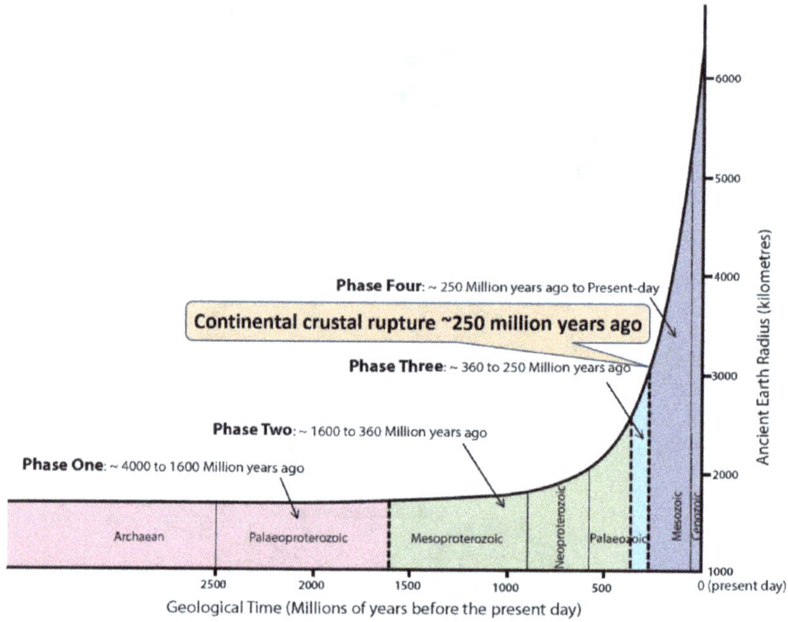

Figure 7.3. Four phases of Earth development history. Phase 1): an early phase of minimal increase in Earth radius extending from 4,000 to 1,600 Ma; Phase 2): a phase of steady to rapidly increasing Earth radius extending from 1,600 to 360 Ma; Phase 3): a short, late-Palaeozoic transitional phase extending from 360 to 250 Ma where continental crustal stretching was exceeded and crustal rupture was initiated; Phase 4): a post-Permian phase extending from 250 Ma to the present-day where increase in radius was predominantly manifested within the modern oceans. [Ma = millions of years before the present]. Maxlow, 2005.

The radius curve in Figure 7.3 shows that crustal development on an increasing radius plate tectonic Earth commenced with an extremely long phase of slow to steadily increasing continental crustal extension. This phase was followed by a phase of steady to rapidly increasing crustal extension leading to crustal rupture, and finally to breakup of the ancient supercontinental crust and formation of the modern continents and oceans during the late-Permian. This entire process was extremely slow and protracted, matched only by the extremely long span of geological time available since the formation and stabilization of the Earth some 4,000 million years ago.

The four phases of continental and seafloor crustal development include:

a) **Phase 1**: An Archaean to Mid-Proterozoic—times older than 1,600 million years ago—phase of minimal increase in Earth radius. The increase in surface area during this phase was readily absorbed during the first beginnings of continental crustal extension and was marked by early sedimentary basin formation and intrusion of granite and volcanic rocks. Deposition of sediments eroded from the exposed lands was confined to a global network of continental sedimentary basins which coincided with relatively shallow continental seas. Breakup and subtle jostling of each of the ancient cratons during changes in surface curvature was first initiated during this phase, in particular within the established network of sedimentary basins, giving rise to long linear zones of crustal weakness. Because of the prolonged period of time involved in this phase, the continental crust may have had a subdued, featureless topography for much of the time and continental seas were relatively shallow. Single to multi-celled organisms first became established and evolved during this phase.

b) **Phase 2**: A mid-Proterozoic to Carboniferous—from 1,600 to 360 million years ago, a rapidly increasing Earth surface area, crustal extension, and changing surface curvature. The increasing surface area during this time was absorbed within the continental crusts as on-going crustal extension, accompanied by magmatic and volcanic intrusions. Crustal extension was again mainly confined within a coincident network of crustal weakness, within sedimentary platform basins, and shallow seas. Over time, the network of sedimentary basins and seas continued to increase their surface areas throughout the early- to mid-Palaeozoic Era—540 to 360 million years ago. Movement of the ancient cratons and orogens during adjustments to changing surface curvature resulted in multiple and often reoccurring phases of crustal jostling and tectonic activity—orogenesis, followed by renewed basin extension as crustal extension continued. The topography contrast between the low-lying sedimentary basins and exposed continents steadily increased during this phase, in particular during periods of crustal interaction and early mountain building. Hard shells and invertebrate life forms and fishes were well established during this phase.

c) **Phase 3**: A Carboniferous through to late-Permian—360 to 250 million years ago—transitional phase, where continental crustal extension was exceeded and extension changed to crustal rupture, rifting, and initiation of continental breakup and opening of the modern oceans. Continental seas commenced draining, and deposition of sediments progressively shifted away from the established continental sedimentary basins into newly opening marine basins. Evidence for this early marine basin phase is now preserved seawards of the continental shelf margins and as remnant ocean plateau within many of the modern oceans, such as the Lord Howe Rise in the Tasman Sea. During this phase reptiles, dinosaurs, plants and mammals evolved to eventually dominate the lands.

d) **Phase 4**: A post-Permian—250 million years ago to the present-day—phase of relatively rapid increase in Earth surface area and opening of the modern oceans. During this time, crustal extension was mainly confined to the modern oceans along well-defined mid-ocean-ridge spreading zones. Over time, the increase in surface area and volcanic activity within the mid-ocean-ridge zones gradually exceeded the input of sediments from the continents. By the early-Jurassic Period the seafloor volcanic rocks were then exposed and preserved as modern seafloor crust within each of the oceans. Two sub-phases are recognized during this time, with each phase corresponding to the distribution of Mesozoic and Cenozoic seafloor crusts respectively. During this phase dinosaurs flourished before becoming extinct and were superseded by the mammals.

As shown on these plate tectonic small Earth models (Figure 7.4) each of the four phases of crustal development are a direct result of an increase in Earth surface area and increasing changes to surface curvature, which is fundamentally governed by the rate of increase in Earth radius. Of note is the interval of time represented by the single, Phase 1 small Earth model, which represents about 60 percent of all Earth history—50 percent more time than the rest of the models combined.

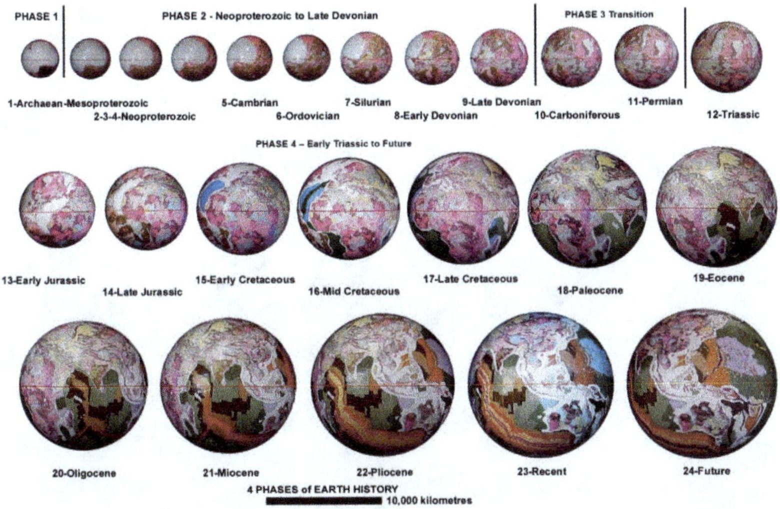

Figure 7.4. Four phases of increasing radius plate tectonic Earth development shown in relation to small Earth models extending from the early-Archaean to 5 million years into the Future. Maxlow, 2001.

This unique crustal development and breakup process will now be elaborated on still further by investigating the development of each of the ancient supercontinents and modern continents. The following observations will be, out of necessity, global scale and no attempt will be made to incorporate complex local geology or tectonic detail. The outcomes of these observations will be further discussed in later chapters when dealing specifically with additional evidence from geology, geophysics, ancient geography, ancient climate, ancient biogeography, and the distribution of natural metallic and fossil fuel-based products.

Chapter 8

Ancient Supercontinents

"*The fit of the continents on a smaller Earth appears to be too good to be due to coincidence and requires explaining.*" Creer, 1965.

Modelling the continental crust on increasing radius plate tectonic small Earth models shows that the ancient supercontinents existed as a complete continental crustal shell encompassing the entire Earth for the first 94 percent of geologic history. This supercontinental phase lasted for some 3,750 million years and culminated during breakup of the Pangaean supercontinent approximately 250 million years ago.

During pre-breakup times, continental crust covered the entire Earth and exposed supercontinental lands were defined by a superimposed network of continental seas. The origin and changing configuration of the supercontinents during these times then involved a progressive, evolutionary crustal process during a prolonged period of crustal stretching, along with changes to both Earth surface area and surface curvature through time.

In contrast, in conventional plate tectonic studies, the ancient supercontinents represent the assemblage of fragmented remains of more ancient supercontinents. The present continents, in turn, represent the assemblage of more recent fragmented crustal remains.

On these conventional tectonic reconstructions there are a multitude of identified ancient supercontinents, most notably Pangaea, Gondwana, Laurentia, Baltica, Laurussia, and Rodinia, each with their own separate crustal assemblage, breakup, dispersal, and subsequent collisional histories. These assemblage processes are depicted as being random and non-predictable and crustal assemblages are often portrayed as having a multitude of plate-fit options and ill-defined plate histories.

On all increasing radius plate tectonic small Earth models, it will be shown that there is no requirement for random, non-predictable, or multiple plate-fit options or ill-defined plate histories. Nor is there a requirement for extensive, largely hypothetical ancient oceans to comply with a constant surface area premise or for the fragmentation of any of the modern continents to comply with palaeomagnetic apparent-polar-wander studies.

Instead, these ancient supercontinental crusts tightly wrap around and fully enclose the ancient Earth. These crusts then assemble against continents that were otherwise not previously considered to be related to conventional assemblages. All crustal elements making up the ancient supercontinents will then be shown to retain an intimate relationship throughout Earth history, lasting until crustal breakup was initiated and the modern oceans began to open during the late-Permian Period.

In the following sections, discussion will start with the most ancient Archaean supercontinent and then move forward in time through to Rodinia, Gondwana, and finally to Pangaea. There are many lesser supercontinents described in conventional literature; however, usage will be restricted to the better-known supercontinents.

The term supercontinent, as used here, will differ somewhat from conventional usage where it is used to imply an assemblage of ancient crustal fragments. The term supercontinent is used here simply to distinguish between the exposed ancient land surfaces on a pre-breakup plate tectonic small Earth and the post-breakup modern continents. As such, it is iterated that on an increasing radius plate tectonic Earth it is the distribution of continental seas that defines the outline of ancient supercontinents, not an arbitrary assemblage of fragmented crusts.

8.1. Archaean Supercontinent

The oldest known rocks found on Earth today represent the period when the ancient crusts first formed and, more importantly, the period when conditions on Earth were stable enough for the rocks to be preserved to the present day. Evidence from minerals preserved in the ancient metamorphic and igneous rocks suggest that global crustal temperatures during the pre- to early-Archaean were extremely high. This suggests that the Earth may have been potentially molten during the pre-Archaean Hadean Eon, prior to cooling and stabilization of the crusts during the early-Archaean and later times.

The primordial Archaean crustal assemblage (Figure 8.1), as constructed during plate tectonic small Earth modelling studies, is shown along with location of the Archaean equator and poles. This figure represents an assemblage of the most ancient cratonic and orogenic crusts that are currently preserved on the present-day Earth. Establishing the location of the equator and poles was achieved via palaeomagnetic studies and will be discussed later. The named remnants of the present-day modern continents are also highlighted in this figure as black outlines.

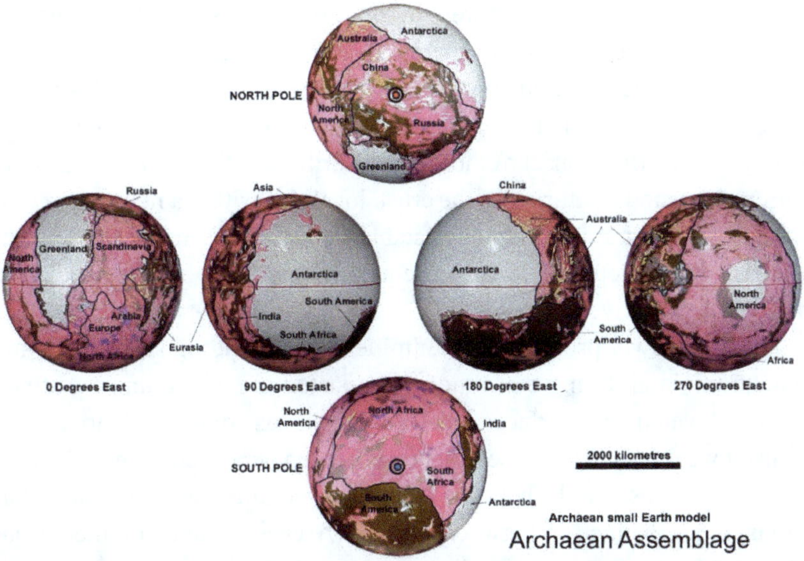

Figure 8.1. The primordial Archaean small Earth model showing the assemblage of ancient early-Precambrian continental crusts. Cratons are shown as pink and red, Proterozoic orogenic rocks are shown as khaki and the location of the present-day Antarctic and Greenland ice-sheets (covering Precambrian crustal rocks) are shown as off white. The ancient equator is shown as a horizontal red line, and the poles are shown as red and blue dots. Ancient remnants of each of the present-day continents are outlined in black. Maxlow, 2001.

The size of this Archaean supercontinent is estimated to be 1,700 kilometres radius, around 27 percent of the present Earth radius and about the size of the present Moon. Because of the extremely low rates of increase in radius existing during Precambrian times, this Archaean plate tectonic small Earth model represents the period extending from the early-Archaean through to the mid-Proterozoic Era. This one model then represents about 2,400 million years, or two thirds of all known geological history.

Modern geological mapping of the Archaean crusts shows that most of these ancient crusts comprise granite and volcanic rocks, with lesser sedimentary rocks. These crusts are often referred to as granite-greenstone terranes, where the term greenstone refers to the distinctive dark green colour of the volcanic rocks. The sediments are also often distinctively different from the later, more modern sediments and are loosely referred to as greywacke, comprising rock particles and volcanic detritus material. This sedimentary evidence suggests that the atmospheric conditions existing during early-Archaean times were very different to those of today, suggesting it may have been a relatively reduced atmosphere devoid of oxygen.

On an increasing radius plate tectonic Earth, during the early-Archaean times, it is envisaged that once the primitive granite and volcanic crust had cooled and solidified, onset of increase in Earth radius was initiated as global-scale cracking and fracturing of the crust, localized within a network of crustal weakness surrounding relatively stable lands. This network of crustal weakness and fracturing was subsequently intruded by renewed granite activity and primitive volcanic lava.

As time slowly progressed, the influence of changing surface curvature, although minimal during that time, also assisted in maintaining a distinction between elevated land surfaces and low-lying areas coincided with a network of crustal weakness. Eroded sediments were then deposited in these low-lying areas. It is considered that this elevation contrast was instrumental in establishing a distinction between the first Archaean supercontinental lands and the first ancient seas and sedimentary basins.

It is envisaged that the emergent Archaean land surface may have been a barren, windswept rocky landscape devoid of all forms of life and exposed to erosion of rock and dust particles by winds and reduced atmospheric rain. During that time, chemical weathering and erosion of the rocks may have been prevalent, erosion was caused by hot and potentially acidic waters from volcanic eruptions localized along the newly established network of crustal weakness. This weathering and erosion may have then given rise to deposition of the first sediments, deposited in low-lying sedimentary basins, along with extruded volcanic lava.

In Figure 8.1, remnants of this network of low-lying sedimentary basins can be seen by the distribution of khaki-coloured Precambrian sedimentary rocks, and the ancient cratonic crusts are shown as areas of pink and red. These early sedimentary basins may have initially been numerous, small, and isolated, becoming progressively more extensive over time.

This early granite-greenstone and sedimentary crust now represent the oldest preserved crustal remnants on Earth today. The prevailing reduced atmosphere and waters also enabled metals rich in sulphur, as well as elemental carbon, to accumulate within the sedimentary basins which, in turn, may have formed the basis for primitive life forms to emerge and evolve during the latter Precambrian times.

Geological evidence suggests that, during the latter part of this Archaean supercontinent phase, sedimentary basins were slowly filling to capacity over an extremely long period of time, operative during these ancient times, to form extensive platform sedimentary basins. The Proterozoic is characterized by the presence of very large stable sedimentary platform basins, forming relatively shallow seas with a very low elevation contrast between the dry lands and the seas.

Any further erosion of the lands and deposition of sediments was then limited to mainly chemically precipitated siliceous chert, carbonate, and banded iron formation rocks within these shallow seas. Remnants of the chemically precipitated rocks, along with the aerially extensive granite, volcanic, and associated sedimentary rocks making up the ancient lands are now preserved on many of the modern continents.

During this extended period, the ancient North and South Poles were located within what are now Northern China and West Africa respectively (Figure 8.1). Similarly, the ancient equator is shown to have passed through what is now North America, East Antarctica, Australia, Greenland, and Scandinavia. Over this extended period the primitive Archaean supercontinent also underwent subtle changes to its network of seaways and sedimentary basins.

These geological changes eventually evolved imperceptibly—via the early-Proterozoic Columbia supercontinent—into the better known Rodinia supercontinent some 1,000 million years ago. It is again emphasized that on an increasing radius plate tectonic Earth it is the changes to sea-levels and distributions of the network of sedimentary basins that define the progression from one supercontinent to the next.

8.2. Rodinia Supercontinent

The conventional plate tectonic Rodinia supercontinental assemblage relies heavily on palaeomagnetic evidence to constrain a multitude of plate-fit options that are currently published in the literature. These assemblages are further supported by ancient climate, fossil, and geographic data as required. These assemblages are, however, seriously hindered by the longitudinal limitations imposed by palaeomagnetic constraints.

Beyond these limitations, the presence of largely theoretical ancient oceans also requires that the continental crustal fragments making up Rodinia must be sourced from remote and equally theoretical continents, which have in turn also been arbitrarily fragmented and dispersed. This fragmentation and dispersal process results in a confused multitude of assemblage options for each supercontinental fragment, usually where fragmentation is not supported by local geological or geographical evidence.

Rodinia on an increasing radius plate tectonic Earth is represented by the late-Proterozoic plate tectonic small Earth model—at about 800 million years ago (Figure 8.2). The assemblage and distribution of the remnant modern continental outlines are shown as black outlines, along with the Proterozoic magnetic poles and equator.

Because there is no published ancient coastal information available for the Precambrian times to properly define the distribution of ancient continental seas, outline of the Rodinia supercontinent and seaways on an increasing radius plate tectonic Earth is inferred from the distribution of Precambrian continental sedimentary basins—shown as khaki colours.

Comparison of the Rodinia plate tectonic small Earth model in Figure 8.2 with the Archaean plate tectonic model in Figure 8.1 shows that there is not a lot of difference between these two models. The location of the poles and equator are essentially the same and the distribution of ancient cratons is also much the same. The final phase of the Archaean supercontinental development history was the formation of very large, stable sedimentary platform basins and shallow seas with a low elevation contrast between the lands and seas. This basin formation, in essence, is the only difference between the Archaean and late-Proterozoic Rodinia supercontinents.

The distinguishing feature of the Rodinia supercontinental history is that it coincides with initiation of the second phase of steadily increasing surface area and Earth radius (shown graphically in Figure 7.3). As a result of this steadily increasing Earth radius, changes to surface curvature were starting to take effect, which began to elevate the exposed lands and, in turn, increase the elevation contrast between the lands and seas. This elevation contrast initiated an increased rate of erosion of the exposed lands, which in turn influenced changes to the distribution of seas and corresponding coastal outlines of the supercontinents.

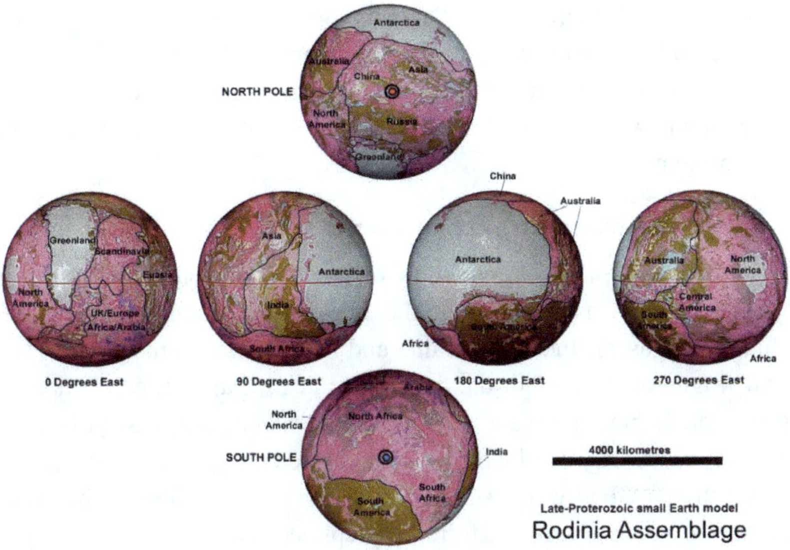

Figure 8.2. Late-Proterozoic Rodinia plate tectonic small Earth supercontinent assemblage. The model shows the distribution of Precambrian cratons (pink and red) inferred to represent exposed land, and a network of orogens and basins (khaki) inferred to represent continental seas. The black lines represent outlines of remnant present-day continents. Maxlow, 2001.

Another distinguishing feature of the Rodinian times was the changing atmospheric conditions. These conditions changed from reducing conditions throughout much of the earlier Archaean times, to the accumulation of atmospheric oxygen during the Proterozoic. This transition was marked by an increasing accumulation of banded iron formation rocks and chemically precipitated calcium and magnesium carbonate rocks. The transition also coincided with development and preservation of the earliest life forms, soon to explode in diversity during the following Palaeozoic Gondwana supercontinental times.

During late-Proterozoic Rodinian times, the ancient North and South Poles continued to be located within what is now Northern China and West Africa respectively. The ancient equator continued to pass through what is now North America, East Antarctica, Australia, Greenland, and Scandinavia, as well as Europe, Asia, and India—essentially the same as during Archaean times.

On increasing radius plate tectonic small Earth models, the Rodinia assemblage represents a supercontinent transition to the better known

Gondwana, and ultimately to the Pangaea supercontinental configurations. As shown on both the Archaean and Rodinia plate tectonic small Earth models, the change from one supercontinent to another is progressive and evolutionary and this theme will continue to follow through to the Gondwana and Pangaea supercontinents.

8.3. Gondwana Supercontinent

The conventional plate tectonic Gondwana supercontinental assemblage is made up of an assemblage of the present-day South America, Africa, Arabia, Madagascar, India, Australia, and Antarctician continents, as well as crustal fragments from Florida, southern and central Europe, Turkey, Iran, Afghanistan, Tibet, and New Zealand. This conventional assemblage shows that these present-day continents were assembled into large, separate, northern and southern Gondwana supercontinents plus smaller sub-continents separated by several largely inferred ancient oceans.

This assemblage is said to have formed during a Pan-African tectonic event. From its initial assemblage, the conventional Gondwana assemblage remained intact throughout the Palaeozoic to Early Mesozoic times. As several researchers have noted though, *...past movements of the Gondwana supercontinent, based on ancient magnetic, biogeographic and climatic evidence, are both equivocal and contentious, and this contention still remains despite the addition of new data from Africa and Australia.*

Gondwana on an increasing radius plate tectonic Earth is reconstructed on the Ordovician small Earth model—at about 460 million years ago (Figure 8.3). In this figure, the distribution of ancient coastal outlines is shown as blue lines and the ancient continental seas as shaded blue areas (data after Scotese, 1994, and Smith *et al,* 1994). The various continental seas and supercontinents are also located and named.

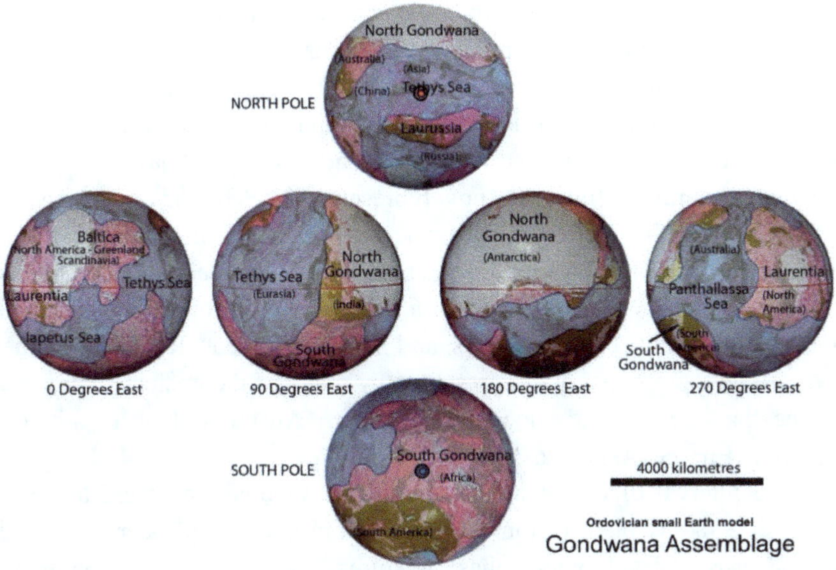

Figure 8.3. Ordovician Gondwana plate tectonic small Earth assemblage. The model shows the ancient coastline distribution (blue lines) defining North and South Gondwana in relation to Laurentia, Baltica, and Laurussia. The ancient Tethys, Iapetus, and Panthalassa Seas form part of a global network of continental seas (shaded blue areas) surrounding each of the exposed supercontinents. Coastline data after Scotese, 1994, and Smith *et al*, 1994.

On an increasing radius plate tectonic Earth, during Gondwanan times, the surface of the Earth underwent a steady to rapid increase in radius and surface area and was also fast approaching crustal rupture and breakup. The Gondwanan crustal assemblage (Figure 8.3) retained the same configuration of cratons, orogens, and basins as seen on earlier Rodinia and Archaean supercontinent models. The only difference being the greater surface area of surrounding sedimentary basins and hence aerial distribution of continental seas. This crustal assemblage is retained still further in time until initiation of crustal rupture and breakup of the pan-global Pangaean continental crust during the late-Permian.

The coastal information on the Ordovician plate tectonic small Earth model shows that, at that time there were distinct elevated ancient land surfaces—supercontinents. These land surfaces were in turn surrounded by a network of equally distinct, relatively shallow continental seas.

Gondwana on this model was subdivided into a North and South Gondwana, separated in part by an early Panthalassa Sea—the precursor to the modern Pacific Ocean. North Gondwana was made up of Australia, East and West Antarctica, and India and may have also included Tibet and Afghanistan. South Gondwana, joined in Madagascar, comprised Africa, Arabia, and South America. The ancient sub-continents Laurentia and Baltica were made up of North America, Greenland, and Scandinavia, as well as smaller Precambrian fragments now represented by Scotland and Ireland.

Similarly, the ancient Laurussia sub-continent was centred on the Precambrian regions of Mongolia and northern Russia. Each of these exposed Gondwanan land surfaces were in turn surrounded and interconnected by the ancient continental Tethys, Iapetus, and Panthalassa Seas. Remnants of these seas are now preserved and represented by many of the ancient sedimentary basins that are currently located in Eastern Australia, North and South America, Europe, Asia, and Africa.

The interval of time that the Gondwana supercontinent existed on an increasing radius plate tectonic Earth coincides with the later part of the second crustal development phase (Figures 7.3 and 7.4). This phase was characterized by a rapidly accelerating rate of increasing surface area and accompanying changes in surface curvature. These accelerating changes gave rise to a marked increase in erosion and deposition of sediments and marked changes to the distribution of continental seas and coastal outlines.

The increasing changes in surface curvature also initiated localized orogenic activity within geosynclinal sedimentary basins to form long linear fold mountain belts, further disrupting established seaways. These times also coincided with the rapid development and evolution of all life forms on Earth. It is considered that the degree of crustal change during this interval of time then had the capacity to markedly influence evolutionary change in all life forms.

During Gondwanan times, the small Earth South Pole was located within central West Africa in what was then South Gondwana. The North Pole was located within Northern China in what was part of the Tethys Sea. The ancient equator passed through East Antarctica, central Australia, North America, central Eurasia, and India, through what was then North Gondwana. This geographic configuration approximates conventional plate tectonic reconstructions in part but differs substantially in the South Pacific region where conventional plate tectonics requires the presence of a wide expansive Panthalassa Ocean.

These Gondwanan configurations are substantiated by existing physical and geological similarities between the Australian and South American continents. These similarities include the distribution of marine and terrestrial plants and animals, which link Australia and New Zealand across to Central and Southern America, and also rock and fossil types that link Australia directly to South America. The changing coastal outlines also show that continental land connections existed between north Australia and North America and between north Africa-Arabia and Scandinavia.

The existence of these land connections formed important migration routes or barriers for terrestrial and marine life forms. A change in configuration of the various continental seas may have also resulted in periods of relatively rapid sea-level changes in these areas, with disastrous consequences for plant and animal species existing at the time.

The late-Palaeozoic to early Mesozoic times eventually coincided with breakup of the Pangaean supercontinent, which was accompanied by draining of the continental Tethys, Panthalassa, and Iapetus Seas as the modern oceans began to open. As a result of this draining of the seas each of the Gondwana and related supercontinents geographically merged with the smaller Laurentia, Baltica, and Laurussia sub-continents to form the more familiar Pangaea supercontinent.

On an increasing radius plate tectonic Earth this evolution of continents and seaways was again reflected in the progressive changes to coastal outlines and sea levels, as displayed in each of the small Earth models.

8.4. Laurentia, Baltica, Laurussia Sub-continents

Laurentia, Baltica, Laurussia and numerous smaller fragments, such as Kazakhstan and fragments making up Southeast Asia are described in conventional plate tectonic literature as sub-continents, or simply continental fragments. These are routinely described in the literature as fragments of pre-existing supercontinents that periodically dispersed and amalgamated between the larger supercontinents during continental breakup and subsequent reassemblage of the various continental fragments.

On the Ordovician increasing radius plate tectonic small Earth model (Figure 8.3), each of these sub-continents were geographically connected to the supercontinents. Their presence and coastal outlines depended primarily on the variation in sea levels existing at the time and the sub-continents were often separated geographically from the main supercontinents by the global network of continental seas.

As the continental crusts began to rupture during the late-Permian, the established continental seas then began to drain into newly formed marine sedimentary basins as the modern oceans opened. The previously submerged seafloor surrounding each of these sub-continents was then exposed as dry lands and the surface areas of the sub-continents progressively increased to eventually form part of the more familiar Pangaea supercontinent.

8.5. Pangaea Supercontinent

The assemblage of a large Pangaea supercontinent was promoted by Alfred Wegener in 1915. In his proposal, he presented geologic evidence to suggest that the large continental areas of the modern world were originally united late in the Palaeozoic Era prior to subsequent breakup and dispersal of the continental fragments. This proposal has since been adopted in conventional plate tectonic studies, whereby Pangaea is said to have assembled from crustal fragments originating from the breakup of Gondwana.

During the late-Palaeozoic Era the Pangaean supercontinent then comprised two contrasting geographical provinces: a southern Gondwana province and a northern Laurasian province. The Pangaea supercontinent was then inferred in conventional Plate Tectonic studies to have fragmented within the Tethyan region, where continental breakup began in earnest during mid-Jurassic times.

Pangaea on an increasing radius plate tectonic Earth is shown on the Permian small Earth model in Figure 8.4. This figure shows the coastal outlines as blue lines and continental seas as shaded blue areas. The various supercontinents and intervening seas existing at the time are also named. The Pangaean assemblage coincides with the Carboniferous to Permian transitional crustal development phase 3 (Figures 7.3 and 7.4). In these previous figures, increasing Earth surface area had progressed from advanced continental extension to true crustal rupture prior to continental crustal breakup and dispersal as the modern continents.

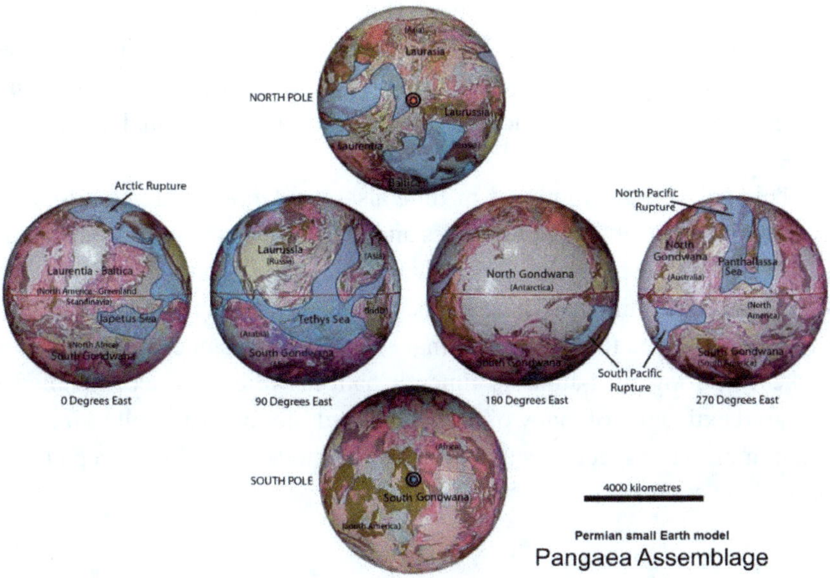

Figure 8.4. Permian Pangaea plate tectonic small Earth crustal assemblage. The model shows the ancient coastline distribution (blue lines) as well as the ancient Tethys, Iapetus, and Panthalassa Seas (blue shaded areas) forming part of a global network of continental seas. The figure also shows the locations of continental rupture commencing in the north and south Pacific and Arctic Ocean regions to form the modern oceans. Coastline data after Scotese, 1994, and Smith *et al*, 1994.

The presence of Pangaea on an increasing radius plate tectonic Earth represents a simple, progressive, although relatively rapidly changing evolution of the Gondwanan coastal outlines prior to continental rupture and formation of the modern continents and oceans. During this Pangaean time, rupture and breakup of the continental crusts had initiated draining of the continental seas which was in turn accompanied by a shift in where eroded sediments were being deposited.

This shift changed from sediments being deposited within an existing network of continental sedimentary basins, to being deposited within newly opening marine basins and along the continental shelf margins of the newly formed modern continents. This influx of sediment, along with intrusion of new volcanic and magmatic rocks, is now commonly preserved within submerged marine plateau as well as continental shelf settings surrounding most of the modern continents.

Following rupture of the supercontinental crust, Pangaea on an increasing radius plate tectonic Earth broke up during the late-Permian and dispersed as the modern continents during Triassic to present-day times. The subsequent migration history of the modern continents and seafloor crustal history is now preserved by the intruded volcanic seafloor lava existing throughout all of the modern oceans.

This post-Pangaea interval of time also saw large apparent shifts in the location of the North and South Poles and equator, occurring as a direct result of opening of the modern oceans and an apparent shift in the location of each of the modern continents. These shifts, as well as geographic isolation during breakup, in turn led to plant and animal evolutionary changes which increased markedly during post-Jurassic times. Changes were also accompanied by increased extinction of many of the established life forms as well as increased isolation of many species as a number of the modern continents separated to form island continents.

Chapter 9

Modern Continents

"...terrestrial expansion brought about the splitting and gradual dispersal of continents as they moved radially outwards during geological time." Vogel, 1983.

On an increasing radius plate tectonic Earth, the modern continents have only existed in their current form since breakup of the ancient Pangaean super-continental crust commenced some 250 million years ago. These modern continents represent the fragmented remains of the ancient Pangaea supercontinental crustal shell. Fragmentation and breakup of Pangaea occurred because the ability of the supercontinental crusts to continue to stretch and extend during on-going increases in Earth surface area was exceeded during late-Permian times. Once crustal stretching was exceeded, the Pangaean supercontinental crust then ruptured, broke apart and fragmented to form the modern continents and opened to form the intervening modern oceans.

When describing the development and subsequent history of the modern continents in this and subsequent chapters, geographical orientations relative to the ancient equator will be given in lower case, for example, the long axis of Australia was orientated north-south throughout the Precambrian—as distinct from its current east-west orientation. In contrast, current geographical descriptors and names of continents relative to the present-day equator will be given in upper case, for example North America.

9.1. Australia

The Australian continent (Figure 9.1) has large areas of ancient crusts preserved throughout what is now Central, Northern, and Western Australia. The rest of present-day Australia is made up of relatively young sedimentary rocks which were originally linked to similar sedimentary basins now located in China and North and South America. The older parts of the Australian continent had their beginnings as part of an ancient Precambrian supercontinental crustal assemblage. This ancient assemblage later extended its surface area as new sedimentary basins opened around its margins.

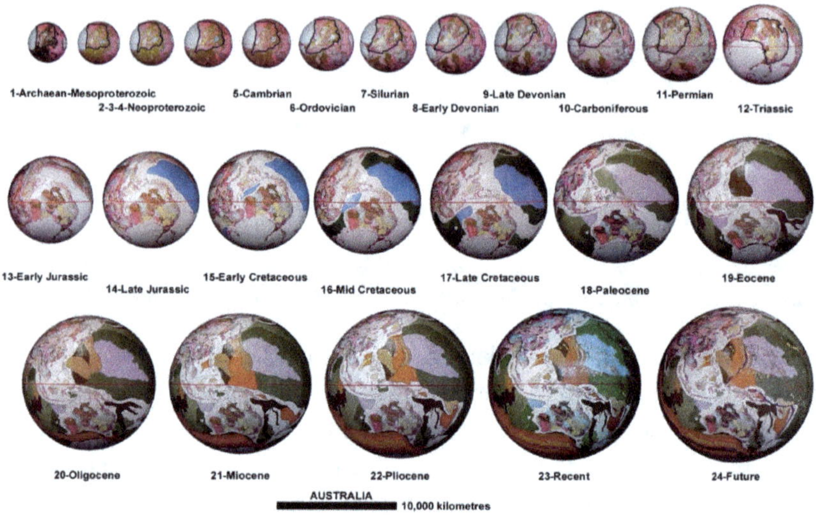

Figure 9.1. Continental crustal development of plate tectonic Australia. The outline of the pre-Jurassic Australian crust is shown as a black line. During the Precambrian to late Palaeozoic Eras Australia was orientated north-south prior to rotating counterclockwise to the present east-west orientation. This occurred during the opening of the modern Pacific Ocean. The horizontal red line represents the location of the ancient equator. Maxlow, 2001.

Throughout the various Precambrian and Palaeozoic Eras, most of ancient Australia was located in the northern hemisphere and the long axis of the primitive Australian continent was orientated north-south relative to the ancient equator. Once the ancient Pangaea supercontinent started to breakup and the modern Pacific Ocean commenced opening during the Mesozoic Era, the newly formed Australian continent then rotated counterclockwise to its present east-west orientation and migrated south into its present location in the southern hemisphere.

During these ancient times, the Australian crusts abutted directly against crusts from China to the north, Canada and North America to the east, New Zealand to the south, and East Antarctica to the west. At that time, ancient Australia was located within mid- to high-northern latitudes relative to the ancient equator and the ancient equator passed through what is now Central and Northern Australia.

During this pre-breakup supercontinental time, the Proterozoic sedimentary basins of Northern and Central Australia formed part of an extensive global network of sedimentary basins. These basins extended north into the Proterozoic basins of Alaska, Canada, Northern Russia, and Asia, and to the east and south these basins were also linked to the Proterozoic basins of North America, Central America, and South America. Deposition of sediments within these ancient Australian sedimentary basins was most active to the south—within what is now Eastern Australia—and this deposition extended into adjoining regions in New Zealand, South America, North America, and Antarctica.

The breakup of this extensive sedimentary basin occurred during Permian times during the initial opening of the South Pacific Ocean. Remnants of this basin are now preserved in East Australia, New Zealand, Central and South America, North America, and Antarctica

Crustal movement during this time was accompanied by ancient mountain building events, mainly within the network of sedimentary basins in what is now Central and Eastern Australia. Remnants of these same mountain building events are also preserved as the Andean Mountain events of South America, the Appalachian and Grenville Mountain events of Eastern North America, and the Cordilleran Mountain event of Western North America and Canada.

Events in Central and Northern Australia were also associated with periods of crustal movement and jostling of the various ancient crusts located between Australia and North America. This jostling occurred as each of the ancient crusts adjusted for changing surface curvature.

Continental crustal rupture to form the modern Australian continent first commenced during Permian times in areas located to the northeast of Australia, relative to the ancient equator, adjacent to what is now the Pilbara and Kimberley regions of Western Australia. Opening also occurred in the south, adjacent to Eastern Australia, separating Australia from New Zealand.

The outline of the modern Australian continent then began to take shape in these areas once the North and South Pacific and similarly the Indian Ocean began to open. After slowly opening during the Permian and Triassic Periods, the early North and South Pacific Oceans then started to rapidly extend, separating Australia from North and South America, while initially retaining a brief land link between Queensland in East Australia and California in North America. These previously separate North and South Pacific Oceans then merged during the Jurassic Period, finally separating Australia and New Zealand from North and South America.

Similarly, the Indian Ocean commenced opening during the Jurassic Period and was located adjacent to what is now Northwest Australia. As a result, Australia separated from China and Southeast Asia as the Indian Ocean continued to open and extend southwards along the west coast of Western Australia. At that time lands connecting Australia with adjoining continents remained attached to Southeast Asia to the north and East Antarctica to the south allowing faunal and floral species migration between each of these continents.

Rifting between Australia and East Antarctica commenced during the Paleocene—about 65 million years ago—during opening of the Southern Ocean. Australia and Antarctica are continuing to migrate away from each other leaving both continents as separate island continents. During this rifting phase, Australia migrated south from mid-northern latitudes crossing the ancient equator into its present-day mid-southern latitude location.

Rocks exposed throughout Australia now reflect this geographic migration history, with extensive coral reef deposits located along the full length of Eastern Australia reflecting its original equatorial location throughout the Palaeozoic Era. Migration was followed by a prolonged period of tropical weathering as Australia rotated and crossed the equator, which is marked by extensive deposits of laterite rocks—rocks that have undergone deep chemical weathering from tropical rains—throughout present-day West and Northeast Australia. This weathering was then followed by a progressive drying and desertification of the landscape to the present-day as Australia continued to migrate south away from the equator.

Once the Southern Ocean began to open, Australia then remained geographically isolated from its neighboring continents and has continued to progressively migrate further south into low- and mid-southern latitudes.

This southern migration of Australia is at odds with conventional plate tectonic studies whereby Australia is instead said to be migrating north, out of more temperate to polar climate zones, to collide with Southeast Asia. This interpretation is based primarily on palaeomagnetic studies and cannot be reconciled on an increasing radius Earth.

9.2. Africa and Arabia

Africa and Arabia contain extensive areas of ancient Precambrian crusts. Both continents originated from an Archaean supercontinental crustal assemblage located adjacent to the Archaean and Proterozoic regions of primitive South America, North America, Central Europe, Scandinavia, India, and East Antarctica. This early continental crustal configuration (Figure 9.2) has since remained spatially intact throughout Earth history relative to each of the surrounding continents.

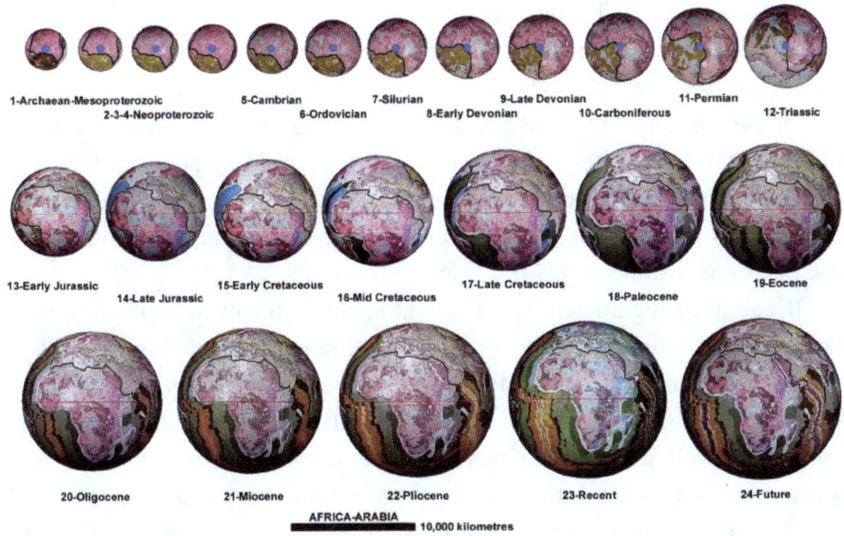

Figure 9.2. Continental crustal development of plate tectonic Africa and Arabia. In each of the Precambrian and Palaeozoic models North Africa is shown located south of the ancient equator and the South Pole is located in west Central Africa. The Archaean to Triassic models is centred on the South Pole (blue dots). The horizontal red line represents the location of the ancient equator, and the black lines represent the African continental crustal outlines. Note: models 1 to 12 are centred over the South Pole and models 13 to 24 are centred on the equator. Maxlow, 2001.

Throughout Precambrian and early-Palaeozoic supercontinental times, the ancient South Pole was located within what is now Central West Africa and during that time a south-polar icesheet extended periodically from there into Arabia and South America. Once crustal breakup commenced, this ancient South Pole made an apparent migration south along the present West African coastline during the Palaeozoic Era as the African continent migrated north (Figure 9.2).

The crustal development of primitive Africa and Arabia during Precambrian and Palaeozoic supercontinent times involved extensive continental crustal stretching and formation of intracratonic sedimentary basins. Each of these areas were, in turn, accompanied by crustal jostling between each of the South, East, and West African cratons during changing

Earth surface curvature.

Throughout North Africa, crustal stretching and deposition of sediments continued into the Palaeozoic and later eras in conjunction with similar crustal stretching and jostling within what are now the Mediterranean, Central European, and Asian Tethys regions.

Precambrian and Palaeozoic crustal movements and mountain building occurred along what is now the present East African coast, associated with similar events in primitive Madagascar and India, as well as along the West African coast associated with the Grenville, Appalachian, and Hercynide Mountain events in Eastern North America and Europe. Remnants of events associated with the younger Alpine and Himalaya Mountain building episodes also occurred within what is now North Africa.

Breakup of the ancient Gondwanan supercontinent to form the modern African and Arabian continents was first initiated between West Africa and North America during the Permian and Triassic Periods during opening of the North Atlantic Ocean. Rifting and separation between South America and Africa, as well as between Africa, Antarctica, Madagascar, and India, commenced during the Jurassic adjacent to what is now South Africa.

This rift continued north, within the opening South Atlantic Ocean, to eventually join with the North Atlantic Ocean. Similarly, rifting continued north along the East coast of Africa during opening of the Indian Ocean which eventually gave rise to the modern African continental outline.

Rifting and separation of Africa from Arabia, and similarly Africa from the Mediterranean and Middle East regions, commenced during the Cretaceous Period during opening of the Mediterranean Sea and rifting continues to the present-day within the opening Red Sea.

During the Mesozoic and Cenozoic Eras, Africa and Arabia continued to slowly migrate north, relative to the South Pole, in conjunction with opening of the Atlantic and Indian Oceans. From its central West Africa location throughout Precambrian to early-Palaeozoic times, the ancient South Pole is shown to have made an apparent migration south along the west coast of South Africa. It then crossed the newly opened South Atlantic Ocean during the Mesozoic Era and continued onto the Antarctic continent during the latter Cenozoic Era.

This apparent polar migration severely disrupted existing plant and animal species development within Africa as the polar climate zone also experienced an apparent migration south across the land surface. Once the ancient South Pole had crossed the Atlantic Ocean, much of Africa and Arabia maintained a centrally located equatorial position through to the present-day with the ancient equator experiencing a slow apparent migration from a North African position to its current Central African position.

On an increasing radius plate tectonic Earth, a direct geographical connection between Europe and Asia is maintained throughout Earth history and only relatively recently are the continents shown to be separating and rifting within what are now the Mediterranean and Red Seas.

This connection is at odds with conventional plate tectonic assemblages where Africa and Arabia are said to be migrating north and colliding with Europe and Asia to form the Alpine Mountain belt. Conventional migration and collision, however, cannot explain the opening of the Mediterranean Sea during the late-Cretaceous times without invoking subduction of pre-existing Tethyan crusts, nor can it explain the present opening of the Red Sea and rifting between Arabia and Africa.

9.3. Antarctica

The continental development of both East and West Antarctica is speculative because of the masking effect of present-day ice coverage. This icesheet, as well as the icesheet covering Greenland, is part of the Geological Map of the World base map used to construct the small Earth models and could not be removed during model construction. The Antarctic icecap is known to be about 33 million years old, representing a considerable part of the Cenozoic Era. Reconstructions and limited published field evidence indicate that the East Antarctica continent comprises mainly Precambrian crusts, and this crust has remained relatively intact throughout Earth history.

The continental reconstruction of Antarctica (Figure 9.3) shows that, unlike present-day East Antarctica, during the Precambrian Eras West Antarctica was made up of small remnants of Proterozoic crusts. The reconstructions also show that, during Precambrian and early-Palaeozoic times, the primitive crusts making up both East and West Antarctica straddled the ancient equator. These continental crusts were located adjacent to primitive Australia, South America, South Africa, India, and Central and Southeast Asian crusts.

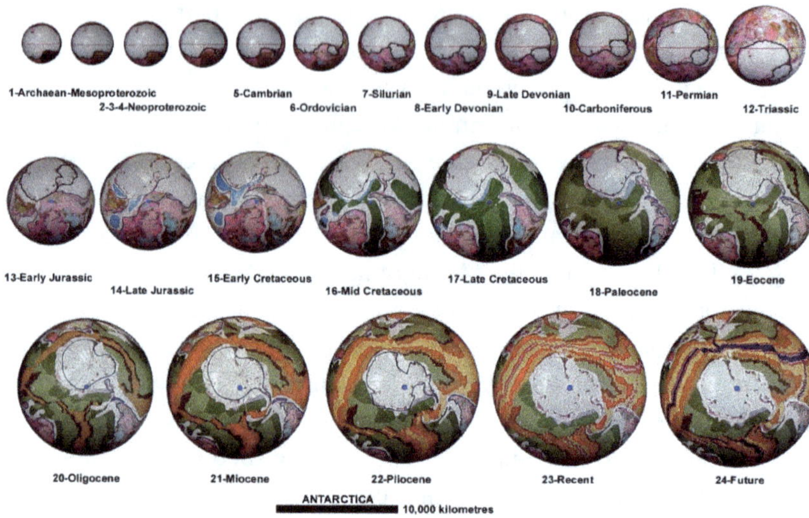

Figure 9.3. Continental crustal development of plate tectonic East and West Antarctica (shown as pale blue—the present-day south-polar ice shelf). During the Archaean to late-Palaeozoic Eras Antarctica straddled the ancient equator. The continent then rapidly migrated south, as the modern oceans opened, to its present location at the South Pole. The horizontal red line represents the location of the ancient equator and blue lines represent the ancient coastlines. Note: models 13 to 24 are centred on the South Pole (shown as blue dots). Maxlow, 2001.

Ancient Proterozoic and Palaeozoic sedimentary basins located in what are now India, Central and Southeast Asia, Australia, and South America surrounded the ancient Antarctican continent and extended into West and East Antarctica beneath the present icesheet. At that time, mountainous regions in West and East Antarctica also formed extensions of similar mountainous regions located in Eastern Australia, as well as the Andean regions of South America.

Breakup of the former Gondwana supercontinent to form the modern Antarctican continent commenced during early-Permian times during a period of crustal rupture and opening of the South Pacific Ocean—the ancient South Panthalassa Sea.

By the late-Jurassic, opening of the Indian and South Atlantic Oceans had also commenced leaving a land connection between Australia and East Antarctica and similarly between South America and West Antarctica. Final breakup, rifting, and separation of modern Antarctica from Australia and East Antarctica commenced during the Paleocene, some 65 million years ago, which will be shown later to coincide with the end-Cretaceous extinction event.

On an increasing radius plate tectonic Earth, the development of West Antarctica relative to East Antarctica was closely related to the opening of the South Pacific Ocean. By the late-Cretaceous, West Antarctica had separated from East Antarctica and had begun to rotate clockwise, relative to East Antarctica, as the South Pacific Ocean continued to widen and open. During that time the West Antarctican Peninsular remained joined to the southern South American Peninsular. This peninsula continued to remain joined until separation and rifting between the two continents began during the Miocene, leaving modern East and West Antarctica as separate and isolated continents through to the present-day.

Throughout Precambrian and Palaeozoic times the Antarctican continent straddled the ancient equator. Once breakup was initiated, the combined East and West Antarctican continent is then shown to have commenced an apparent migration south, relative to the ancient South Pole. It then continued migrating further south to its present south-polar location throughout Mesozoic and Cenozoic times, establishing and preserving the present south-polar icesheet 33 million years ago.

This reconstruction and migration history shows that Antarctica evolved throughout most of its history within a warm tropical environment, prior to migrating south into high southern and then into south polar regions. The Oligocene to present-day small Earth models also show that Antarctica is currently not stationary over the South Pole but is continuing to migrate slowly across the south-polar region. Seafloor mapping also shows that the surrounding Southern Ocean contains a single, continuous mid-ocean-rift zone. All of the surrounding continents are migrating north away from Antarctica as new seafloor crust continues to be intruded along this Southern Ocean mid-ocean-rift zone.

9.4. Europe, Russia and Asia

The European and Asian continent, inclusive of Russia, is a vast area and constitutes the largest land mass on Earth today. The continental crustal development of Europe and Asia is shown to have a complex and prolonged history of Precambrian to recent crustal stretching and extension, sedimentary basin formation, mountain building, and intrusion of magmatic rocks. Much of the present European and Asian continent now represents the uplifted and exposed seafloor of the former Tethys Sea region, with relatively minor areas of more ancient crusts.

In conventional plate tectonic studies, a large Tethys Ocean is depicted as being located between fragmented remains of the present European and Asian continent and is inferred to have covered many small crustal plates, as well as Cretaceous island-arcs and smaller continents. Parts of Central and Eastern Europe are also said to have been covered by a northern branch of the Tethys Ocean. This branch was separated from the Tethys by the formation of the Alps, Carpathians, Dinarides, Taurus, and Elburz Mountains, before it gradually disappeared during the late-Miocene Period, becoming an isolated inland sea.

In contrast, on an increasing radius plate tectonic Earth, during Archaean and early-Proterozoic times, the European and Asian continent was largely made up of a network of Precambrian sedimentary basins—the early Tethys Sea basin in particular. It also included small fragments of Archaean crusts that are now dispersed throughout Europe, as well as the larger ancient Mongolian and Chinese crusts. These crusts were once united as part of the ancient Archaean supercontinent. After fragmenting during the Precambrian, these Archaean crusts now form small remnants embedded within the much larger Tethys sedimentary basins.

Throughout the Precambrian and into the following Palaeozoic times, the Northern Asian region of the ancient European and Asian Tethys Sea was centred over the ancient North Pole. Similarly, the European region, in what is now the Mediterranean region, was centred over the ancient equator. East Antarctica was located along the eastern margin, Australia was located to the north, Canada to the northwest, Greenland to the west, Scandinavia and Arabia to the southwest and India was located to the south (Figure 9.4).

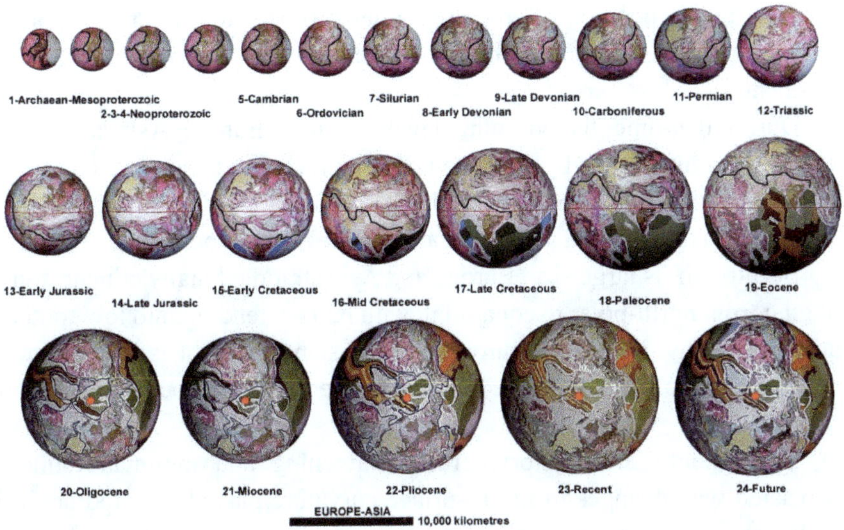

Figure 9.4. Continental crustal development of plate tectonic Europe and Asia. Models 20 to 24 are centred on the North Pole (shown as red dots). The horizontal red line represents the location of the ancient equator, blue lines represent the ancient coastlines, and the black line represents the outlines of the ancient European and Asian continent. Maxlow, 2001.

During Precambrian and Palaeozoic times, sediments were being eroded from each of the exposed surrounding continental land surfaces and deposited within the extensive Tethys Sea basin. This north-south orientated Tethys Sea basin, in turn, formed part of a much more extensive global network of continental seas. In this context, the ancient Tethys Sea also included Precambrian sedimentary basins that are now exposed in Central Australia and India and further afield also included ancient basins located within South America.

Breakup of the Pangaea supercontinental crust first began in the ancient Arctic, North Atlantic, and North Pacific Oceans during the late-Permian which then initiated breakup and separation of the combined European and Asian continent. During that time, as each of the surrounding modern continents began to rift apart and the modern oceans began to open, the existing Tethys continental sea was then disrupted and began to progressively drain.

During the Mesozoic, the ancient Tethys Sea was then exposed as dry lands as the waters slowly drained into the opening modern oceans. Deposition of sediments within the ancient Tethys Sea region was then disrupted and deposition of eroded sediments shifted into the newly formed marine basins, now located around the margins of many of the modern continents.

During that time, the on-going development of Europe, Asia, and Russia was strongly influenced by ever changing sea levels and changing coastlines. The Tethys Sea was completely drained during the Cenozoic, exposing Europe, Asia, and Russia as the elevated continent it is today.

Because of its large size, Europe and Asia straddled many climate zones, ranging from north-polar to equatorial, with parts extending into low southern latitudes. Today, the entire European and Asian continent is located in the northern hemisphere but still extends from the equator through to the North Polar Region.

Throughout Earth history, crustal stretching and mountain building associated with changes in Earth surface curvature played an important role in shaping the European and Asian continent. Precambrian and Palaeozoic events in Western Europe were associated with the ancient Grenville, Appalachian, and Hercynide Mountain building events now preserved within Eastern North America and Scandinavia.

Similarly, during the Mesozoic and Cenozoic Eras, the Alpine and Himalaya Mountain belts were formed during opening of the Mediterranean Sea and were accompanied by renewed stretching and crustal extension between Europe and Asia relative to Africa.

This plate tectonic increasing radius continental crustal history differs markedly from conventional plate tectonic reconstructions. On an increasing radius plate tectonic Earth, fragmentation of former supercontinents and inclusion of an extensive ancient Tethys Ocean is not necessary in order to close off the North Atlantic Ocean, or to conform to seafloor bedrock mapping data.

9.5. India

The Indian continent is traditionally shown on conventional plate tectonic reconstructions to be an island continent migrating north during the Mesozoic Era, moving across a vast pre-existing Tethys Ocean until it collided with Asia during the Cenozoic. Collision with Asia is then said to have resulted in formation of the Himalaya Mountain belt.

In contrast, crustal development of the ancient Indian continent on an increasing radius plate tectonic Earth (Figure 9.5) during Precambrian times initially formed a southern extension of the European and Asian Tethys Sea basin. During that time, and extending into later Palaeozoic times, India was located adjacent to East Antarctica to the northeast, Madagascar and South Africa to the southwest, and Arabia to the west. The ancient crust making up the present Indian continent was originally located within mid-southern to equatorial latitudes throughout the Precambrian and Palaeozoic times.

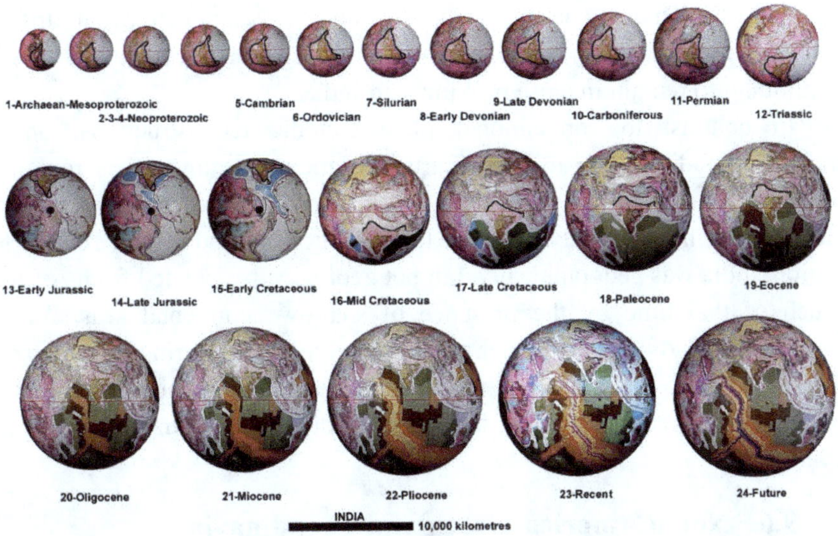

Figure 9.5. Continental crustal development of plate tectonic India. Each of the early Jurassic to early Cretaceous models (models 13 to 15) are centred over the South Pole (shown as black dots). The horizontal red line represents the location of the ancient equator, blue lines represent the ancient coastlines, and the black line represents the Indian continental crustal outline. Maxlow, 2001.

During the Mesozoic, India briefly migrated into mid-southern latitudes before returning to equatorial and low-northern latitudes during the latter Cenozoic Era. This apparent migration of India was related to its proximity to the rapidly extending European and Asian Tethys Sea basin. It was also influenced by the migration of adjoining continents away from the ancient South Pole, which was then located in central West Africa, as well as the newly opening Indian Ocean.

Crustal extension between the ancient north and south Indian crustal regions occurred during the Proterozoic Era in conjunction with related crustal development in the Tethys region. This period of crustal extension continued into the Mesozoic Era. Crustal movement and mountain building in India was associated with crustal motion relative to Antarctica, Madagascar, and Africa, as well as a number of Palaeozoic to Cenozoic Mountain building events along the northern Himalaya contact with Europe and Asia. In this context, the Himalaya Mountain chain was intimately associated with changes to surface curvature focused along this northern Himalaya contact.

Continental separation and rifting of India, Madagascar, and Sri Lanka from Antarctica and Africa commenced during the Jurassic with initial opening of the Indian Ocean. Madagascar and Sri Lanka then began drifting away from India during the early- to mid-Cretaceous, with Sri Lanka continuing to remain in close proximity to India.

In contrast to conventional plate tectonic reconstructions, on an increasing radius plate tectonic Earth the Indian continent has remained geologically attached to the Asian continent throughout all Earth history. Because of the proximity of India to the European and Asian Tethys Sea region, India was geographically, but not geologically, isolated from Asia for much of that time by the presence of shallow continental seas. As the European and Asian Tethys Sea progressively drained during the Cenozoic Era, India and Asia were then fully exposed as one continuous continental plate with no requirement for a separate Indian sub-continent or collision event.

9.6. North America-Greenland-Scandinavia

The development of the North American continent on an increasing radius plate tectonic Earth (Figure 9.6) was intimately related to the ancient Archaean supercontinent crustal assemblage and its ultimate breakup to form the ancient Canadian, Greenland, and Scandinavian crusts. This cluster of ancient continental crusts is discussed together and is here referred to as the North American cluster.

On conventional plate tectonic reconstructions, this cluster is referred to as the Laurentia and Baltica supercontinents and these have been extensively studied and referenced throughout North American literature.

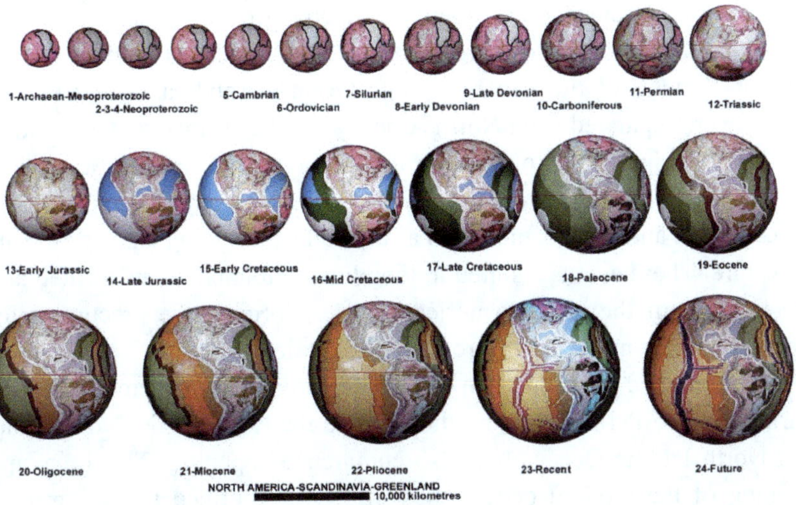

Figure 9.6. Continental crustal development of plate tectonic North America-Greenland-Scandinavia. The horizontal red line represents the location of the ancient equator and black lines represent the continental outlines. Maxlow, 2001.

On increasing radius plate tectonic small Earth models (Figure 9.6) the ancient North American cluster of crustal fragments remained intact throughout Palaeozoic and Precambrian times. This cluster is shown to straddle the ancient equator, extending from mid-southern to high-northern latitudes. The ancient cluster was, in turn, assembled against the European and Asian Tethys region to the north, the Australian Proterozoic basins to the west, South America to the southwest, and West Africa to the south and southeast, relative to the ancient equator. Small Precambrian crustal fragments now located within Europe were also clustered adjacent to and southeast of Greenland and Scandinavia.

In effect, this ancient North American cluster formed a nucleus for surrounding crustal development. The cluster represented exposed elevated land throughout ancient times and supplied eroded sediments to surrounding sedimentary basins. The Precambrian and Palaeozoic development of the North American cluster then involved crustal extension and basin sedimentation within a surrounding network of continental sedimentary basins. These basins included links to the Tethys Sea basin to the southeast, as well as basins linked to what are now Russia, China, Australia, and South America to the north, west, and southwest.

The North American cluster remained intact throughout the Precambrian and Palaeozoic times until the Pangaea supercontinental breakup and opening of the North Atlantic Ocean commenced during the late-Permian Period. During opening of the North Atlantic Ocean, Scandinavia and the Baltic region were separated from North America and have since remained attached to Europe during further opening of the North Atlantic and Arctic Oceans.

Crustal jostling and mountain building events occurred during Precambrian and Palaeozoic times as a result of on-going changes to surface curvature. These events resulted in long linear mountain belts located around the margins of the North American cluster, forming the precursors to the Cordilleran, Grenville, and Hercynide Mountain belts.

Breakup and fragmentation of the ancient Pangaea supercontinent began during the late-Permian and by the Triassic the early Arctic, North Atlantic, and North Pacific Oceans had also commenced opening. This breakup and opening of the modern oceans then effectively defined the modern North American continental outline.

By the early-Jurassic, breakup had continued to extend into the Arctic and North Atlantic Oceans, and also into the Caribbean and Labrador Seas. Greenland was then separated from Canada and has remained in close proximity. Similarly, South America began an apparent migration away from North America. Fragmentation of the Northern Canadian Islands also occurred during the Jurassic Period, which was intimately related to rifting between Canada and Greenland. Further rifting within the Northern Canadian region has continued to the present-day.

During the breakup of Pangaea and the opening of the Pacific, Atlantic, and Arctic Oceans, each of the established Precambrian, Palaeozoic, and Mesozoic Mountain belts were then fragmented. Remnants of these mountain belts are now separated as far away as Australia, Africa, South America, Russia, and Europe. Much of the Grenville and Appalachian fold-mountain belts remained attached to Eastern North America and the Cordilleran Mountain belt remained attached to Western North America.

The Hercynides remained attached to Europe, the Caledonides to Scandinavia, and the Mauritanides to West Africa. The northern extension of the Cordilleran Mountain belt continued via Alaska into Asia and continued as the Andean Mountain belt into South America. Fragments of this Andean belt also include the New England fold belt of Eastern Australia and remnants can also be seen in New Zealand.

Throughout late-Jurassic to Miocene times, north-south stretching of the

Cordilleran and Andean Mountain belt maintained a continuous continental link extending from South America through to North America and Siberia. The Siberian connection was then severed during the Pliocene Epoch during opening of the Bering Strait, and the Central American link still remains attached today.

From its original equatorial location, the North American continent and continental cluster slowly rotated clockwise as a result of crustal breakup and opening of the Atlantic, Pacific, and Arctic Oceans. Each of the North American, Scandinavian, and Baltic continents has since migrated north into mid- and high-northern latitudes, relative to the ancient North Pole.

9.7. South America

Development of the South American continent on an increasing radius plate tectonic Earth (Figure 9.7) is closely associated with the development of Africa. The South America to Africa assemblage has long been recognized in all plate tectonic reconstructions. Closing of the Atlantic Ocean and assemblage of the American and African plates also forms the basis of the Gondwana supercontinental assemblage, as well as the basis for both continental drift and conventional plate tectonic theory.

In increasing radius plate tectonic studies, the South American crusts originally formed part of an ancient Archaean supercontinent, which in turn formed part of an extensive network of ancient Precambrian sedimentary basins. East Antarctica and Precambrian remnants of West Antarctica and New Zealand were located to the northwest, Australia was located to the north, North America to the northeast, and Africa to the east, south, and west. Subsequent development of South America involved an extended period of crustal extension and fragmentation of the ancient crusts during the Proterozoic Eon and Palaeozoic Era. This development occurred in conjunction with similar Pan-African events in South and West Africa and was also associated with events in ancient Australia and New Zealand.

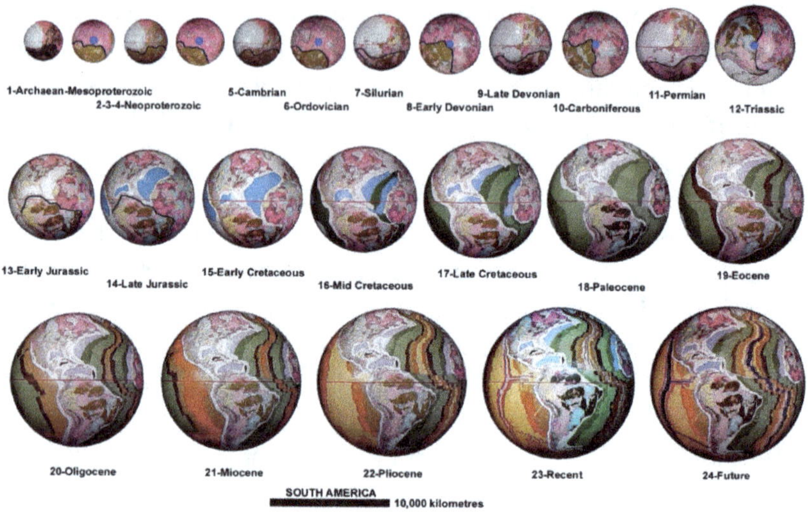

Figure 9.7. Continental crustal development of plate tectonic South America. The horizontal red line represents the location of the ancient equator, blue dots represent the South Pole, and black lines represent the South American continental outline. Maxlow, 2001.

During Precambrian and Palaeozoic times, the ancient South American continent extended from low equatorial to high south polar latitudes. During the late-Palaeozoic to mid-Cretaceous, as the Atlantic and Indian Oceans progressively opened, the South American continent then slowly migrated north, in conjunction with Africa, relative to the ancient South Pole.

From its Precambrian and Palaeozoic southern hemisphere location, South America then rotated clockwise, in sympathy with opening of the Atlantic Ocean, and migrated north to straddle the present-day equator. During the Palaeozoic, crustal extension and basin formation also gave rise to extensive deposition of basin sediments located between what are now South America, Antarctica, Australia, New Zealand, and North America.

Continental breakup between South America, New Zealand, Australia and Antarctica commenced during late-Permian to Triassic times. This breakup occurred in conjunction with opening of the South Pacific Ocean, with New Zealand retaining a brief link to Mexico.

During the Jurassic, opening of the South Atlantic Ocean commenced in the south and this opening then continued north to merge with the opening North Atlantic Ocean. During that time, South America separated from North America during opening of the North Atlantic Ocean and Caribbean Sea. A land connection with North America still remains along the South American and Central American peninsulas. A land connection between the West Antarctican Peninsular and southern South America also remained until final separation during the Miocene during opening of the Southern Ocean.

Mountains developed during the Mesozoic and Cenozoic as long linear belts along the West Coast of South America. This occurred in conjunction with further opening of the South Pacific Ocean and formed a Southern extension of the Cordilleran event in North America. Fragments of these mountain events also occur in New Zealand and Eastern Australia.

This crustal history of South America is in strong contrast with conventional plate tectonic assemblages where opening of the South Atlantic Ocean is compensated for by subduction of the South Pacific plate beneath the west coast of South America to form the Andean Mountain belt. Closing of the Pacific plates by subduction against the Americas is not required on an increasing radius plate tectonic Earth.

9.8. The Future

In current cosmological thinking, the eventual demise of planet Earth involves a fiery end, related to a steady expansion and decay of the Sun in the far distant future. As the Sun expands and decays, scientists tell us that the Sun will eventually envelope and consume each of the planets in turn to form a red giant star.

On increasing radius plate tectonic small Earth models, projection into the future is readily achieved by simply calculating a future Earth radius at any moment in time and extrapolating opening of the mid-ocean-ridges throughout all of the oceans. When the rate of increase in Earth radius established from the small Earth models is projected forward in time, both Earth surface area and radius are shown to increase to the size of the planets Jupiter and Saturn by about 500 million years into the future.

While this scenario is a conjecture based on extrapolation of the past, it is envisaged that one of two things may happen to the Earth during that time. Firstly, if it is at all possible, the Earth may fragment and disintegrate to form a second asteroid belt or, secondly, it may simply continue increasing in size to become another giant gaseous planet.

Because the present Earth is essentially a wet planet, when compared to the dry inner rocky planets of the Solar System, the giant planet scenario seems the most likely outcome. As the Earth's core-mantle continues to increase in size, it is envisaged that entrapped fluid and gas will continue to be expelled from the mantle to form a dense gaseous atmosphere in the far distant future. This expulsion may then extend in time to form planetary ring structures as lighter gases are progressively lost into space.

The reconstruction of an increasing radius plate tectonic Earth five million years into the future (Figure 9.8) is readily achieved by simply calculating the predicted Earth radius from the established radius formula and then adding new seafloor crust along each of the mid-ocean-ridge axes. Apart from the increased distances between the various continents and subtle changes to the coastlines, on this model the distribution of continents and oceans on the future Earth is shown to be essentially the same as it is today.

During this geologically short interval of time, it is calculated that Earth radius will increase by 107 kilometres to 102 percent of the present radius. The series of images in Figure 10.8 show that the increase in Earth radius to 5 million years into the future is consistent with a continued increase in surface area of each of the oceans and lengthening of each of the present-day mid-ocean-ridge spreading axes.

This process of lengthening of the mid-ocean-ridge spreading axes is a direct result of an increase in circumference of the Earth during increase in Earth radius. The mid-ocean-ridge lengthening process then represents an important mechanism for future crustal development during on-going increases in Earth radius.

On an increasing radius plate tectonic Earth, as the mid-ocean-ridge spreading axes lengthen, and the ridges open, they can be visualized as being large propagating cracks in the seafloor—which is precisely what they are. As the cracks propagate and lengthen, they continue to break up the continental and seafloor plate components into ever smaller fragments.

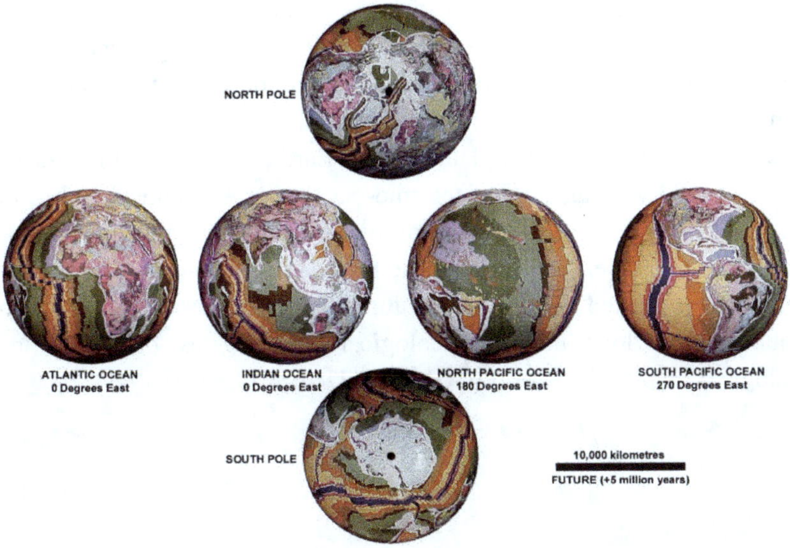

Figure 9.8. The increasing radius plate tectonic Earth projected to 5 million years into the future. The dark blue spreading ridge represents spreading along each of the mid-ocean-ridges for the next 5 million years. The model shows an extension of all mid-ocean-ridge spreading axes into seismically active areas such as Turkey, Japan, California, and New Zealand. Maxlow, 2001.

This breakup is currently occurring within each of the major earthquake prone areas of the world today. In these earthquake prone areas seismic and earthquake activity occurs as a result of tensional cracking and breakup of the crusts. This breakup is also accompanied by intrusion of volcanic lava, elevated heat flow, and expulsion of new water and gases from the mantle.

Lengthening of the East Pacific mid-ocean-ridge spreading axis on an increasing radius plate tectonic Earth is currently occurring as a northward extension of the spreading ridge passing through the Gulf of California. This gulf will eventually rift and separate the Californian Peninsula from North America to form an island.

A northward extension of the Red Sea Rift zone through the Gulf of Aqaba and Dead Sea region into Turkey will eventually result in rifting and separation of the Sinai Peninsula from Arabia. A northern extension of the Marianas spreading ridge is shown to be continuing towards Japan and a southern extension of the Tongan spreading ridge is also continuing through New Zealand.

Elongation of the mid-ocean-ridge spreading zones within these areas contrast strongly with the conventional plate tectonic requirement for plate convergence, continental collision, and subduction of vast areas of oceanic crust.

On an increasing radius plate tectonic Earth, an increase in surface area accompanied by elongation of the mid-ocean-ridges is considered to better represent the breakup and separation of the continents and opening of the existing oceans when moving into the future. This mid-ocean-ridge opening process, as projected well into the future, also continues to comply with the established seafloor bedrock geological mapping as portrayed in the Geological Map of the World (CGMW & UNESCO, 1990).

Chapter 10

Modelling Summary

"*It is not a task of the geologist to explain problems beyond their discipline. Their task is to see and correctly explain all geological facts.*" Stefan Cwojdzinski (in Maxlow 2005).

The bedrock Geological Map of the World was first published in 1990. This map demonstrates conclusively the indisputable presence, configuration, and distribution of all identified plates, which quantifies the underlying basis of plate tectonics. This map also demonstrates conclusively that there is no evidence for the randomness of plate motion insisted on by continental drift theory.

While completion and publication of this geological map was originally intended to provide a key role in defining and quantifying plate movement in our oceans, it is unfortunate that, because it fails to quantify continental drift, it remains underutilized and ignored in all modern plate tectonic studies.

The essence of these past chapters has been to test and quantify the proposal that modern published geological mapping of the oceans and continents can be used to constrain crustal plate assemblages on models of ancient Earth. Heavy reliance has been made on using the published bedrock Geological Map of the World map (1990) to constrain assemblage of both the oceanic and continental plates.

In order to achieve this aim, all preconceptions about Earth's size were simply ignored in order to both measure the potential size of the ancient Earth and to establish a formula for determining ancient Earth radii at any moment in time.

This bedrock mapping and measured ancient radius formula were then used to construct spherical small Earth models of the Earth, referred to here as plate tectonic small Earth models. These models extend from the early-Archaean, some 4,000 million years ago, to the present-day, plus one model extended to 5 million years into the future. In all cases, this plate assemblage exercise, constrained by the geological mapping, achieved a series of small Earth models where all crustal plates are shown to assemble together precisely, with only one unique plate-fit option.

It is emphasized that, although conventional plate tectonic reconstructions of individual regions on a constant sized Earth can achieve a localized high degree of plate fit-together, in most cases the use of palaeomagnetic apparent-polar-wander to constrain assemblages results in multiple plate-fit options. These reconstructions obscure the fact that crustal assemblage, development, and displacement in one region of the Earth globally affect all other areas.

Modern bedrock geological mapping also shows that the continents are not rigid plates, as is often portrayed in conventional tectonic studies, but are instead a complex array of crustal domains. These domains each have their own complex crustal histories, which must be taken into consideration, and strictly adhered to, when undertaking any modelling or crustal assemblage studies.

The modelling studies presented here demonstrate conclusively that seafloor crustal plates, when reconstructed on plate tectonic small Earth models, coincide fully with the seafloor spreading and geological data and accord precisely with the derived ancient Earth radii for each model constructed. This coincidence applies not only to the more traditional oceans, such as the Atlantic Ocean, where conventional reconstructions agree in principle, but also to the Pacific Ocean where the necessity for subduction of all, or part of the seafloor crusts generated at spreading ridges is refuted.

The plate tectonic small Earth models presented here demonstrate that the mechanism of seafloor spreading and distribution of crustal plates, as highlighted by the seafloor geological mapping, provides a definitive means to accurately constrain crustal assemblages on small Earth models, with only one plate-fit option.

By progressively removing age-dated seafloor volcanic crust from each of the small Earth models in turn, it is shown that the global plate fit-together along each of the mid-ocean-ridge plate margins achieves a better than 99% fit, for each post-Triassic model constructed.

This unique plate fit-together of the seafloor crusts is considered to empirically demonstrate that post-Triassic plate tectonic small Earth geological modelling is indeed a viable process and, from this demonstration, it is therefore justifiable to consider extending continental crustal modelling studies back to the Archaean.

This empirical study further demonstrates that all remaining continental crusts assemble as a complete Pangaean Earth at approximately 50 percent of the present Earth radius during the late-Permian—around 250 million years ago.

Quantification of an increasing radius plate tectonic Earth back to the early-Archaean requires an extension of the fundamental cumulative seafloor crust premise to include continental crusts. Continental crust was reconstructed on pre-Triassic small Earth models by considering the primary crustal elements cratons, orogens, and basins.

In order to assemble pre-Triassic plate tectonic small Earth models, consideration was given to an increase in Earth surface area occurring as a result of crustal stretching and extension within an identified network of continental sedimentary basins.

Moving back in time, this crustal extension was progressively restored to a pre-extension, pre-stretching, or pre-rift crustal configuration by simply removing young sedimentary and intruded magmatic rocks and reducing the surface areas of each of the sedimentary basins in turn. Consistent with the geological mapping shown on the Geological Map of the World.

During this process, the spacial integrity of all existing ancient cratons and orogens was retained until restoration to a pre-orogenic configuration was required. By removing basin sediments and magmatic rocks, as well as progressively reducing the surface area of each sedimentary basin in turn, a potential ancient primordial small Earth with a radius of approximately 1,700 kilometres was achieved during the early-Archaean.

This primordial Earth, representing 27 percent of the present Earth radius, is comprised of an assemblage of the most ancient Archaean cratons and Proterozoic basement rocks. All other rocks were simply returned to their places of origin, albeit back to the mantle or back to the ancient lands as required.

Crustal assemblages on each of the plate tectonic small Earth models also show that large Panthalassa, Tethys, and Iapetus Oceans are not required during model construction. These oceans are instead replaced by lesser continental Panthalassa, Iapetus, and Tethys Seas, which represent precursors to the modern Pacific and Atlantic Oceans, as well as ancient sedimentary basins located on many of the present-day continents.

Similarly, emergent land surfaces during the Precambrian and Palaeozoic Eras are shown to equate to the conventional Rodinia, Gondwana, and Pangaea supercontinents and smaller sub-continents.

On each plate tectonic small Earth model, supercontinental development is shown to be progressive and evolutionary, which is related to crustal extension within an established network of continental sedimentary basins, changes to Earth surface curvature, and changes to sea levels.

At this stage, it is considered that the outcomes of this geologically-constrained small Earth crustal modelling exercise are more than adequate to quantify the validity of adopting an increasing radius crustal process as the primary mechanism for plate tectonic modelling.

This exercise also demonstrates that there is indeed a *crisis situation* arising in what we are currently taught in conventional continental drift-based plate tectonics, which prompts a need for a paradigm shift in plate tectonic thinking.

This crustal modelling exercise, however, is not the problem that people see. The fundamental problem that scientists, and the general public, have is comprehending where did the huge volume of material making up the seafloor crusts and underlying mantle go to when moving back in time in order to reassemble the continents? And, more importantly, where does this huge volume of material come from when moving forward in time?

From this perceived problem, it would seem that it doesn't matter how convincing any of the constructed empirical models or outcomes are, if an explanation for these observations cannot be given to the satisfaction of scientists and the general public alike, then all increasing Earth considerations must remain rejected.

It is emphasized in this geological modelling summary that, even if a mechanism for an increasing Earth mass and radius were previously not fully known or understood, the extensive analysis presented here is based on readily available modern bedrock mapping data.

The outcomes of this analysis are further based on readily reproducible empirical geological modelling studies, and it is important to emphasize that a prior lack of a fully comprehended mechanism does not invalidate the need to at least test the concept of an increasing radius plate tectonic Earth model.

As Cwojdzinski wrote in 2005 (pers. comm.), "*The insinuation that we still do not know a physical process responsible for an accelerated expansion of the Earth is not a scientific counterargument.*" He further commented that, "*It is not a task of the geologist to explain problems beyond their discipline. Their task is to see and correctly explain all geological facts.*"

It is fair to then ask the very pertinent question that, if an acceptable causal mechanism is proposed, as palaeomagnetics did for the rejected continental drift theory during the 1950s, do we seriously consider this mechanism, test the new proposal in light of modern global observational data, accept the empirical evidence in support of this proposal, and revise the current plate tectonic theory? Or do we continue to reject the observational data and acceptable mechanism and instead remain supportive of an outdated theory based on a pre-assumed constant Earth radius premise?

Part Two

Empirical Global Data Modelling

"*Ultimately world reconstructions must be congruent not only with the data from geology and geophysics, but also with palaeobiogeography, palaeoclimatology, and palaeogeography.*" Shields, 1997.

Chapter 11

Proposed Causal Mechanism

"*...it may be fundamentally wrong to attempt to extrapolate the laws of physics as we know them today to times of the order of the age of the Earth and of the Universe.*" Creer, 1965.

Irrespective of empirical evidence, the theory of Earth Expansion was rejected during the mid-twentieth century based mainly on the lack of a suitable and convincing mechanism to explain the necessary increase in Earth mass and radius over time. This rejection was further quantified by the conclusions of palaeomagneticians, based on measurements of ancient Earth radius from palaeomagnetic pole data. This palaeomagnetic pole data was also used earlier for quantifying the movement of continents in the continental drift theory.

Since this rejection, a considerable amount of new scientific data and evidence has become available for study, in particular space-based observational data, over the past decades. This modern evidence seriously challenges conventional insistence that Earth mass and radius has remained constant and instead may offer a plausible causal mechanism for an Earth increasing in mass and radius over time.

The detailed knowledge and influence of charged solar wind-related particles emanating from the Sun on the near-Earth environment has been available since the Cluster II satellites were launched by the European Space Agency in year 2000. The new space-based observational data subsequently collected has highlighted the introduction of large quantities of solar wind-related electrons and protons into the Earth, propelled by the Earth's magnetic field. This observational data then begs the question as to what is happening to the particles—the building blocks of all matter on Earth—once they enter the Earth?

11.1. Plasma Transfer from the Sun

The near-Earth observations presented here are based on a suggestion put forward by Eichler in 2011. Eichler posed the question, "*Does plasma from the Sun cause the Earth to increase in size?*" and by presenting a new argument based on known physical observations, he suggested that the answer to this question might indeed be the case. Eichler elaborated with his statement that:

> "*To assume that the Earth is gaining matter and that this may be due to nucleosynthesis within the Earth seems to fly in the face of conventional wisdom—and it does. Based on empirical geologic evidence which strongly indicates that... [an increase in Earth size] ...is indeed valid, the task confronted is to formulate a viable mechanism whereby this occurs. In a plasma universe, the Earth is under constant bombardment from space, with all the necessary components to reconstitute matter from its component parts deep within the Earth not requiring theoretical constructs which have never been experimentally observed. The Earth, having a magnetic field strong enough to interact with impinging particles, gathers more than sufficient fundamental particles, namely electrons and protons, to account for a slow increase in matter internally over hundreds of millions of years. There is therefore no lack of component particles to create new matter deep within the body of the Earth. The exact process by which this occurs is complex in nature and, like the interior of the Earth itself, involves speculation as to its dynamics. It is argued that the avenue of approach proposed here is plausible and warrants further serious scientific investigation. If new matter has been added to the interior of the Earth, there must be an answer to the riddle of the dynamics of the process.*"

This proposal also incorporates the observations of Kremp who, in 1992, suggested that new geophysical evidence indicates that the Earth has been growing rapidly in the past 200 million years. Kremp indicated that seismologists have located the existence of a zone, about 200 to 300 kilometres thick, located at the base of the mantle directly above the core-mantle boundary, designated the D" region. Yuen and Peltier in 1980, as well as Boss and Sacks in 1985, had earlier postulated the existence of a substantial flow of heat across the core-mantle boundary and concluded that if whole-mantle convection—a plate tectonic requirement for convection within the mantle—were to occur in the Earth's mantle, this D" region should be the lowest thermal boundary layer of the whole-mantle system. With the temperature of the outer core of the core-mantle boundary estimated to be about 800 degrees higher than the D" layer of the mantle, or perhaps even 1,500 degrees higher, Kremp concluded that this thermal increase in the outer core may be a fairly recent process forcing a rapid increase in Earth radius.

The proposed causal mechanism for matter transfer from the Sun is summarized by Eichler in the following salient points:

1. If the Earth is undergoing an increase in radius, as geological studies presented here strongly indicate, there must be a mechanism whereby the Earth increases in mass with increasing time.
2. Previously proposed mechanisms for mass increase are not considered feasible for various reasons, primarily because of the lack of current principles of known physics or esoteric theories lacking experimental verification.
3. It is conservatively estimated that the Sun ejects sufficient quantities of matter in the form of electrons and protons to account for an increase in mass of the Earth over the geological periods of time being discussed here.
4. Individual atoms and ions are generally of sufficient size—of the order of a nanometre (one-billionth of a metre)—that penetration of such particles into solid matter to any significant depth is, in general, highly unlikely.
5. On the other hand, individual electrons and protons are of roughly comparable size—of the order of one-millionth of a nanometre or less—and small enough to freely move through solid matter by the process known as conduction, although the nature of conduction varies considerably for each particle. Proton conduction has been extensively investigated, in particular through forms of matter constituting the generally accepted geological composition of the Earth.
6. Any new matter formed requires the presence of both electrons and protons, without which atom formation cannot occur.
7. Extensive investigation of LENR (low energy nuclear reactions) has been carried out over the past quarter century and continues today. A number of international researchers have obtained experimental

results strongly indicative that LENR occurs, although the conditions under which they occur are intermittent and not well understood. This investigation also applies to comprehensive theoretical understanding of mechanisms involved. A number of scientific books and articles have been written on the subject, and a number of theories have been put forth, although much of this material is based on experimental evidence and is a poorly understood area of science.

8. Since the movement of both electrons and protons—as for all electrically-charged particles—is dependent on the presence of magnetic fields, celestial bodies not having magnetic fields like that of the Earth, such as the rocky planets and the Moon, most likely do not exhibit similar mass-increase via this mechanism.
9. Exactly where within the Earth electrons and protons combine to form new atomic structures is open to conjecture. The D" layer is a prime candidate because seismic-detectable activity appears to be occurring at that level, although it may be happening elsewhere such as within the mantle or even perhaps the upper core itself. It has even been suggested that the existence of a molten mantle may be necessary for mass gain to occur.
10. Although poorly understood—and desperately begging more theoretical studies, this mechanism appears to be a viable solution to the problem of apparent mass increase of the Earth over geological time.

11.2. Earth and Solar System

The Sun is the Solar System's only star and is by far its main physical component. Its large mass, estimated to be 332,900 Earth masses, produces temperatures and densities in its core that are high enough to sustain nuclear fusion. These nuclear fusions release enormous amounts of energy, mostly radiated into space as electromagnetic radiation, peaking in the 400 to 700 nanometre band of visible light, but is by no means confined to the visible spectrum only.

Along with light, the Sun also radiates a continuous stream of plasma—known as the solar wind—which represents a stream of charged particles released from the upper atmosphere of the Sun consisting mostly of electrons, protons, and other ions. This stream of charged particles spreads outwards at roughly 1.5 million kilometres per hour, creating a tenuous atmosphere called the heliosphere that permeates the Solar System out to well beyond the orbit of Neptune. This atmosphere is also known as the interplanetary medium. Activity on the Sun's surface, including solar flares and coronal mass ejections, disturbs the heliosphere, creating a form of space weather, which in turn generates geomagnetic storms.

Coronal mass ejections and other similar events, in turn, produce a magnetic field where huge quantities of charged material are ejected from the surface of the Sun. The presence of the Earth's own magnetic field stops its atmosphere from being stripped away by the solar wind. This magnetic field contrasts with Venus and Mars, for instance, that do not have well developed fields and, as a result, it is speculated by others that the solar wind has caused their atmospheres, if present, to gradually bleed away into space.

Interaction of the solar magnetic field and ejected plasma with the Earth's own magnetic field funnels charged particles into the Earth's upper atmosphere, where it forms auroras—a natural light display in the sky seen near both of the magnetic poles. The Earth's magnetic field, in effect, makes the capture cross section many times larger than the cross section of the Earth itself.

The total number of magnetized particles carried away from the Sun by the solar wind is now estimated to be about 1.3×10^{36} per second. From this estimate, the total mass loss of the Sun each year is further estimated by others to be about 4 to 6 billion tonnes per hour, or 35 to 53 trillion tonnes per year. This mass loss is equivalent to losing a mass equal to the Earth every 150 million years or conversely providing a mechanism to double Earth radius in the past 250 million years.

11.2.1. Effects on the Present-Day Earth

In strong contrast to what was available during the mid-twentieth century, modern space technologies now show that the Earth and other planetary bodies are constantly immersed in a solar wind. This solar wind travels to Earth with a velocity of around 400 kilometres per second and a density of around 5 ions per cubic centimetre. During solar magnetic storms, the flow of plasma-related ions can be several times faster and the interplanetary magnetic field may also be much stronger.

The Earth's magnetosphere has now been shown to be full of trapped plasma emanating from the solar wind as it passes the Earth. This flow of plasma into the magnetosphere increases with increase in solar wind density and speed, as well as increases in turbulence in the solar wind during solar storms.

In addition to moving perpendicular to the Earth's magnetic field, it is shown that magnetospheric plasma travels down along the Earth's magnetic field lines within the auroral zones. New Cluster II satellite research by the European Space Agency suggests that this process may be more common than previously thought and possibly represents a means for the constant penetration of solar wind-related plasma into the terrestrial environments.

It was also shown by the European Space Agency that the existence of certain waves in the solar wind enables incoming plasma to breach the magnetopause, suggesting to scientists that the magnetosphere responds more as a filter rather than a continuous barrier. Of particular importance here is that these discoveries were considered by the European Space Agency's project scientists to be of great importance because it showed how the Earth's magnetosphere can be penetrated by solar particles under specific interplanetary magnetic field circumstances.

Beyond the known observations about solar wind-related plasma flow into the Earth, all else remains speculative once it enters the Earth. Again, as Eichler states, "*There is therefore no lack of component particles to create new matter deep within the body of the Earth. The exact process by which this occurs is complex in nature and, like the interior of the Earth itself, involves speculation as to its dynamics. It is argued that the avenue of approach proposed here is plausible and warrants further serious scientific investigation. If new matter has been added to the interior of the Earth, there must be an answer to the riddle of the dynamics of the process.*"

The suggested effects of this penetration on present-day Earth are schematically summarized in Figure 11.1. The Earth is known to have a strong iron-related magnetic field that extends from the interior to the outer magnetosphere. The Earth's magnetic field approximates that of a magnetic dipole and is currently tilted at an angle of about 10 degrees with respect to the Earth's rotational axis.

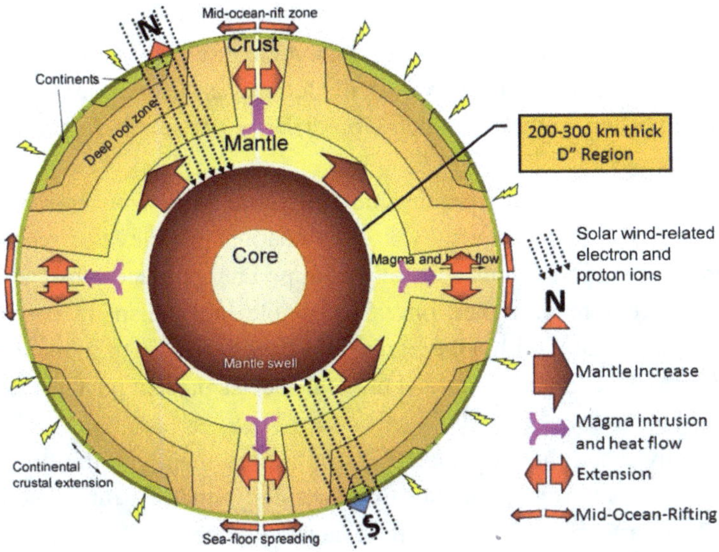

Figure 11.1. A schematic cross-section of the present-day Earth highlighting the influence of charged electrons and protons entering the Earth resulting in increase in mass and radius over time.

This magnetic field contrasts with the inner rocky planets of the solar system, as well as the Earth's Moon, which have non to very weak magnetism, and the giant planets which all have very strong to exceptionally strong magnetic fields. This contrast between the essentially non-magnetic inner rocky planets and the strong magnetic giant planets may suggest a common theme for increase in mass and radius of the giant planets, as distinct from no to very little increase in mass or radius of the smaller planets and moons.

It is envisaged that magnetically charged electrons and protons enter the Earth's magnetosphere and lower terrestrial layers, primarily at the polar auroral zones, and as random lightning strikes during electrical storms. These magnetically charged particles are further attracted by conduction to the strongly magnetic core-mantle region of the Earth.

The elevated core-mantle temperatures and pressures present enable the particles to dissipate and recombine via nucleosynthesis as new matter within the upper core or lower mantle regions, in particular the 200 to 300 kilometres thick D" region located at the base of the mantle, directly above the core-mantle boundary. This nucleosynthesis is precisely as Kremp suggested in 1992: "*...this thermal increase in the outer core may be a fairly recent process forcing a rapid expansion of the Earth.*"

The difference between what Kremp envisaged and what is presented here is that this matter formation represents a mantle growth process—an increase in mass and volume—focused within the lower mantle to upper core regions, not an expansion process—which would imply phase changes with no net increase in mass.

New matter formation requires not only pure energy, but the presence of both electrons and protons. The combination of elevated core-mantle temperatures, and the abundance of incoming charged electrons and protons within this region, may then provide a viable mechanism to continuously synthesis new matter within the Earth. This new matter, in turn, represents the building blocks of all elements and mineral species present on and in the Earth and Universe today.

It is envisaged that new matter is synthesized mainly within the reactive upper core or D" region of the lower mantle, which in turn results in an increase in Earth mass. This growth of new matter causes the core and mantle to increase in volume. This increase in volume is then transferred to the Earth's outer surface crust via two primary mechanisms. Firstly, as an increase in Earth radius, and secondly, as laterally-directed crustal extension which is presently occurring on the surface of the Earth as extension along the full length of the mid-ocean-rift zones, within continental sedimentary basins, and within more localized mantle plume and large igneous complex regions.

Extension within the upper mantle and surface crustal regions is highlighted by paired red arrows, shown at each mid-ocean-rift spreading zone in Figure 11.1. This mantle and crustal extension process enables newly formed magma to be squeezed from deep within the Earth, where it travels by convective flow up to the surface, as indicated by the upward facing magenta-coloured arrows.

The surface expression of this magma and high heat flow process is focused along the full length of the centrally located mid-ocean-rift zones, plus leakage along known hot spots, volcanic centres, and large and small igneous provinces. This process is also accompanied by intrusion and extrusion of new basaltic mantle-derived lava and granite-related magma, as well as expulsion of new water and atmospheric gases at the surface.

On an Earth increasing its mass and radius over time, all geological observations now seen at surface, such as basin formation, folding, faulting, orogenesis, magmatic intrusion, mountain formation, and so on, are considered to be related to this crustal extension and continental breakup processes. These observations are particularly related to changes in relief of surface curvature of continental and seafloor crusts, as a direct result of an increase in Earth radius and surface area over time.

11.2.2. Effects on the Ancient Earth

The potential effects of this causal process on the ancient Earth are summarized in Figure 11.2. This figure is an extension of Figure 11.1 and shows a schematic cross section of three small Earth models: the Cambrian (around 540 million years ago), the Permian (around 250 million years ago), plus the present-day model repeated.

These models are all on the same scale and are based on the geological plate tectonic small Earth modelling studies. The full range of small Earth geological models extend back to the early-Archaean, as well as providing infill on the schematic models shown in Figure 11.2 plus one model extended to 5 million years into the future.

The schematic models shown in Figure 11.2 have been chosen because, on an increasing radius plate tectonic Earth, these models mark times of significant changes to crustal development. The changes basically include changes to the aerial distribution of ancient sedimentary basins and continental seas during ongoing crustal extension throughout the Palaeozoic Era, plus breakup of the Pangaean supercontinent and formation of the modern continents and oceans during the late-Permian. These, and additional changes, are discussed in detail throughout this book.

Prior to the Cambrian model shown in Figure 11.2, the 3.5-billion-year interval of time known as the Precambrian Eon, where it has been shown that the rate of change in Earth mass, surface area, and radius during this extended period of time was extremely small. During that time, it is envisaged that matter formation within the primitive lower mantle and upper core was at an early stage of development, and the increase in Earth radius amounted to microns increasing to millimetres per annum.

It is envisaged that the limited size of the Precambrian Earth, along with the possibility of a much-reduced magnetic field, may have limited the amount of solar particles entering the Earth, hence limited the rate of increase in mass and radius over time.

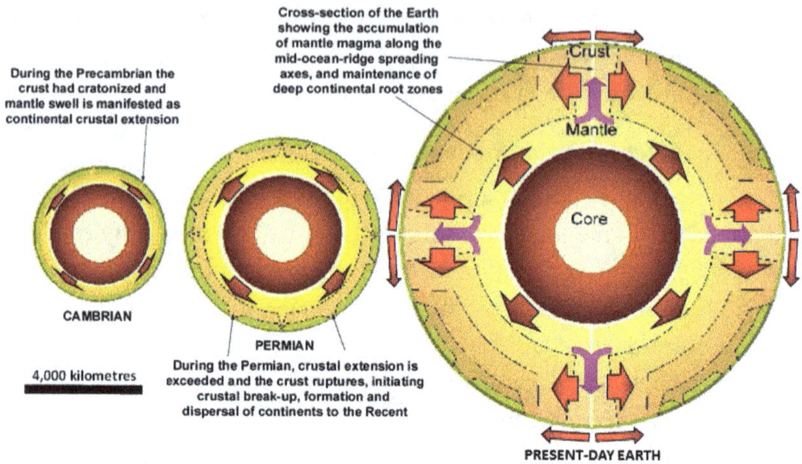

Figure 11.2. A schematic cross-section of the Cambrian, Permian and present-day Earth highlighting the effects of an Earth increasing its mass and radius over time. Core and mantle dimensions for the Permian and Cambrian plate tectonic models remain speculative only.

The approximately 310-million-year time interval between the beginning of the Cambrian to the end of the Permian Periods is the Palaeozoic Era and, on an increasing radius plate tectonic Earth, it is shown that this era represents a time when Earth mass, radius and surface area gradually increased from a steady to steadily accelerating rate of change. Again, this increasing rate of change in mass and radius may have been the result of an increasing Earth magnetic field intensity, as well as increased changes to output of charged solar particles from the Sun.

The end of the Permian Period is marked by a breakup of the Pangaean supercontinent to form the modern continents and opening of the modern oceans. The following Mesozoic and Cenozoic Eras represent a time of dispersion of the modern continents, ongoing opening of the modern oceans, and a rapidly accelerating rate of change in Earth mass, radius, and surface curvature.

On plate tectonic small Earth models, for the first 3,750 million years of Earth history—to the end of the Permian Period some 250 million years ago—it has been shown that all continental crusts remained united to form a single supercontinental landmass encompassing the entire ancient Earth. This supercontinental landmass was, in turn, partly covered by a network of relatively shallow continental seas. During this extended period of time, the overall crustal extension and increase in surface area amounted to microns, increasing to tens of millimetres per annum, spread over the entire surface area of the ancient Earth.

During these pre-Permian times, volumetric mantle growth was manifested at surface as an increase in surface area within an interconnected network of crustal weakness—sedimentary basins and orogenic zones. This network was first established during the early-Archaean Eon and progressively extended in surficial area over time.

Continental crustal extension was focused within this network of crustal weakness, which initiated formation of low-lying sedimentary basins and accumulation of relatively shallow seas. This crustal extension continued unabated until the late-Permian Period where the ability of the crust to continue to stretch and extend was physically exceeded.

During the late-Permian Period, the ability of the crusts to stretch and extend within the Pangaean supercontinental crusts—shown as green coloured crusts on the Permian model (Figure 11.2)—was exceeded, and rupturing of the Pangaean supercontinent was initiated.

The Pangaean supercontinental crusts then ruptured and broke apart to form the modern continents and the modern oceans. The outcomes and growth history of this breakup and opening process, extending from early-Jurassic to the present-day, is now preserved within the seafloor crustal geology shown in modern publications, such as the Geological Map of the World (CGMW & UNESCO, 1990).

On a plate tectonic increasing Earth model, all geological observations now seen at surface, such as basin formation, folding, faulting, orogenesis, magmatic intrusion, mountain formation, crustal extension, continental break-up and so on, are considered to be ultimately caused by the volumetric growth in Earth mass and volume over time. These observations are particularly related to changes in relief of surface curvature of continental and seafloor crusts as a direct result of the subsequent increase in Earth radius and surface area over time.

Irrespective of the infancy of this solar wind-related observational data, and causal mechanism for an increase in Earth mass and radius, it is considered that there is more than enough justification to at least consider that the mass and radius of the Earth may in fact be increasing over time. This observation then begs the question as to what would happen to global tectonics on an increasing radius Earth model?

We can no longer justify rejection of an increasing Earth mass and radius proposal in favour of a constant radius Earth model, without at least scientifically testing this new proposal using modern global observational data. Validation of this new proposal would then constitute a paradigm shift in the way we currently promote conventional plate tectonics and hence justify science looking beyond conventional continental drift-based plate tectonics to see what else this modern data has to offer.

Chapter 12

A Paradigm Shift in Palaeomagnetics

"*The many geophysical and geological paradoxes that have accumulated during the past two or three decades are apparently the consequences of forcing observational data into an inadequate tectonic model.*" Storetvedt, 1992.

Small Earth geological modelling studies have now reached a point where additional palaeomagnetic evidence can be introduced in order to physically locate the ancient magnetic poles and corresponding equators on each small Earth model. Once the magnetic poles and equator are located, this information will then be used to establish a geographical latitude and longitude grid for each model, which will be used in later chapters as a platform for modelling additional global observational data.

These pole and equator locations have previously been shown on plate tectonic small Earth models in previous chapters.

Palaeomagnetics is the study of remnant magnetism preserved in rocks and is credited with leading to the revival of the theory of continental drift, as well as its subsequent transformation into plate tectonics during the mid-1960s.

The primary role of palaeomagnetic studies carried out on samples of magnetized rock is to provide measurements to determine the ancient latitude of a sample site, and to measure a direction and angular distance to the ancient magnetic pole. Similarly, measurements taken from widely separated sample sites are also used in conventional studies to determine apparent-polar-wander-paths—a plate tectonic concept referring to the apparent movement of an ancient magnetic pole location—for ancient continental fragments.

Palaeomagnetics is traditionally considered the cornerstone of conventional plate tectonic studies and has been used to supply a large amount of data about past locations of continents and crustal plates. This data has provided evidence about motion histories of the various continental crusts, continental growth, mountain belt formation, as well as measurements of the Earth's ancient radius.

It was concluded in 1975 from the palaeomagnetic studies of McElhinny and Brock that "*...within the limits of confidence, theses of exponential Earth expansion, or even moderate expansion of the Earth are contradicted by the palaeomagnetic evidence.*" This conclusion therefore led McElhinny and Brock to further conclude, "*...there has been no significant change in the ancient radius of the Earth with time.*" With this in mind, we will now look at palaeomagnetics in more detail to try and resolve this dilemma.

12.1. Remnant Magnetism

Remnant magnetism is preserved in rocks that contain magnetic minerals, in particular volcanic rocks containing the iron-rich mineral magnetite. Over time, these rocks acquire a weak but permanent magnetism during crystallization, deposition, or cementation of the magnetic minerals. The orientation and polarity of this preserved remnant magnetism have been shown elsewhere to be aligned parallel to the Earth's ancient magnetic field that existed at the time of preservation.

Creer and others first appreciated the application of palaeomagnetics to tectonic studies in 1954. At that time, they observed that remnant magnetism, measured in a series of rock samples from the same general area but of increasing age, gave a progressive change in the remnant magnetic vector and hence position of the derived ancient magnetic pole.

This change in magnetic vector suggested that the ancient magnetic pole had either moved because of changes in the position of the pole, or because of changes in the position of the continent containing the rock sample. A distinction between these two suggestions was made possible by comparing the apparent polar wander paths determined from sample sites located on different continents. Early tests confirmed that the ancient continents had indeed moved relative to each other, rather than the poles moving.

The collection of palaeomagnetic data from sample sites located on all continents is now extensive, and the measurement and statistical treatment of this data has reached a very high degree of precision. A number of limitations to the use of palaeomagnetic data must, however, be clarified prior to interrogating palaeomagnetics and establishing a basis for palaeomagnetic modelling on an increasing radius plate tectonic Earth.

Limitations to the use of palaeomagnetic data include:

1. Measurement of remnant magnetism in a sample of rock can only be used to determine the precise ancient latitude—palaeolatitude—of a sample site, as well as a precise direction—declination—to the ancient magnetic pole—palaeopole.
2. Palaeomagnetics cannot determine, nor can it constrain the ancient longitude—palaeolongitude—of a sample site.
3. Determining an ancient Earth radius using palaeomagnetic data assumes that continents have maintained a constant or near constant surface area over time.
4. Ancient magnetic pole locations, determined from palaeomagnetic data, are derived locations, which are based on conventional formulae that are in turn based on the fundamental assumption that Earth's radius is constant.

These limitations to the use of palaeomagnetic data represent significant constraints during conventional plate tectonic reconstructions of the ancient supercontinents, as well as plate motion studies on a constant radius Earth model.

12.2. Palaeomagnetic Dipole Formula

Palaeomagnetic measurements on samples of rock are based on weak remnant magnetic vectors preserved in rocks that contain small amounts of iron-rich magnetic minerals. To determine both the ancient latitude and ancient pole location from a magnetized rock sample, there are a set of well-established formulae based on known magnetic properties. Derivation of these conventional palaeomagnetic formulae is shown in Appendix B. These formulae are based on the unwavering assumption that Earth's ancient radius is equal to, or approximately equal to, the present Earth radius.

Conventional palaeomagnetic formulae are based on what is referred to as the geocentric axial dipole model of the Earth (Figure 12.1). This model forms the basis for all palaeomagnetic studies and is equally applicable to a magnetized sphere or a planet of any radius.

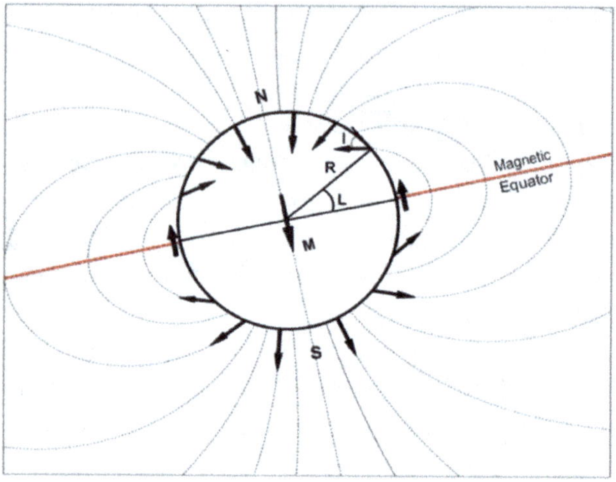

Figure 12.1. Geocentric axial dipole model of the Earth. A magnetic dipole M is located at the centre of the Earth and is aligned with the rotation axis. The geographic latitude is L, and the mean Earth radius is R. The magnetic field directions at the Earth's surface, produced by the geocentric axial dipole, are schematically shown. Inclination I is shown for one location and N is the North magnetic pole. Maxlow, 2001.

In Figure 12.1, the geographic pole and rotational axis of the Earth are coincident with the mean magnetic north and south poles. A single magnetic north and south dipole vector, shown as M, is located at the centre of the Earth and is aligned with the geographic rotation axis. This single dipole produces the entire Earth's magnetic field, shown as grey dotted lines, and its configuration is exactly the same for any radius planet or magnetized object.

Palaeomagnetic formulae use measurements of the natural remnant magnetic vector preserved in a rock sample to calculate the ancient latitude, or ancient colatitude of a sample site (Figure 12.2). Latitude is the angular distance of a sample site north or south of the ancient equator and colatitude is the angular distance from a sample location to the ancient North or South Pole. From this information, in conjunction with measured declination, the location of an ancient magnetic pole can then be calculated for each sample site measured.

The formula used to determine the ancient latitude, or colatitude, of a sample site was defined by Butler in 1992 as:

$$\tan I = 2 \tan L = 2 \cot P \qquad \text{Equation 1}$$

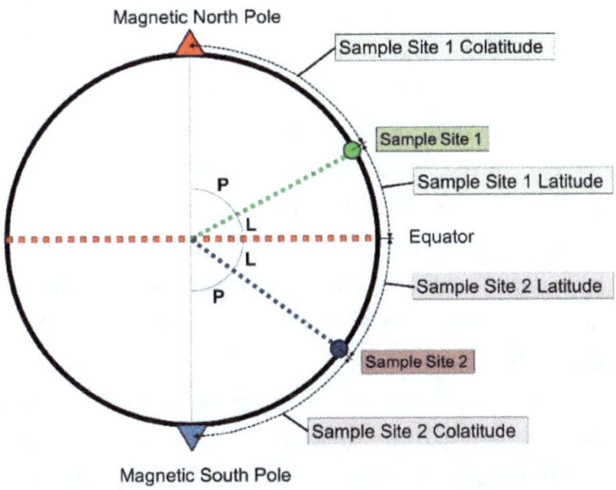

Figure 12.2. Cross section of the Earth showing latitude (L) and colatitude (P) relationships between two sample sites, the magnetic North and South Pole, and the equator.

Rearranging the dipole equation gives colatitude P as:

$P = \cot^{-1} (\tan I/2) = \tan^{-1} (2/\tan I)$ Equation 2

Where:

1. I is the mean angle of inclination of the remnant magnetic field which is preserved in rocks located at a sample site. This angle increases from minus 90 degrees at the magnetic North Pole to plus 90 degrees at the magnetic South Pole.
2. L is the ancient geographic latitude, calculated from the measured angle of inclination I, and it represents the angular distance along a meridian extending from the equator to the ancient sample site.
3. P is the ancient geographic colatitude, calculated from the measured angle of inclination, and represents the angular distance along a meridian extending from the sample site to the ancient magnetic North or South Pole, depending on which hemisphere the sample site is located (Figure 12.2).

To minimize errors during sampling of the magnetized rock under study, and to give a more statistically representative measurement, a series of samples are generally taken for each sample site. These samples are structurally corrected to compensate for any Earth movements that may have occurred to the site since deposition of the rock sample. They are also magnetically screened to remove any overprinting magnetic vectors, such as metamorphism, that may have occurred since initial preservation.

A statistical mean value for the angle of inclination is then determined from the corrected samples, and this mean value is assumed to represent the ancient time-averaged magnetic field present at the time of preservation. This sampling procedure, in effect, compensates for any small annual variations in the magnetic field which may have been caused by wandering of the magnetic pole axis around the geographical axis, as well as helping to mean out any errors in the determinations.

For the geocentric axial dipole field (Figure 12.1), the time-averaged and structurally corrected angle of inclination for the particular sample site is then used to calculate the ancient latitude, or colatitude, of the sample site for when it was first preserved in the rock-record. Similarly, the time-averaged angle of declination—the measured angular direction to the ancient pole—measured from the site sample, is also used to determine the angular direction towards the ancient magnetic pole.

The primary purpose of this sampling and statistical process is to locate the position of the ancient magnetic North Pole, as determined by the site sample on the present-day Earth. The location of the ancient magnetic South Pole can also be determined from sample sites located in the southern hemisphere. Using conventional palaeomagnetic formulae, the angular distance to the ancient magnetic North Pole location is equal to the ancient northern colatitude, measured from the sample site along an ancient meridian in the direction shown by the measured angle of declination.

Using spherical trigonometry, additional formulae, as outlined in Appendix B, are then used to calculate the geographical coordinate location of the ancient magnetic North Pole on the present-day Earth.

By assuming a constant Earth radius, it is also assumed that the ancient geographical coordinate system was identical to the present-day coordinate system. By using these additional formulae, the colatitude and declination vectors of the ancient geographical system are then simply added, using spherical trigonometry, to the present geographical system in order to locate the ancient magnetic pole, and from this to derive apparent-polar-wander paths for the various continents under study.

12.3. Apparent-Polar-Wander

Understanding apparent-polar-wander is crucial in understanding the difference between conventional palaeomagnetics on a constant radius Earth and palaeomagnetics on an increasing radius plate tectonic Earth model. The fundamental difference being that apparent-polar-wander is only applicable to a constant radius Earth model and does not apply to an increasing radius Earth.

Apparent-polar-wander on a conventional plate tectonic Earth model is simply a means of using the ancient magnetic pole locations from a number of continents to assist in reassembling the ancient continents. On an increasing radius plate tectonic Earth, it will be shown that ancient pole locations coincide at a single ancient pole location—as they should—and do not form apparent-polar-wander paths.

Conventional palaeomagnetic studies use measured palaeomagnetic data to plot the locations of the ancient poles for a range of ages for each continent. In Figure 12.3 the red and blue apparent polar wander paths for North America and Eurasia are shown in conjunction with the ancient magnetic declination lines—red and blue dotted lines. Declination is measured at the sample site and is projected from the sample site towards the ancient magnetic pole location. The dotted declination lines shown simply join each of the sample sites to their respective calculated pole locations for each continent.

In Figure 12.3 the blue European apparent-polar-wander path, for the various ages, is shown in relation to the red North American apparent-polar-wander path. By rotating the North American continent with respect to Europe, for instance, until its red apparent-polar-wander-path coincides with the blue European path, the past locations of each of the ancient poles will then coincide. While apparent-polar-wander on a global scale is far more complex than shown here, this fundamental technique has since become the primary means of assembling and constraining past locations of continental plates on a conventional Plate Tectonic Earth.

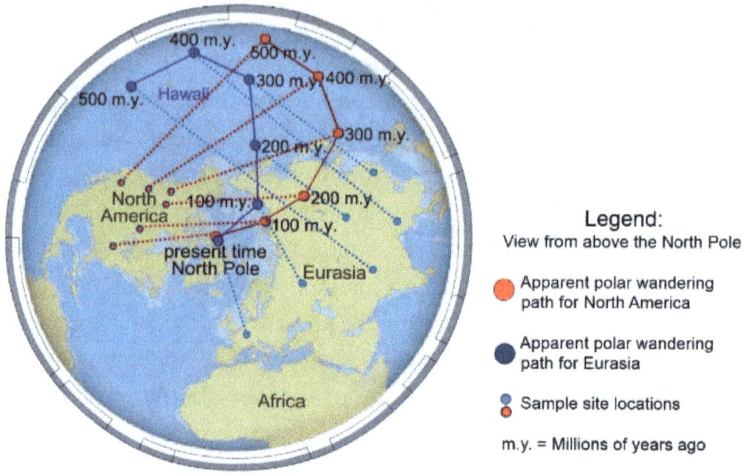

Figure 12.3. Conventional Apparent-Polar-Wander-Paths for North America and Eurasia showing hypothetical sample sites and declination lines to apparent-polar-wander poles.

In order to fully understand the limitations of what apparent polar wander represents, Figure 12.4 is a schematic cross-section of an ancient and present-day Earth crust extending vertically through the geographic and magnetic rotation axis. It is important to appreciate that conventional palaeomagnetic studies have concluded that Earth radius is constant, hence there is no consideration of or provision for palaeomagnetics on an increasing radius plate tectonic Earth model.

Figure 12.4 shows an ancient sample site located on the surface of an ancient Earth—shown as a red dot. By taking palaeomagnetic declination and inclination measurements from this sample site, the geographic location of the ancient magnetic pole can then be determined—shown as a red triangle on the ancient Earth. This red triangle represents the actual physical location of the magnetic pole on the ancient Earth, as determined from the ancient site sample colatitude and declination measurements.

If the Earth increases in radius from the ancient Earth to the present-day Earth radius, the ancient sample site and magnetic pole locations are shown by the dashed red lines projected onto the present-day Earth—again shown as a red dot and red triangle. These symbols represent the actual ancient magnetic pole and sample site located on the present-day Earth as determined from the ancient geographical coordinates.

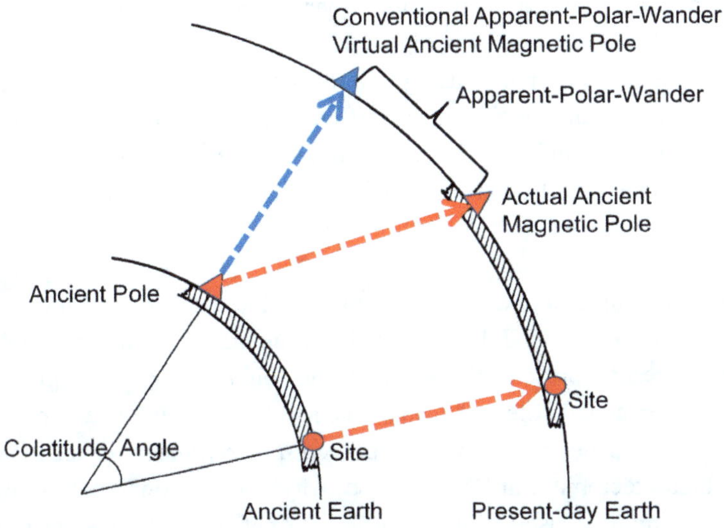

Figure 12.4. Schematic cross-section of an ancient and present-day Earth showing an ancient sample site plus its actual and virtual magnetic pole locations on the present-day Earth.

In order to determine a pole location from this sample site located on the present-day Earth, conventional palaeomagnetic formulae adopted the premise that Earth radius and geographical coordinates have remained constant throughout time. In other words, conventional palaeomagnetics does not consider that there has been an intermediate smaller radius Earth, so Figure 12.4 has no relevance to conventional apparent-polar-wander studies.

Instead, by adopting a constant radius Earth premise, the colatitude, calculated from the ancient sample site measurements located on the present-day Earth, uses the present-day geographic coordinate system to determine the location of the ancient pole and hence the apparent-polar-wander.

Using conventional palaeomagnetic formulae, the calculated colatitude angle is identical to that determined for the ancient site and will locate the ancient magnetic pole at the blue triangle in Figure 12.4. In this example, the colatitude angle is simply projected radially onto the present-day Earth surface.

If there were two or more different ancient sites of the same age located on the ancient Earth crust, their ancient magnetic poles would then plot as a scatter of locations on the present-day Earth. By joining each of these pole locations together, they would then generate a separate apparent-polar-wander-path for each continent. This scenario is equivalent to the blue and red apparent-polar-wander paths in Figure 12.3 for North America and Eurasia.

In contrast, on an increasing radius plate tectonic Earth, the ancient magnetic poles from two or more different sample sites of the same age located on the ancient Earth plot and coincide at the same Ancient Pole site—red triangle in Figure 12.4. This coincident pole location will, in turn, transfer to the same Actual Ancient Pole site—red triangle—as shown on the present-day Earth.

In the example above, the conventional blue triangle pole locations are virtual pole locations, and these can only be regarded as actual pole locations if they represent present-day sample sites on a present-day Earth. This discrepancy in actual pole and virtual pole locations is the fundamental reason why there is no need to consider apparent-polar-wander on an increasing radius plate tectonic Earth. It is also why conventional palaeomagnetic formulae cannot be used in their current format to locate ancient magnetic poles or to constrain crustal assemblages on increasing radius small Earth models.

This scenario is displayed in Figure 12.5 by plotting the locations of actual ancient magnetic pole locations—green and magenta dots—along each declination line.

The actual pole locations plot much closer to the original ancient sample site than the conventional poles, and their geographical colatitude distances will vary for each time period. Moving back in time, by moving each continent closer together, each actual ancient pole will then plot as a coincident pole—in reality as a cluster—which will also have a diametrically opposed South Pole equivalent cluster. Each of the continents will also merge on a smaller radius Earth to form a Pangaean supercontinent. Note: this scenario is an idealized example only.

In the example shown in Figure 12.5, the actual colatitude distance—the actual physical distance measured from the ancient sample site to the ancient magnetic pole—is what is used in plate tectonic small Earth studies to locate the ancient magnetic pole position on a present-day Earth model. Only by using the actual pole locations can all poles from each sample site cluster as a single magnetic pole location on an ancient smaller radius plate tectonic Earth model.

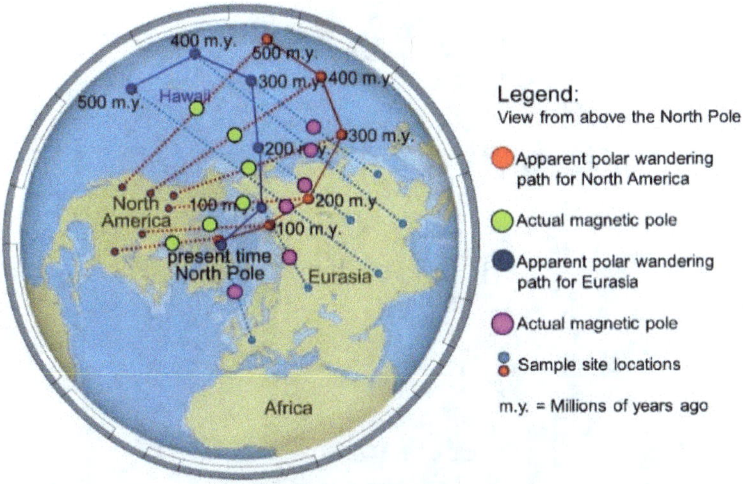

Figure 12.5. Location of actual ancient magnetic pole locations on the present-day Earth in relation to conventional Apparent-Polar-Wander-Paths and magnetic poles for North America and Eurasia.

12.4. Present-day Palaeomagnetic Poles

In conventional palaeomagnetic studies, the measured and structurally corrected angle of declination determined from an ancient site sample is projected, using spherical trigonometry, from the sample site to the ancient pole location on present-day Earth.

In Figure 12.6, the locations of the present-day palaeomagnetic poles for each sample site recorded in the International Palaeomagnetic Database (McElhinny & Lock, 1996) are shown as red dots. Each of these point locations also have an error factor ascribed to them, but these are not shown in this figure. This error factor is plotted as an ellipse around each pole, with dimensions dependent on the accuracy of the determination—generally from a few to tens of degrees radius.

In Figure 12.6, the distribution of present-day palaeomagnetic pole locations determined from the various sample sites is shown to have an overall scatter of between 25 to 35 degrees radius, centred on the present-day geographic North Pole. This scatter is referred to as secular variation of the magnetic pole, and the scatter is largely due to errors introduced during measurement and structural correction of the site samples as well as the presence of annual variations in the Earth's magnetic field.

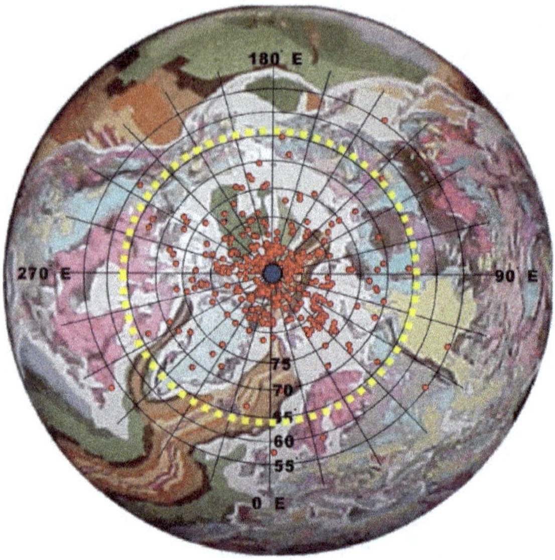

Figure 12.6. Present-day North Pole and magnetic pole locations. The magnetic pole locations are shown as red dots, and a 25-degree general scatter of poles is highlighted as a dashed yellow circle. These poles are plotted from the published international palaeomagnetic database of McElhinny & Lock, 1996. Maxlow, 2001.

The distribution of present-day palaeomagnetic pole data, where structural correction and magnetic interference should be at a minimum, are shown clustered around the geographic North Pole. In conventional palaeomagnetic studies, in order to determine a pole location from present-day palaeomagnetic data, the data are statistically treated to establish a mean, time-averaged, geocentric magnetic pole. If there is enough data, this magnetic pole will coincide with the Earth's geographical pole, as shown by the central blue dot in Figure 12.6.

For magnetic poles other than the present-day, conventional palaeomagnetic pole data from each continent, or selected portions of a continent, are statistically meaned to establish separate palaeomagnetic pole locations for each continent. These pole locations are then used in conventional studies to establish an apparent-polar-wander path for each continent, or part continent. Similarly, statistically treated poles from selected sample sites within a continent are also used to investigate potential small-scale crustal displacements, referred to as displaced terranes, between the various sample site locations.

12.5. Palaeomagnetics on an Increasing Radius Earth

The application of palaeomagnetic formulae to an increasing radius plate tectonic Earth is crucial in understanding both the limitations of conventional palaeomagnetics on a constant radius Earth model and its application on plate tectonic small Earth models. In both cases, measurements of the remnant magnetism preserved in rocks determine a direction to the ancient magnetic pole plus an estimate of the ancient latitude or colatitude of the sample site. The difference between conventional palaeomagnetics and palaeomagnetics on an increasing radius Earth is in the mathematical treatment of the raw field observational data plus interpretation of the measured data.

The application of palaeomagnetic formulae to an increasing radius plate tectonic Earth is detailed in Appendix B where the conventional formulae have been modified by reconsidering the palaeomagnetic dipole formula in conjunction with both an exponential increase in Earth radius and geological time. Modified palaeomagnetic formulae used to determine the geographic location of an actual ancient magnetic pole on the present-day Earth are also presented in this appendix. It must be appreciated, however, that modifications to the Earth's crust during relief of surface curvature over time will influence these formulae in determining actual ancient pole locations.

The mathematical relationship for an exponential increase in the Earth's ancient radius, derived from empirical measurements of seafloor and continental surface area data, was previously found to be (derived in detail in Appendix A):

$$R_a = (R_0 - R_p)\, e^{kt} + R_p \qquad \text{Equation 3}$$

where R_a = ancient radius of the Earth, R_0 = present mean radius of the Earth = 6370.8km, R_p = primordial Earth radius = approx. 1,700 km, e = exponential, t = time before the present-day (= negative), k = a constant = 4.5366×10^{-9}/yr

By incorporating this formula into the conventional magnetic dipole formula introduced previously (equation 2), the actual ancient colatitude, extending from the sample site to the ancient palaeopole position on an Earth of present-day radius, is then equal to:

$$P = ((R_0 - R_p)\, e^{kt} + R_p))\, (\tan^{-1} (2/\tan I))/R_0 \qquad \text{Equation 4}$$

where I is the mean inclination of the ancient magnetic field determined from the magnetic site data and P is the actual ancient colatitude determined from

I and is expressed in degrees.

The application of this modified dipole formula to palaeomagnetic site sample data enables the ancient colatitude to be converted from the ancient geographical grid system to the present-day geographical grid system. This formula converts, and will correctly locate, the ancient magnetic pole position on the present-day Earth surface.

What this conversion means on the present-day Earth is that the actual ancient magnetic pole (Figure 12.4) is always located closer to the sample site than the same pole determined using conventional palaeomagnetic formula. The location of an actual ancient magnetic pole on the ancient Earth cannot be calculated using conventional palaeomagnetic formulae. Conventional formulae simply give a dimensionless angular value for its geographical location, which is correct only if the ancient Earth radius equals the present-day Earth radius.

The unforeseen limitations placed on magnetic pole locations using conventional palaeomagnetic formulae, and similar ancient radius determinations outlined later in this chapter, are shown schematically in cross-sections of an ancient and present-day Earth in Figures 12.7 and 12.8.

In Figure 12.7, a cross-section of a fragment of continental crust is shown as pink crust on an ancient Earth. This fragment contains three ancient sample sites, labelled Sample Site 1, 2, and 3—red, green and yellow dots respectively. The ancient magnetic pole, as determined from these sample sites, is shown by the blue triangle on the ancient crust and this pole location is coincident for each of the ancient sample sites.

As the ancient Earth radius increases radially to the present radius, the ancient continent is shown to fragment into three cratons—pink crusts—which are separated by younger sedimentary basins—green crusts. In this figure, the ancient Earth undergoes a simple idealized radial increase in radius to the present-day Earth radius and the distances between each sample site also increase.

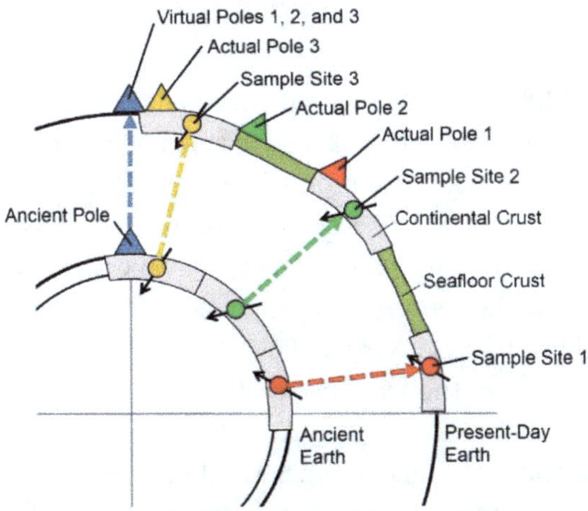

Figure 12.7. Schematic cross-sections of ancient and present-day Earth showing sample site and pole locations determined from conventional palaeomagnetic formulae and modified increasing radius formulae.

On the present-day Earth, the ancient magnetic pole locations, as determined using conventional palaeomagnetic formulae, are shown to remain coincident at the blue triangle labelled Virtual Poles 1, 2, and 3, which also coincides with the present-day magnetic pole location. This coincidence occurs because there has been an increase in separation of each of the sample sites on the present-day crusts during formation of the two green sedimentary basins.

A conventional palaeomagnetic interpretation of this scenario would insist that, because each of the ancient magnetic pole locations are shown to be coincident, the original ancient continent must have remained intact, inclusive of the sedimentary basins, during the interval of time involved. This example would insist that there was no apparent polar wander of the continental fragments and there is no indication of any increase in Earth radius.

In contrast, the actual magnetic pole locations for each site (Figure 12.7), calculated from the modified increasing radius palaeomagnetic formulae, are shown dispersed at the Actual Poles 1, 2 and 3—red, green and yellow triangles on the present-day Earth. Because there is prior knowledge of the increase in Earth radius involved in this hypothetical scenario, it can be appreciated that the ancient continent has, in fact, physically fragmented and these crustal fragments have migrated and formed intervening sedimentary basins. When these crusts are reassembled on a smaller radius Earth model each of the actual ancient poles will then coincide at the correct ancient Earth radius.

In Figure 12.8, the same hypothetical ancient continental fragment, containing ancient sample sites 1, 2, and 3, again breaks-up and undergoes an idealized radial increase in Earth radius from the ancient radius to the present-day radius. In this example, one craton is shown to break up and separate during mid-ocean-rifting, while the remaining two cratons remain intact.

The locations of the conventional magnetic poles on the present-day Earth are shown as blue triangle Virtual Poles. The virtual magnetic poles for sample sites 1 and 3 are shown to be coincident, whereas the virtual pole for sample site 2 is shown to be separated. For a constant radius Earth, the locations of the ancient magnetic poles would suggest that the two separate cratons have either undergone apparent-polar-wander relative to sample site 2, or they have remained intact relative to sample sites 1 and 3.

Conventional plate tectonic theory is now less rigid in its interpretation of continental crust than those of the 1960s when van Hilten first calculated Earth radius. Plate tectonic theory would now suggest that the ancient virtual poles 1 and 2 (Figure 12.8) originated from the same continent and there has been either a displacement of the cratons relative to each sample site, or again, an apparent-polar-wander of poles has occurred relative to site 2.

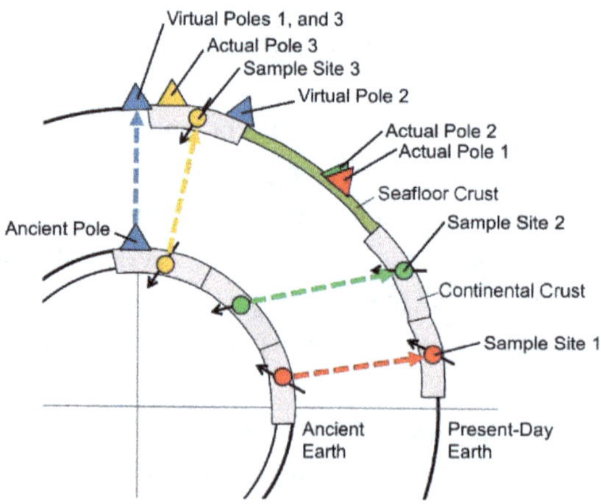

Figure 12.8. Schematic cross-sections of ancient and present-day Earth showing sample site and pole locations determined from conventional palaeomagnetic formulae and modified increasing radius formulae.

In contrast, on an increasing radius plate tectonic Earth, the positions of the actual ancient magnetic poles 1 and 2 on the present-day Earth coincide—

red and green triangles on the present-day Earth. These poles require no displacement of the cratons and the isolation of the actual magnetic pole 3 confirms that there has indeed been migration and continental breakup of the ancient continent. All three ancient poles again cluster as coincident magnetic poles on the ancient Earth.

While complex, the hypothetical examples shown in these figures serve to illustrate the potentially unforeseen and unimagined limitations imposed on conventional plate tectonic interpretations of palaeomagnetic pole data. They also serve to illustrate the ramifications incurred in assuming continental crustal stability on a constant radius Earth when attempting to calculate an ancient Earth radius.

The global palaeomagnetic database has increased considerably since early attempts at determining apparent-polar-wander-paths and ancient Earth radii were first made during the 1960s. Because of this increased dataset, what conventional palaeomagnetic studies now show is a marked increase in the complexity of apparent-polar-wander between continents, as well as an increasingly complex crustal displacement history between individual fragments of continental crust.

12.6. An Alternative Palaeopole Method

To overcome the problem of excessive scatter of palaeomagnetic poles—as per Figure 12.6—an alternative method of plotting palaeomagnetic pole data is to re-plot the data as small circle arcs centred on each sample site. In this method the actual colatitude is drawn as an arc extending from the known sample site in the general direction of the magnetic pole rather than using spherical trigonometry. For conventional palaeomagnetic studies, this method is only applicable to present-day palaeopole data.

Because of the complexity of the often hundreds of overlying arcs involved in this drawing method, in Figure 12.9 the full spectrum of individual arcs plotted from the published present-day data is shown summarized as coloured swaths. Each swath represents the lateral distribution of individual arcs plotted for each labelled continent shown on the figure and may contain tens to hundreds of individual arcs. The swaths of colatitude arcs in this figure typically range from 20 to 30 degrees wide. This width is consistent with the error factors ascribed to each conventional palaeopole location. From these swaths, a mean magnetic pole is readily located at the intersection of each of the coloured swaths.

This non-conventional method, using actual colatitude arcs instead of single points, is presented here as an alternative method for displaying the variation in magnetic pole locations for each continent. In this example each swath, or multiple swaths, defines the location of each continent relative to the present-day magnetic pole and each swath straddles the coincident magnetic pole location.

In Figure 12.9, the intersection of swaths collectively determines a mean magnetic North Pole, independent of statistical treatment, which is coincident with the geographical North Pole. In addition, the equator and magnetic South Pole can also be confidently plotted by measuring 90 and 180 degrees of latitude respectively.

Figure 12.9. Swaths of present-day north palaeomagnetic pole locations plotted as small circle arcs on a present-day Earth model. These swaths represent clusters of arcs plotted for each continent highlighted.

12.7. Ancient Palaeomagnetic Poles

Locating the ancient magnetic poles on each plate tectonic small Earth model uses a combination of actual magnetic pole locations, as determined using equation 4 and detailed in Appendix B, and the colatitude small circle arc method. To locate ancient palaeomagnetic poles on each small Earth model, the known published sample site locations are first manually transferred from the present-day Earth model to each small Earth model in turn, using the detailed bedrock geology as a guide to locate each sample site. The actual colatitude, as calculated for each site sample using the modified palaeomagnetic formula (equation 4), is then manually drawn as an arc on the small Earth model in the direction of the measured declination.

Figure 12.10 is an example of the distribution of North and South Pole colatitude arcs, as well as the derived mean pole locations for the primordial Archaean plate tectonic small Earth model. In this example the Archaean small Earth model is an extreme example of the colatitude arc method of plotting ancient poles. This Archaean example has the dual role of showing how to locate the ancient magnetic poles on plate tectonic small Earth models as well as quantifying assemblage of the ancient crusts on the primordial Archaean small Earth model.

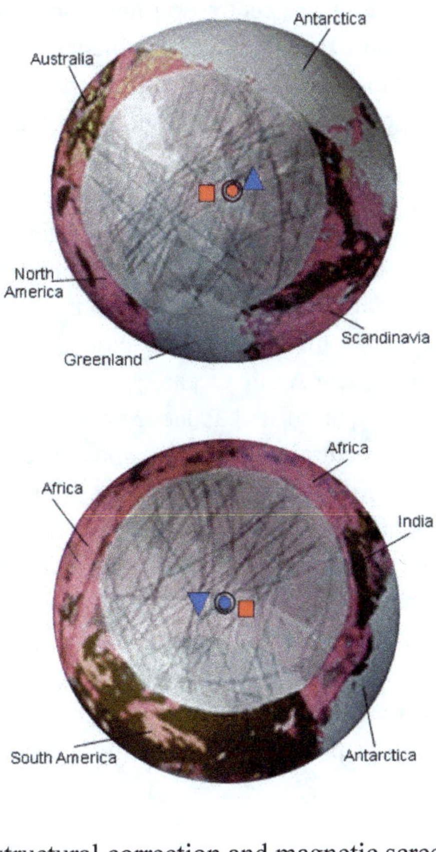

(Left) Figure 12.10. Archaean to mid-Proterozoic north and south palaeomagnetic pole colatitude arcs. Red squares are the mean and projected north poles established from the colatitude arcs. Blue triangles are the mean and projected south poles. Red bullseye is the location of the meaned North Pole, and the blue bullseye is the meaned South Pole. Maxlow, 2001.

By using this method to plot ancient magnetic poles, each small Earth model consistently shows that approximately 95 percent of the colatitude arcs plot within twenty-five degrees radius of each magnetic pole location. This plot compares favourably with the distribution of magnetic poles shown for the present-day North Pole palaeomagnetic data (Figure 12.6). The remaining arcs—less than five percent—are assumed to contain sporadic, site-related structural correction and magnetic screening errors and were discarded.

For each plate tectonic small Earth model, the magnetic pole positions are located by visually averaging the clusters of North and South Pole arcs. Each pole location was then transferred 180 degrees to the opposite pole and again averaged to determine a mean North and South magnetic pole location. From these meaned pole locations, an equator can be established for each model by scaling through 90 degrees of latitude, measured from each of the established poles. Longitude can also be established by adopting the ancient location of Greenwich, England, as zero degrees longitude and scaling east and west from Greenwich along the ancient equator as per conventional geographic usage.

The locations of ancient north and south magnetic poles established for each of the plate tectonic small Earth models are shown in Figures 12.11 and 12.12 respectively. These models extend from the Archaean to the present-day, plus one model extended to 5 million years into the future. The small Earth models shown in each of these figures are centred on the North and South Poles respectively. These images show the crustal development of each of the continents as they migrate, relative to the ancient magnetic poles, during crustal extension, breakup, and opening of the modern oceans to the present-day.

In Figure 12.11, the Precambrian through to the Palaeozoic magnetic North Pole is shown located in eastern Mongolia. As the Pangaea supercontinent ruptured during the late-Permian, and the various modern continents slowly migrated south, the distribution of pole locations shows there was an apparent northward migration of the magnetic pole through Siberia to its present location within the present Arctic Ocean.

Similarly, in Figure 12.12, the Precambrian and Palaeozoic magnetic South Pole is shown located in west central Africa. As the Pangaea supercontinent ruptured and the various modern continents slowly migrated north, the distribution of pole locations shows there was an apparent southward migration of the pole along the South American and West African coastlines to its present location in Antarctica.

The significance of both Figures 12.11 and 12.12 is that the distribution of North and South Poles, plotted independently on each plate tectonic small Earth model, confirms that each pole remains stationary throughout Earth history and, more significantly, each pole plots as diametrically opposed North and South Poles.

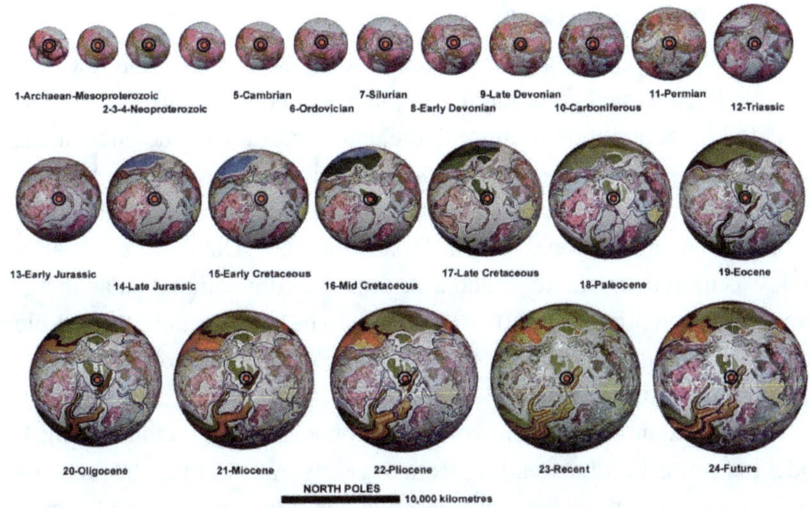

Figure 12.11. Small Earth Archaean to Future magnetic North Poles (red dots). Maxlow, 2001.

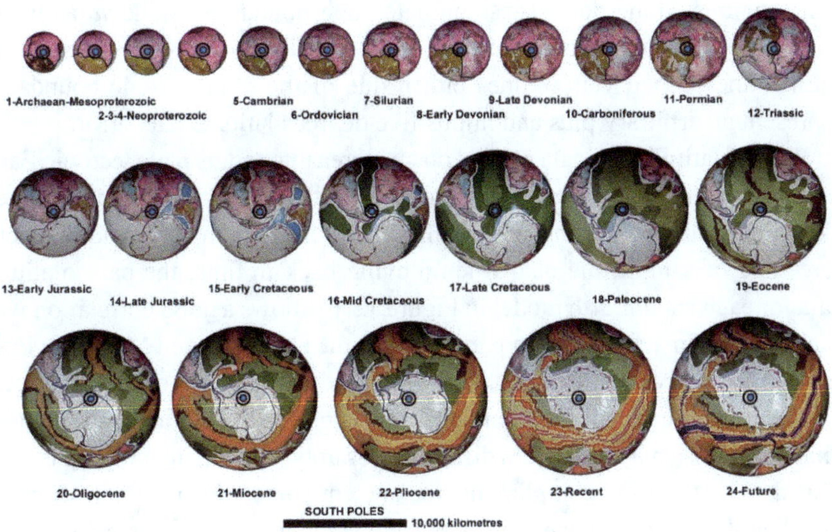

Figure 12.12. Small Earth Archaean to Future magnetic South Poles (blue dots). Maxlow, 2001.

In addition, the distribution of these poles confirms an evolutionary history of development of the supercontinents followed by subsequent breakup and dispersal of the continents throughout history.

12.8. Ancient Latitude

In addition to plotting palaeomagnetic pole data, ancient latitude can also be plotted from each measured sample site. This data is unique in that calculated palaeolatitude represents the actual latitude of the ancient sample site and, unlike palaeopole data, does not require projection beyond the sample site. Structurally corrected and magnetically screened paleomagnetic data from the International Global Palaeomagnetic Database of Pisarevsky, 2004, was used to calculate palaeolatitude using conventional palaeomagnetic formulae. Conventional formulae were able to be used because the sample site represents the actual ancient site measured in degrees of latitude, not a mathematically derived pole site.

Palaeolatitude data are shown plotted on each plate tectonic small Earth model in Figure 12.13, extending from the early-Archaean to the present-day. In this figure the calculated palaeolatitude data are colour coded to represent data located within the north and south equatorial climate zones—red dots, the north and south temperate zones—green dots, and the north and south polar regions—blue dots. These climate zone boundaries—shown as heavy dashed yellow lines—are based on zonal distributions on the present-day Earth. Fine dashed yellow lines either side of the climate zone boundaries represent an arbitrary plus and minus five degrees latitude data error.

Palaeolatitude site data plotted on an increasing radius plate tectonic Earth represent actual latitudes which must accord with actual climate zones. Considering the increasing uncertainty in structural correction and magnetic screening of sample site data when moving back in time, the palaeolatitude data for each small Earth model in Figure 12.13 shows a good correlation with each of the climate zones—in particular for the Cenozoic and Mesozoic Eras.

An increased spread of equatorial zone red dots on each of the Palaeozoic and Precambrian small Earth models may suggest that the ancient magnetic dipole response may have been different, possibly weaker, than what it is now. This data distribution compliments the ancient palaeomagnetic pole data and together they quantify the use of palaeomagnetics on an increasing radius plate tectonic Earth model.

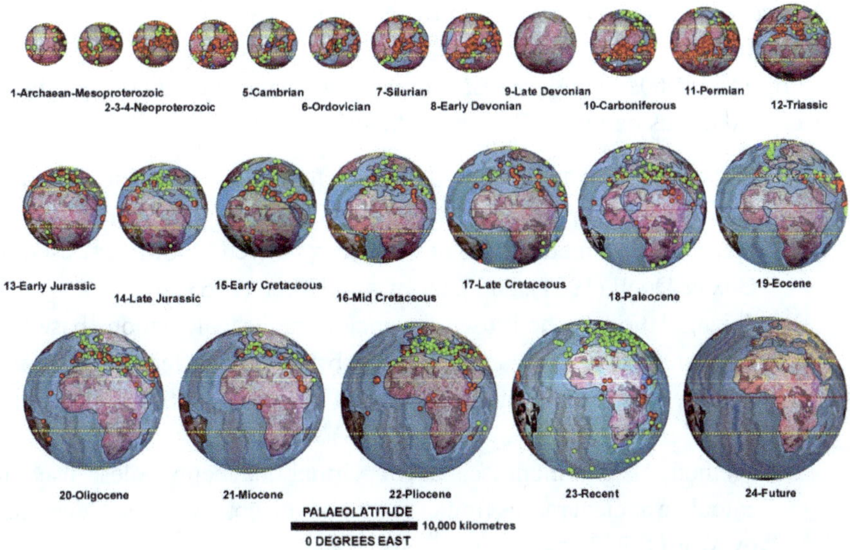

Figure 12.13. Archaean to present-day palaeolatitude sample site data centred on zero degrees east longitude. Red dots represent calculated data located in the equatorial climate zones, green dots represent data located in the north and south temperate zones, and blue dots represent data located in the north and south Polar Regions. Heavy dashed yellow lines represent climate zone boundaries and fine dashed yellow lines represent an arbitrary plus and minus 5 degrees data error. Note: no data is shown for the late-Devonian model.

12.9. Palaeoradius Using Palaeomagnetics

Palaeomagnetic measurements were first used during the 1960s and early 1970s in an attempt to determine a potential ancient Earth radius. At that time there was considerable debate amongst scientists as to whether the Earth radius was or was not increasing, and, if so, by how much. Earth radius calculated from palaeomagnetic measurements was then used in an attempt to resolve this debate.

Palaeomagnetic site data from Europe, Siberia, North America, and Africa were used in various studies during the 1960s, with mixed and generally inconclusive results. Published estimates of ancient Earth radius made from these studies varied from large changes in Earth radius, comparable to rates derived from expanding Earth modelling studies to negligible changes in radius. Based on results from the African evidence in particular, McElhinny and Brock in 1975 concluded that, "*...for the past 400 million years the amount of potential Earth expansion is limited to less than*

0.8 percent of the present Earth radius." No further estimates have been made since then.

From the 1960s to mid-1970s, determining an ancient radius of the Earth was carried out using three different methods:

1. Palaeomeridian method: used for ancient radius calculations based on palaeomagnetic data situated approximately on the same ancient meridian. The method was developed by Egyed (1960), first used by Cox & Doell (1961), and later by van Hilten (1963).
2. Triangulation method: used for ancient radius calculations based on palaeomagnetic data situated on substantially different meridians. The method was originally developed by Egyed (1960) and used in a modified format by van Hilten (1963).
3. Method of minimum scatter of virtual magnetic poles: used to calculate ancient radius from a scatter of site observations. Developed by Ward (1963).

For each of these ancient radius methods, the distances between the various palaeomagnetic sample sites were typically in excess of 5,000 kilometres. Small Earth modelling studies now show that these distances represent a significant proportion of ancient Earth circumference. Figure 12.14 emphasizes the significance of this observation where a 5,000-kilometre-wide crustal fragment located on an ancient Pangaean Earth—typical of Europe—represents approximately 25 percent of the prevailing Earth circumference. This contrasts with the same crust on present-day Earth which represents only 10 to 12 percent of the present circumference.

Modern geological studies also show that, over this 5,000-kilometre distance, continents have undergone considerable amounts of both crustal extensions to form basins or rift zones, as well as contractions during periods of orogenesis. This modern crustal evidence is contrary to the premises placed on the European ancient radius data of Egyed in 1960, and similarly Cox and Doell in 1961 and hence negates both their determinations and conclusions.

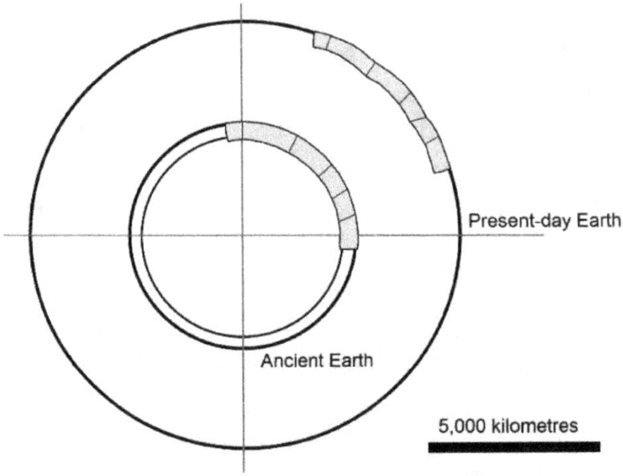

Figure 12.14. Schematic cross section of a 5,000-kilometre-wide crustal fragment (shown as pink) on a Permian and present-day Earth. Maxlow, 2001.

The mechanics behind conventional plate tectonic continental breakup and dispersal are also well established, as demonstrated by both seafloor mapping and apparent-polar-wander studies. The seafloor mapping, in particular, is contrary to the premises imposed on the European and North American ancient radius data of van Hilten in 1963 and again negates both the determinations and conclusions. Similarly, rotation and displacement of parts of continental crust are also recognized within continents, and these displacements are contrary, in part, to the premises imposed on the African palaeomagnetic and ancient radius determinations of McElhinny and Brock in 1975.

12.10. African Palaeoradius Determinations

In 1975, McElhinny and Brock published a paper outlining new palaeomagnetic results from Triassic and Cretaceous sample sites located in north, east, and South Africa. These results were used to calculate an ancient radius of the Earth for the Mesozoic Era in order to resolve criticism of previous, inconclusive estimates of radius determined from sample sites located in Europe, Siberia, and North America.

Africa was considered by McElhinny and Brock to be ideally suited for determining an ancient Earth radius because of its implied crustal stability. These crustal conditions fitted the palaeomagnetic requirement prevailing at the time that the surface area of continents must remain near constant, and continents must act as rigid plates over time. In their paper, McElhinny and Brock went to great lengths to emphasize the quality of both the palaeomagnetic results and age dating of samples taken from the various sample sites in Africa. In addition, their sample sites were shown to be located on approximately the same ancient meridian. This enabled McElhinny and Brock to use the palaeomeridian calculation method to determine an ancient Earth radius.

McElhinny and Brocks' published sample site locations are shown in Figure 12.15 as red squares and their calculated magnetic pole positions are shown plotted as yellow stars on the present-day Earth.

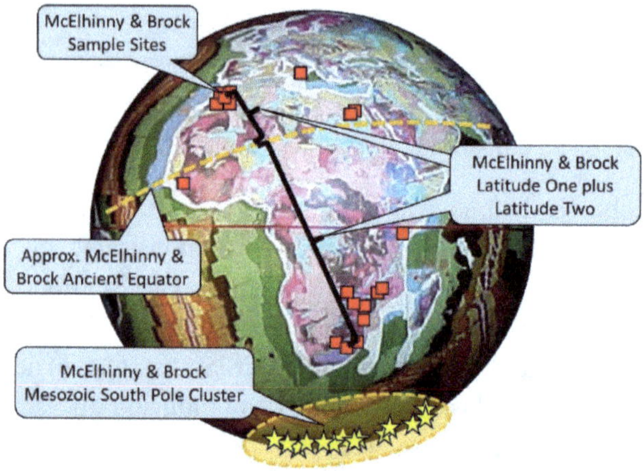

Figure 12.15. Published palaeomagnetic sample site locations of McElhinny & Brock 1975 for Africa (red squares), in relation to calculated conventional magnetic pole locations (yellow stars), and inferred equator (curved yellow dashed line). Modified after Maxlow, 2001.

In this figure, the locations of the ancient magnetic South Poles, as calculated using conventional palaeomagnetic formulae, form a broad cluster for each age grouping. This clustering of ancient pole positions suggested to McElhinny and Brock that the ancient magnetic South Pole had remained essentially the same position throughout the Triassic and Cretaceous Periods, with respect to Africa. This observation is based on a similar scatter of poles as shown for the present-day magnetic poles (Figure 12.6).

Because the ancient magnetic poles were considered to cluster together, McElhinny and Brock assumed that the ancient equator passed between the northern and southern African sample sites—dashed yellow line in Figure 12.15. To determine past geographic separation of the sample sites the calculated ancient latitudes from the various sample sites located north and south of the ancient equator were then simply added.

Ancient radius was then determined by McElhinny and Brock by comparing the ancient geographic separation with the present-day geographic separation between combinations of any two sample sites located on the same ancient meridian. In each case, and for each age group measured, the calculated ancient Earth radius was found to be very similar to the present Earth radius. From this, McElhinny and Brock concluded that Earth radius has not changed by more than 0.8 percent during the past 400 million years of Earth history. They further concluded that "*...hypotheses of Earth expansion were therefore very difficult to sustain.*"

In contrast, in addition to McElhinny and Brocks' sample site data, the precise locations and migration path of the small Earth magnetic South Poles are shown plotted as yellow stars along the west coast of Africa on the present-day Earth model in Figure 12.16. These pole locations were transferred directly from each of the small Earth models in Figure 12.12. Each of these pole locations were established using published palaeomagnetic pole data, where the poles were shown to plot as diametrically opposed North and South Poles for each plate tectonic small Earth model.

The palaeopoles and shaded red migration path for the Palaeozoic to present-day south magnetic pole (Figure 12.16) shows that the ancient South Pole was originally located in central West Africa throughout the Precambrian and Palaeozoic Eras. This was followed by an apparent migration of the South Pole along the then joined coastlines of South Africa and South America, prior to migration across the opening Atlantic Ocean to its present location within East Antarctica as each of the continents broke up and progressively migrated north.

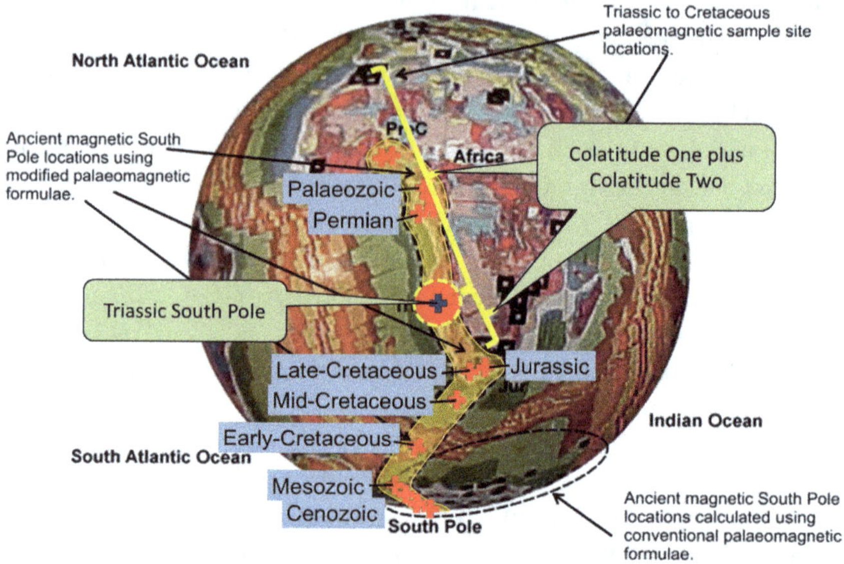

Figure 12.16. Published palaeomagnetic sample site locations of McElhinny & Brock 1975 for Africa (red squares) shown in relation to increasing radius small Earth magnetic pole locations (yellow stars and highlighted polar migration path). Modified after Maxlow, 2001.

In this figure, the small Earth magnetic poles are located much closer to the sample sites than those of McElhinny and Brock determined using conventional palaeomagnetic formulae. Also, in this figure, it is only the Triassic sample data, located in north and South Africa that lie along a common ancient meridian. As such, only these data can be used to calculate an ancient Earth radius on an increasing radius Earth.

In Figure 12.16 the Triassic plate tectonic small Earth magnetic South Pole, highlighted in red, was located between the north and South African sample sites. The ancient equator was similarly located around the perimeter of the small Earth model shown, well beyond each of the sample sites. To determine the past angular separation of the Triassic sample sites, the ancient colatitude from each site must then be added, not ancient latitude as McElhinny and Brock had done. Of note is that on an increasing radius plate tectonic Earth it is the actual ancient colatitudes that are added, calculated using modified palaeomagnetic formulae, not the virtual colatitudes as calculated using conventional palaeomagnetic formulae.

When compared to the present angular separation of the sites, the ancient radius of the increasing radius plate tectonic Earth existing during the Triassic Period then calculates as 52 percent of the present Earth radius. This value is the same as the Triassic small Earth radius established from measurements of seafloor surface area data.

While McElhinny and Brock went to great lengths to present quality data and sound methodology, at that time there was very little agreement as to what a potential increase in Earth radius may or may not have been. Previous ancient Earth radius determinations from Europe and North America were variable and inconclusive and there were many conflicting arguments for and against an increasing radius Earth.

What McElhinny and Brock failed to comprehend was the significance of the magnetic pole locations, as determined from their conventional palaeomagnetic formulae. The poles that McElhinny and Brock plotted were virtual pole locations and present-day geographic angular measurements were used in their calculations, not actual pole locations or ancient geographic measurements. From this, they then concluded that the ancient equator passed between the various sample sites and therefore added latitude, rather than colatitude, to determine an ancient Earth radius.

The requirement to establish pole and equatorial positions on plate tectonic small Earth models that correspond to observed evidence is a prime objective of this research. On an increasing radius plate tectonic Earth, it is maintained that the palaeomagnetic pole and latitude data located on models presented here more closely conform to what is observed than those on a fixed radius Earth.

Chapter 13

A Paradigm Shift in Space Geodetics

Proof of any theory comes through direct observation or by direct measurement.

Both palaeomagnetic and space geodetic measurement techniques are now routinely used in conventional plate tectonic studies for determining past and present-day plate motions and plate assemblages on a constant radius Earth. In addition, the outcomes of these techniques are used as confirmatory evidence in support of a constant radius plate tectonic Earth model. It should be appreciated, however, that the evidence presented by both of these disciplines are derived mathematical entities, and the established formulae used are constrained to, and must adhere to, a number of applied constancy assumptions prior to calculation.

Space geodetics is a relatively new and sophisticated physical science that uses a network of radio telescopes, satellites, and ground-based receiver and transmitter stations from around the world to routinely measure the precise dimensions and continental plate motions of the Earth. This technology also forms the framework for modern GPS technology which is more familiar today. Because of its implied sophistication and complexity, it is therefore very important at this stage to investigate space geodetics in the context of a potentially increasing radius plate tectonic Earth model.

Space geodetic measuring techniques developed to measure the dimensions of the Earth stem from the early 1970s and include Very Long Baseline Interferometry (VLBI), Satellite Laser Ranging (SLR), Lunar Laser Ranging (LLR), Global Positioning Systems (GPS), and Doppler Orbitography and Radiopositioning Integrated by Satellite (DORIS). Observational data are now routinely recorded from each of these measurement techniques. The mathematically and statistically treated data from all receiver stations are combined and used to calculate a solution to the global geodetic network—a three-dimensional measurement framework of the Earth.

In 1993, when Robaudo and Harrison first combined SLR data, including all global geodetic data from 1976 to the beginning of 1991, as well as VLBI data containing data up to the end of 1990, they allowed all observational stations to have three independent "*X-Y-Z motion velocities*"—vector motions of the ground-based observational stations in three-dimensional coordinates. When these motion, velocities were used to establish a "*...global geodetic network*" they calculated "*...a Root Mean Squared value of up-down* [variation in Earth radius] *motions of over 18 mm/year.*" In other words, the radius of the Earth was found to increase by up to 18 millimetres per year.

At that time, Robaudo and Harrison considered this up-down motion to be extremely large. They did not consider any increase in Earth radius when making this judgement but instead compared it to values that were expected from areas undergoing "*crustal rebound*" during glacial melting—the relaxation and rise of a continent after glacial melting—estimated to be less than 10 millimetres per year. It is significant that Robaudo and Harrison "*...expected that most VLBI stations will have up-down motions of only a few mm/year,*" and they went on to recommend that the vertical motion "*...be restricted to zero, because* [they considered that] *this is closer to the true situation than an average motion of 18 mm/year.*"

Robaudo and Harrison were, in fact, faced with a daunting problem. When they calculated the 18 millimetre per year variation in Earth radius from 15 years' worth of observational data they found, but did not acknowledge, or recognize, that Earth radius was actually increasing by 18 millimetres per year. This value is very close to the value of 22 millimetres per year calculated here using published seafloor mapping, especially when Robaudo and Harrison's error margins are taken into consideration.

Instead, this value of 18 millimetres per year was assumed by Robaudo and Harrison to be an accumulation of systematic errors during the collection and mathematical and statistical treatment of the raw observational data. Since then, the mathematical formulae and applied correction parameters attributed to this data have been extensively refined, which has forced all perceived errors to be statistically meaned out to zero.

The application of advanced space geodetic techniques to study of the Earth has now progressed to the point where estimation of Earth radius and present-day plate motion is quoted to sub-millimetre accuracy. Results of Shen *et al.* in 2011 now show that "*...both geodetic and gravimetric observations support the conclusions that the Earth is expanding at a rate of*

0.2 millimetres per year in recent decades." This is encouraging, but the value is a factor of 100 too low when compared with the current 22 millimetres per year rate of increase in radius based on the seafloor mapping data used in this study.

13.1. Space Geodetic Data

One of the primary limiting factors to the accuracy of measurements in all space geodetic measuring techniques is the systematic errors that come from seasonal atmospheric interference to the signals being measured. This atmospheric interference alters the timing of the optical or radio signals being measured due to refraction of the signal as it passes through the atmosphere. To minimize errors, correction factors are routinely introduced during mathematical treatment of the raw measured data. These correction factors make allowance for the variations in wet and dry atmospheric conditions throughout the year.

For SLR, GPS, and DORIS measurements, additional limiting factors also include satellite tracking and modelling of the Earth's magnetic force field. Force field premises imposed on the mathematics are based on adopting a constant universal gravity G, a constant Earth mass M, and a constant product G·M. This product is then used to calculate Earth's surface gravity and to locate the physical centre of the Earth, which is used in both satellite altimetry control and as the X-Y-Z coordinate reference point. Satellite positioning and altimetry control are also known to be sensitive to both universal time and to the value of G·M.

In 2002, Koziar mentioned that, even though Earth mass and the gravitational constant are assumed to be constant for space geodetic purposes, the incremental change in Earth mass can be deduced from SLR observational data. The precise measurement of G.M began in the late-1970s and in his review Koziar took into consideration measurements that continued into the 1990s. The SLR data were shown to consistently record a slow increase in Earth mass of the order of 3×10^{19}grams/year, which is of a similar order of magnitude of 6×10^{19} grams/year as calculated here using Earth radius derived from seafloor mapping data.

Current known errors in the positioning of the centre of the Earth, and hence the coordinate system used by space geodesists, have been quoted in the literature as plus or minus 50 to 100 millimetres, so constraining Earth mass to a constant value may well be introducing additional unforeseen errors during data processing.

Vertical crustal motion studies carried out by geodesists have also shown that the systematic and random errors in space geodetic measurements are largest in the vertical component of a site measurement which directly affects measurement of Earth radius.

VLBI is also shown to be insensitive to changes in the vertical. Similarly, for SLR the greatest discrepancies between estimates of plate motion and established plate model values are in the vertical component and the largest systematic error source for absolute height determination is the value of G·M.

Regardless of this, in 1997 Larson *et al.* concluded from the observational data that there is "*...no persuasive evidence for any significant net vertical crustal motions*"—increase in Earth radius. Similarly, Soudarin *et al.* in 1999 concluded that "*...with the exception of a few stations, vertical plate motion rates are generally less than 5 millimetres per year.*"

More recently, Shen *et al.* in 2011 gave a brief overview of how raw observational data is currently treated before using it to calculate a variation in Earth radius. In their overview they used spacial geodetic data published in 2008.

In this data "*...there are 1572 readings at present from various stations* (including GPS, VLBI, SLR, and DORIS) *...Due to discontinuities of many recordings, only 930 recordings...are available for present study.*" Also, "*...stations that are very close to each other...are merged to one station...Then, there are 841 stations left...in this study the stations located in active tectonic zones...have been removed from our calculations.*" Furthermore, "*Another concern is that the absolute values of the vertical velocities of some stations are beyond 0.02m/year, and so large vertical movements of such kinds of stations are not related to Earth expansion...Hence, such kinds of stations are not included in our calculations...After removing the stations located in the orogenic zones and the stations whose vertical velocities are greater than 0.02m/yr, there are 625 stations left...Our calculations show that the Earth is expanding at present at an expanding rate of 0.24±0.04mm/yr.*"

From this discourse on data treatment by Shen *et al.* it can be seen that 60 percent of the raw observational data are eliminated before calculating a rate of change in Earth radius. In other words, all data that might otherwise indicate an increase in Earth radius are removed.

This, in effect, smooths out the raw data before making their calculation. By doing this treatment Shen *et al.* then concluded "*...the Earth is expanding at present at an expanding rate of 0.24±0.04mm/yr.*" This smoothed and calculated data is now routinely published and used to quantify a limited increase in Earth radius and to continue to discredit any suggestion of a large increase in Earth radius.

13.2. Space Geodetic Limitations

In many publications, the way the Earth is perceived to increase in size is likened to a balloon, where the balloon is considered to be merely pumped up with a relatively uniform increase in radius each year.

The implications of this pumping up process can be further visualized by imagining the addition of a thin smear of damp clay coating the outside of the balloon in order to simulate a Pangaean supercontinental crust. By merely pumping up the balloon, the damp clay will stretch and distort to emulate the Earth's supercontinental crusts before breaking up. Ultimately, the clay will rupture and break apart simulating breakup of the ancient Pangaean supercontinent to form the modern continents and oceans.

By adopting this simplistic balloon example as representative of the Earth it could then be mistakenly perceived that the increase in radius of the Earth can be measured from a relatively few surface measurements. This is what Shen *et al.* tried to emulate the Earth by eliminating 60 percent of the raw space geodetic observational data before making their increase in radius calculation.

Completion of the Geological Map of the World in 1990 now shows that the Earth's continental and seafloor crusts are not uniform and are far more complex than what this simplistic balloon example implies. The continents in particular comprise a mixture of ancient cratons, orogens, and sedimentary basins, each with their own age and definitional requirements, as well as the relatively modern seafloor crusts, which make up around 70 percent of the surface area of the Earth.

Simulating this complex crustal composition during increase in Earth radius can be further visualized by varying the balloon example to consider these different types of crusts. The cratons, by definition, are crusts that have stabilized before about 2,400 million years ago. These can be simulated by partly drying the coating of clay on the surface of the balloon while leaving the underlying clay still damp.

By again pumping up the balloon, the dried clay outer crust will crack and fragment to accommodate for the increase in balloon radius. By continuing to pump-up the balloon three things will happen: firstly, the dry outer crust will continue to fragment—simulating ongoing fragmentation of the cratons; secondly, the underlying damp clay will stretch—simulating the opening and extension of sedimentary basins; and thirdly, the soft clay located around the margins of the dry fragments will wrinkle and crumple—simulating formation of geosynclines, orogens and fold mountain belts.

In addition to fragmenting, because of its increased rigidity, the dry outer clay coating will retain a partial super-elevation at the centre of each fragment, allowing it to rise in relation to the rest of the balloon surface during change in balloon radius. The outer edges of each dry fragment will also be forced down into the balloon along the respective margins. Ultimately, the exposed damp clay will then rupture and break apart as the balloon continues to be pumped-up, again simulating breakup of the ancient Pangaean supercontinent to form the modern continents and oceans.

In this second example, even though hypothetical, the pumped-up surface of the balloon is not smooth or uniform. The surface will instead reflect the variation in super-elevation and fragmentation of the dried crusts, the depressed margins around these same crusts, and the relatively uneven surface of the stretched, wrinkled, and ruptured damp clay.

In other words, the surface of the crusts will be going both up and down, depending on where the surface radius measurements are taken from. This pumping up process will then reflect the variability of the up-down motions of the various crusts, identical to what was seen by Robaudo and Harrison from early space geodetic measurements of the Earth.

To eliminate 60 percent of the measurements, as Shen *et al.* did before calculate change in Earth radius, especially when the vertical motions are so small, is therefore seen as potentially misleading and erroneous. In addition, the choice, location, and amount of observational stations distributed around the world may also be biasing outcome of the current space geodetic calculation. For example, "*... removing the stations located in the orogenic zones and the stations whose vertical velocities are greater than 0.02m/yr*" will bias the outcome of the variable radius calculation by removing the very measurements that may show an increase in Earth radius.

To overcome these problems, it is considered that substantially more observational stations than currently available around the world would be needed before this bias could be minimized and a meaningful increase in Earth radius calculated from space geodetic data.

13.3. Continental Plate Motion

The primary function of space geodetics is not to measure variation in Earth radius. The primary function is to measure the horizontal plate motion and plate motion directions of each of the continents in order to determine the relative movement of continents constrained to a constant radius plate tectonic Earth model. The horizontal plate motions of each of the continents are then routinely published as horizontal velocity field vectors relative to the centre of the Earth. Figure 13.1 shows the published motion vectors from the International Terrestrial Reference Frame 2008 data.

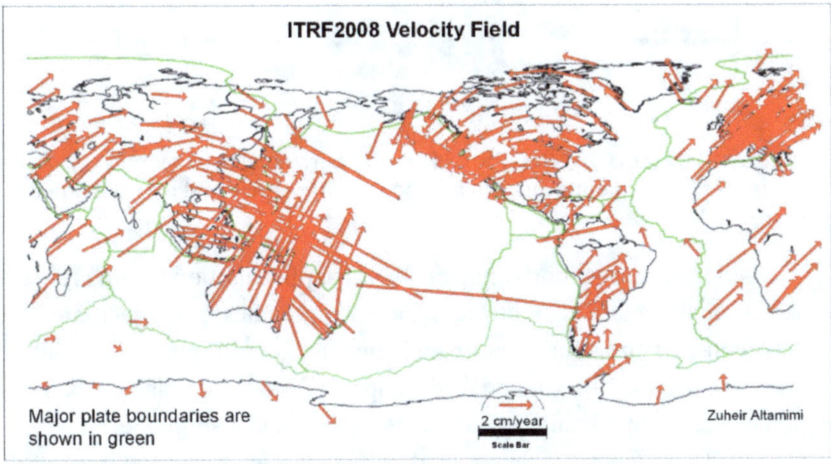

Figure 13.1. International Terrestrial Reference Frame 2008 horizontal velocity field vectors. International Terrestrial Reference Frame, 2008.

Earlier, 1997 versions of this plate motion vector data are shown in spherical format (Figure 13.2). In this second figure the motion vector data are shown in relation to vectors for the mid-ocean-ridge spreading zones. The mid-ocean-ridge spreading zones are shown as pink and red Pleistocene and Pliocene stripes respectively and the outlines of present-day continents are shown as blue lines.

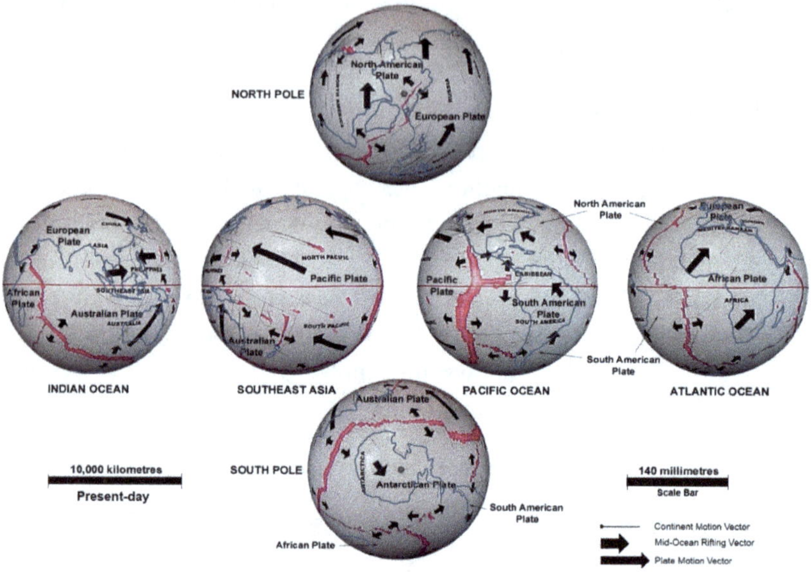

Figure 13.2. 1997 VLBI, SLR, GPS, and DORIS horizontal plate motion vector plots. Shown is present-day continents and mid-ocean-rifting (pink and red). Modified after Maxlow, 2001.

To understand what horizontal and vertical plate motion vectors represent on conventional plate tectonic and increasing radius plate tectonic Earth models, consider the simple cases in Figure 13.3 and Figure 13.4 where an ocean basin, such as the Atlantic Ocean, opens. Both figures show a part cross section of the Earth containing continental crusts and a single ocean basin.

Figure 13.3 shows a conventional plate tectonic cross section of part of the Earth where, because Earth radius is assumed to be constant, as the ocean basin opens during mid-ocean-rifting only the dark blue horizontal plate motion vector is required. Note: these vectors are grossly exaggerated and are only meant to represent centimetres of crustal movement per year as distinct from hundreds of kilometres of ocean basin opening.

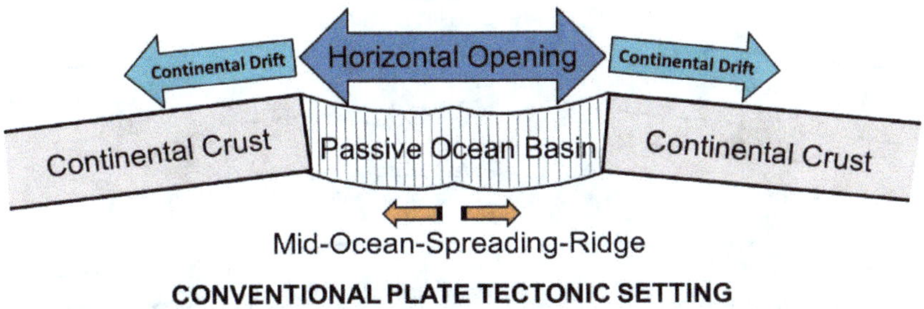

Figure 13.3. A cross-section of a conventional plate tectonic Earth showing a horizontal plate opening vector arrow (grossly exaggerated) and associated continental drift for a passive ocean basin such as the Atlantic Ocean and two adjoining continents. Crustal thicknesses are not to scale.

In this simple example, as the ocean widens, both continents located to the left and right of the ocean must then migrate away from the mid-ocean-spreading-ridge by continental drift. Furthermore, in order to maintain a constant Earth surface area, an equivalent amount of pre-existing crust must also be disposed of elsewhere.

In contrast, Figure 13.4 shows an increasing radius plate tectonic Earth cross section of part of the crust where, because Earth radius increases, horizontal, vertical, and actual plate motion vectors are required. Again, the scale of these motion vectors is grossly exaggerated and represents centimetres of crustal movement per year. In this example, Earth radius has increased from the lower A to the upper B crustal position, simulating a present-day radial increase of 22 millimetres per year.

In this example, the continental crusts are shown to move radially from A to B and, as such, there is no requirement for continental drift. The annual increase in Earth radius equals the green vertical motion vector, and opening of the ocean along the mid-ocean-ridge spreading centre equals the dark blue horizontal vector—again in centimetres per year. For an increasing radius plate tectonic Earth, the sum of half the blue horizontal vector plus the green vertical motion vector equals the red actual plate motion vector for each continental crust.

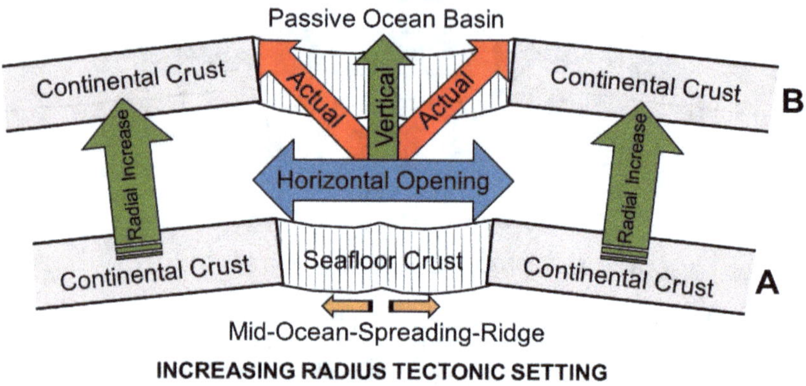

Figure 13.4. A cross-section of an increasing radius plate tectonic Earth showing horizontal, vertical, and actual plate opening vector arrows (grossly exaggerated) in relation to radial increase for a passive ocean basin such as the Atlantic Ocean and two adjoining continents. Crustal thicknesses are not to scale.

The simple example in Figure 13.4 is precisely what happens for all ocean basins on an increasing radius plate tectonic Earth. To further demonstrate this observation, the example is extended in Figure 13.5 to show the same motion vectors in relation to opening of two separate oceans. This example can be further extrapolated to the rest of the Earth.

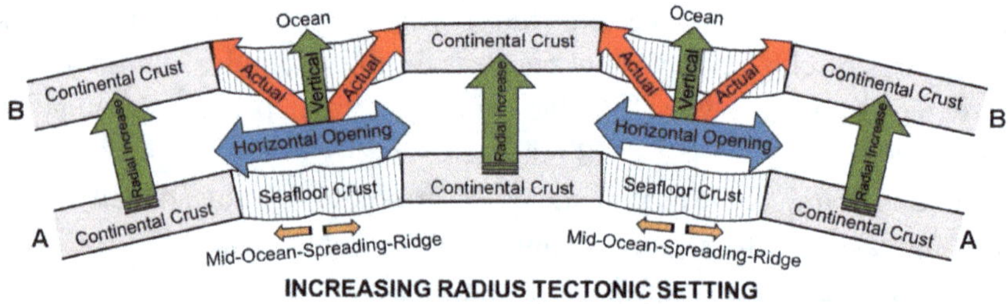

Figure 13.5. A cross-section of the Earth increasing in radius from crustal position A to B, showing horizontal, vertical, and actual plate motion vector arrows (grossly exaggerated) for three continents plus two adjoining ocean basins. Crustal thicknesses are not to scale.

The figure is again shown in cross-section where the Earth increases in radius from the lower crust A to the upper crust B position. On an increasing radius plate tectonic Earth, both of the oceans depicted open and spread apart at the mid-ocean-ridges in sympathy with a radial increase in Earth radius, with no requirement for continental drift.

In this figure, the increase in Earth radius is again represented by the vertical green radial motion vector arrows and opening of each ocean is represented by the dark blue horizontal motion arrows, representing centimetre scales. The net opening of each of the oceans plus the increase in Earth radius is then depicted by the red inclined actual crustal motion vector arrows shown for each continental crust.

In Figures 13.4 and 13.5 the positions of both the mid-ocean-ridge spreading axes and the continents, relative to the centre of the Earth, remain relatively stationary. The primary displacement of the continental crusts and the mid-ocean-ridges is then radial. This situation is unique to an increasing radius plate tectonic Earth, and the figures simulate the actual motions required for each of the continents during an increase in Earth radius over time.

For all oceans, opening along each of the mid-ocean-spreading-centres and accumulation of new seafloor crust is then dictated by an increase in Earth surface area as a direct result of increase in Earth radius. Again, in each of these increasing radius examples there is no requirement for disposal of excess crusts by subduction, and no requirement for continental drift.

In contrast, in conventional space geodetic and plate tectonic studies Earth radius is assumed to be constant and therefore any vertical vectors present are constrained to zero. Because of this constraint, on all conventional plate tectonic models, adjacent ocean basins cannot open without continental drift because the horizontal motion vectors of each continental crust oppose each other. This opposition can be seen in Figure 13.6 where, relative to the fixed centrally located continental crust B, continental crust A and continental crust C must move away from crust B by the amount of horizontal crustal opening shown by the green arrows in each ocean basin.

In this example, opening along each of the mid-ocean-ridge spreading zones will move each adjacent continent, as well as their respective mid-ocean-ridge spreading zones, away from the fixed continent B. This is the fundamental basis of and reason why continental drift is necessary in conventional plate tectonic studies.

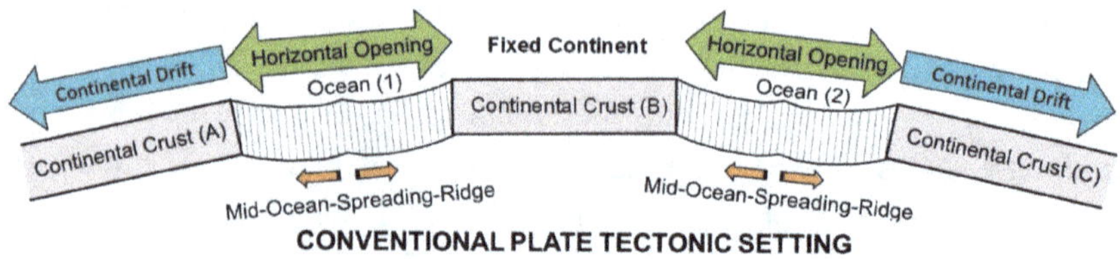

Figure 13.6. A cross-section of a conventional plate tectonic Earth showing the required sideways drift motion of two continental crusts A and C relative to a fixed continent B during mid-ocean-spreading within two separate oceans (vector arrows are grossly exaggerated). Crustal thicknesses are not to scale.

For the horizontal motion vectors in Figure 13.6 the sideways continental drift away from the fixed continental crust B is a purely local phenomenon. At some point around the globe continental drift of the various crustal plates must then interact and the plates will oppose each other. This opposition of crustal plates is the mechanism predicted in conventional plate tectonic studies for both collisions to form mountains and subduction along trenches.

In order to compensate for opening of the oceans on a conventional plate tectonic Earth model, such as the Atlantic Ocean relative to the Pacific Ocean, Plate Tectonics insists that excess crust must be disposed of elsewhere by subduction (Figure 13.7).

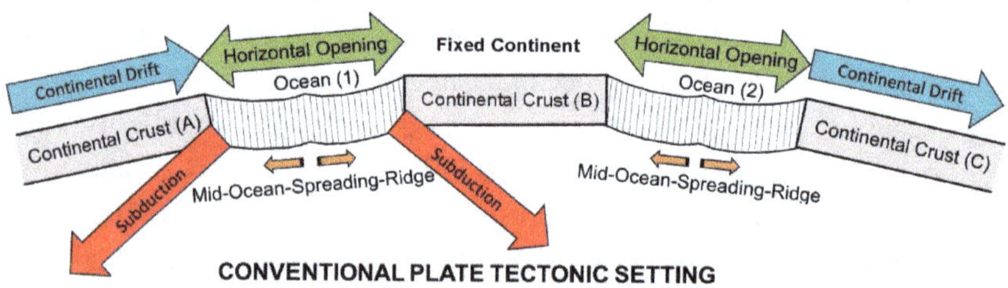

Figure 13.7. A cross section of a conventional plate tectonic Earth showing horizontal plate motion vectors for a number of continental crusts plus two intervening ocean basins. In this example excess crusts must be disposed of by subduction—shown as red vector arrows. Crustal thicknesses are not to scale.

This proposal now forms the fundamental basis of plate tectonic theory whereby excess crusts are disposed of within the Pacific Ocean via an inferred conveyor-belt motion of the continents which is driven by mantle convection. This scenario is depicted in Figure 13.7, where both of the ocean basins open and excess seafloor crust is consumed beneath the continents along subduction zones—shown as red vector arrows.

From the published horizontal plate motion vectors in Figures 13.1 and 13.2, it can be seen that the scenario in Figure 13.7 accords with the published plate motion results. This is because the published horizontal plate motion vectors are constrained to a conventional plate tectonic crustal model in order to maintain a static radius Earth premise.

The published space geodetic vectors in Figures 13.1 and 13.2 are therefore unique to a constant radius plate tectonic Earth model and cannot be reconciled on an increasing radius plate tectonic Earth. These same horizontal plate motion vectors on an increasing radius Earth are, in fact, meaningless because they do not consider vertical motion vectors relative to an increase in Earth radius.

Chapter 14

Geological Implications

"*Why should a body which is expanding develop huge diapiric extensional basins with a very deep source, as testified by the earthquake foci pattern, while elsewhere, very shallow extensional ridges with an associated shallow seismicity implies a great complexity of the global expansion process?*" Scalera, 1990.

By far the single most important scientific contribution to understanding the concept of an increasing radius plate tectonic Earth model has been publication of the bedrock geological map and age dating of all the continental and seafloor crusts during the early 1990s. This geological mapping has enabled assemblage of all crustal plates to be accurately constrained on small Earth models and, for the first time, has enabled modelling studies to be extended back to the earliest Archaean times. This mapping has also provided a means to calculate an increase in Earth radius over time and a means to investigate additional related physical observations.

Each of the plate tectonic small Earth models presented in previous chapters demonstrate that, by using modern bedrock geological mapping to constrain plate assemblage, an Archaean to present-day increasing radius Earth process is indeed viable. The basis of this small Earth modelling study is that volumetric increase of the Earth is simply transferred from the mantle to the outer crusts as continental and seafloor crustal stretching and extension.

It was recognized that this crustal extension process was operative throughout Precambrian and Palaeozoic times prior to supercontinental crustal failure and rupture during the late-Permian. Crustal failure and rupture then gave rise to breakup of the Pangaean supercontinent to form the modern continents and rifting and opening to form the modern oceans.

As Shields astutely wrote in 1997, "*Ultimately world reconstructions must be congruent not only with the data from geology and geophysics, but also with palaeobiogeography, palaeoclimatology, and palaeogeography*". In this context, in the following chapters, the geological implications of what seafloor and continental crustal small Earth modelling are showing will be further investigated by way of additional geological, geographical, and biogeographical evidence. This evidence is preserved in the rock-record of all continents and seafloors throughout the world, for all to see and study, and is collectively referred to here as global observational data.

When correctly interpreted, this global observational data provides an extensive amount of information about the past geological history, past geography, biogeography, and ancient climates, along with additional physical parameters, such as atmospheric conditions, sea levels, metallogeny, location of the ancient equator and poles, Earth kinematics, and so on.

For an increasing radius plate tectonic Earth to be seen as a truly viable process, the evidence preserved in this rock-record must then fully support and substantiate the small Earth crustal assemblages and ancient Earth radii so far investigated.

14.1. Historical Overview

Acceptance of the historical theory of Earth expansion was regarded by many researchers as being thwarted by major obstacles. These obstacles were originally considered by researchers to outnumber the evidence in favour. These concerns, however, stemmed from times dating back to the 1950s, well before the advent of modern-plate tectonic studies and well before modern understanding of these perceived obstacles.

Included in these concerns was the need to explain the existence of two very different extensional structures observed on the seafloors. These structures included the mid-ocean-ridges and the trench-arc and back-arc zones characterized by very different seismic activity and volcanism. Concerns were also raised regarding an explanation for the perceived problem of accumulating ocean waters and an atmosphere on an increasing radius Earth, as well as the adaptation of palaeomagnetics to a constantly changing Earth size.

Brunnschweiler in 1983, in a paper dealing with the evolution of modern tectonic concepts, considered that "*Earth expansion was essentially a radial movement and therefore its tangential plate displacements are only apparent, not real.*" In 1990 Scalera asked, "*...why should a body which is expanding develop huge diapiric extensional basins with a very deep source, as testified by the earthquake foci pattern, while elsewhere, very shallow extensional ridges with an associated shallow seismicity implies a great complexity of the* *global expansion process?*" Similarly, the possibility of orogenesis developing under conditions of radial expansion was discounted by Rickard as early as 1969 because he considered, "*...the necessary vertical movements did not appear to explain the observed compressional features.*"

These early discussions about crustal mechanisms on an increasing radius Earth model relate to times when it was considered that continental crusts acted as rigid bodies and hence, any subsequent breakup of these crusts only involved simple crustal fragmentation. This consideration continues, to some extent, through to present-day thinking where conventional plate tectonic modelling studies continue to treat supercontinental assemblages as amalgamations of previously fragmented and accreted crusts.

Since early times, modern global geological mapping now shows that continental and seafloor crusts have had a long and complex geologic history, a history that must be strictly adhered to during all crustal modelling and theoretical studies.

Bailey and Stewart in 1983 further considered that, "*...for an Earth undergoing expansion with time, the bulk of the oceans would have to be outgassed since the Palaeozoic, requiring fundamental changes in atmosphere, climate, biology, sedimentology and volcanology.*" In 1986 Weijermars considered that, "*...for a pre-Jurassic small Earth with a continuous continental crust, a large expansion process implies that the entire Earth would have been covered by an ocean with an average depth of 6.3 kilometres.*" This implication of Weijermars is contrary to the evidence preserved in sedimentary rocks, hence his concern.

Subsequent literature, such as Smiley in 1992, indicates there is also an increasing awareness of a number of "*perceived problems*" confronting conventional plate tectonics, as modelled on a constant radius Earth. Smiley considered that these perceived problems continue to be simply ignored or are overcome by invention of new "*ad-hoc fixes.*"

The outcomes of empirical small Earth plate tectonic modelling studies detailed in previous chapters lend substance to these concerns that the present concepts of conventional plate tectonics, including continental drift and palaeomagnetic based polar wandering studies, may indeed need to be "*...re-evaluated, revised, or rejected*" as Smiley suggested.

14.2. Changing Earth Surface Curvature

During the 1960s, there was active debate in the literature regarding the perceived incompatibility of the continental crusts during changes in Earth surface curvature over time. This incompatibility reflected the prevailing view that the Earth's continental and seafloor crusts acted as rigid bodies. It was further considered that the continental crusts could not change shape without undue fragmentation or distortion

Other researchers though were a little more imaginative. Some considered that crustal distortion on an increasing radius Earth should be possible because the complex folding, faulting, and shearing that is seen in the present-day continental rocks would lead one to expect continental fragmentation and distortion in any event. Others suggested that the outline of continental margins may have changed in time as a result of accretion of the continents, by loss of continental crust, or both.

All of these perceived problems regarding changing Earth surface curvature revolved around not only a lack of appreciation of what crusts are made of, but a lack of appreciation of the extensive amount of time available during increasing Earth radius. In particular, appreciation of the extremely small annual increases in radius that this research suggests with respect to the overall dimensions of the Earth itself. This is understandable and in the eyes of many, regardless of the potential for an increase in Earth radius, at the time a rigid, constant radius Earth crust was a justifiable conclusion.

It is now known that the Earth's continental crust is primarily made of stabilized ancient cratons, complexly folded and metamorphosed orogens, sedimentary basins, and volcanic seafloor crusts. The ancient cratons comprise mainly granite and volcanic rocks, with lesser sediments and can, by definition, be considered as rigid crusts. The orogens comprise mainly metamorphosed sedimentary rocks—rocks that have been folded and recrystallized as a result of Earth pressures and temperatures—and when originally formed these rocks were essentially flexible crusts. These orogenic rocks may have since increased their rigidity during metamorphism. The orogenic rocks can then be considered as having an intermediary strength between the ancient cratonic rocks and the younger basin sediments. The rocks forming the sedimentary basins and seafloor crusts are, by comparison, truly flexible rocks. Over the extended period of geological time available, the basin sediments in particular will readily flex, fold, fault, shear, stretch, and distort, as highlighted in many rock exposures throughout the world today.

14.3. Relief of Surface Curvature

Relief of surface curvature on an Earth undergoing a progressive increase in radius over time was first considered at length by Rickard in 1969 and was modelled extensively by Koziar during the 1980s and 1990s. The associated implied stress relief was also discussed by Weijermars in 1986. The latter author discounted Earth expansion on the grounds of a perceived "*spatial incompatibility*" resulting from a changing surface curvature.

This spatial incompatibility was previously emphasized by Jeffreys in 1962 when commenting on Barnett's 1962 small Earth model. Jeffreys pointed out that crust on Barnett's 115-millimetre diameter model could not be simply placed on his 76-millimetre diameter model without undue distortion. The reconstruction of continents on a small Earth model was then considered by Jeffreys to be dependent on the distribution of continental distortion.

In 1962 Dennis, in reply to Jeffreys' comment about crustal spatial incompatibility, suggested that the implied crustal distortion could be met because of what he called "*pseudo-viscosity of the crust.*" He further commented that distortion by normal geological processes, such as folding, faulting, and shearing, would lead one to expect continental distortion in any event.

Van Hilten in 1963 considered the problem of crustal distortion and breakup on an increasing radius Earth during his early investigations into palaeomagnetism. Van Hilten suggested that "*...during a potential Earth expansion process, a complete continental crust would distort and eventually be displaced by forming a number of radial tears in the crust.*" He likened this process to the radial peeling of an orange and introduced the term "*orange peel effect*" to describe such a tearing process (Figure 14.1).

This orange peel model, which suggests that the continental outlines were formed by merely pumping up the size of the Earth to split the continental crust like an orange peel, was never seriously accepted by researchers at the time.

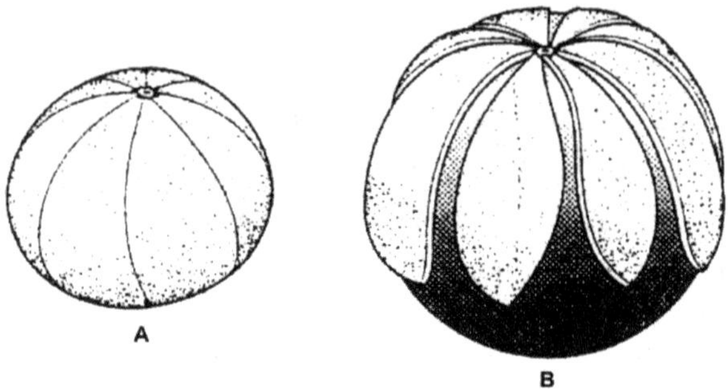

Figure 14.1. Van Hilten's "*orange peel effect*" model for an increasing radius Earth model. Van Hilten suggested that, during increase of the Earth from A to B, the continental crusts would distort and displace by forming radial tears. van Hilten, 1963.

Yet, as shown on the plate tectonic increasing radius small Earth models, once the continental crust first began to rupture during the late-Permian Period, this orange peel model gives a very simple but effective visual impression of how the complete Pangaea supercontinental crust eventually broke-up and how each of the modern oceans opened and spread apart.

This orange peel effect is shown on the Geological Map of the World (Figure 14.2), displayed in Mollweide equal area projection, where the Americas and Africa, along with India, Australia, and Southeast Asia can be likened to four of van Hilten's radial tears. This is also supported by observations of Carey in 1963, where he commented that "*...the greater separation of the southern continents, and general northward migration of all continents, has resulted because of a greater expansion of the southern hemisphere than in the northern hemisphere,*" identical to what van Hilten portrays in Figure 14.1B.

In 1969, Rickard further suggested that relief of surface curvature during an increase in Earth radius was a possible mechanism for both "*geosynclinal*" formation and "*orogenesis.*" The term geosyncline has now been replaced by additional plate tectonic terminology, but it basically refers to a subsiding low-lying linear trough or down-warp in the Earth's crust where sediments accumulate. Geosynclines can be viewed in a similar context to continental crustal extension and development of a network of crustal weakness and sedimentary basins on each of the small Earth models.

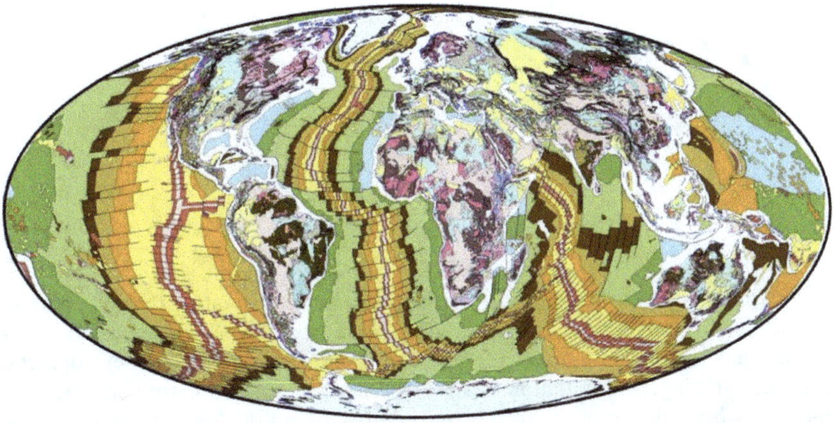

Figure 14.2. The Geological Map of the World displayed in Mollweide equal area projection highlighting both van Hilten's "*orange peel effect*" model for an increasing radius Earth and Carey's comment regarding the greater separation of the southern continents and general northward migration of all continents.

In contrast to comments by others, where it was suggested that continental crust adjusted to the new surface curvature by plastic flow and cracking during increasing Earth radius, Rickard suggested that the Earth's crust may be sufficiently strong to ensure a considerable time lag before complete adjustment for continental curvature was achieved. Rickard argued that, during an increase in Earth radius, geosynclines—sedimentary basins—would initiate as furrows along the margins of continental cratons (Figure 14.3). He considered that these furrows occurred because of a perceived "*differential radial expansion*" operating during opening of the oceans.

The effective increase in slope of continental crusts by a few degrees was then considered to increase the rate of erosion of the exposed lands, giving rise to a rapid accumulation of new sediment within the low-lying marginal basins—note that the "*furrows*" shown along the margins of the continent in Figure 14.3A are analogous to trenches in conventional plate theory.

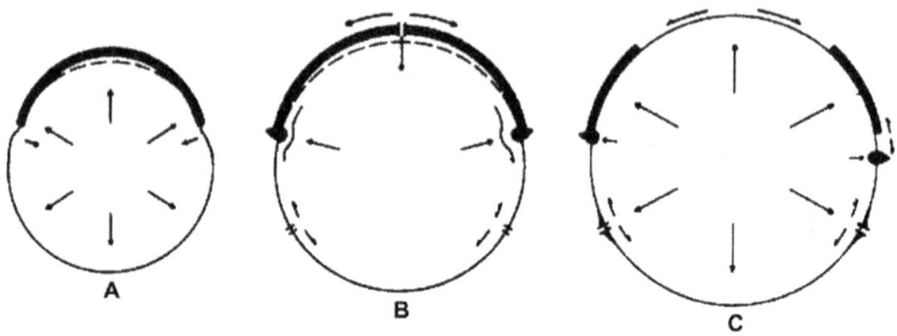

Figure 14.3. Rickard's 1969 model for relief of surface curvature during an increase in Earth radius. Figure A represents an "*initial stage*" of "*marginal geosynclinal furrow*" development; Figure B represents a "*critical stage*" in the development of orogenesis, rifting, and initiation of sea-floor spreading; and Figure C represents the "*relieved stage*" of curvature readjustment, isostatic uplift, block faulting, sea-floor spreading, continental separation, and development of island-arcs. Rickard, 1969.

The delicate state of balance between the opposing forces involved in relief of surface curvature was considered by Rickard as adequate to account for the complex variety of vertical movements required during early geosynclinal activity. Rickard further considered that eventually a critical stage would be reached where magmatic activity and rising heat may accompany orogenesis.

Carey was critical of Rickard's model because of his assumption of a significant "*...enduring strength in the continental crust and an implied super-elevation of the continents.*" Carey instead considered that, because of the "*...rapid adjustment of isostatic inequalities in the lower crust the required super-elevation could never come about.*" Carey appreciated that the central sector of the continental crust must rise "*...because of the megatumour beneath it*" but maintained that it would never depart far from isostatic equilibrium, nor would there be any "*...lateral gravitational force beyond that arising from hydrostatic equilibrium.*"

All this simply means is that, rather than the continental crust doming upwards by super-elevating the central part of the crust, as in Figure 14.3A, the crust would naturally adjust for changing surface curvature by distortion and fragmentation. The continental crust would also retain some superelevation within the central part of the continent due to a limited retention of an enduring strength of the crust, plus a degree of isostatic equilibrium.

Carey further considered that "*...the Earth consists of a crystalline mantle some 3,000 kilometres thick over a fluid core.*" The first result of an increase in Earth radius would be rupture of the whole mantle—plus overlying crust—into polygonal blocks of a few 1,000 kilometres across (Figure 14.4). He considered that these polygonal blocks would then be surrounded by accreted seafloor crust added during the last 160 million years as Earth radius continued to increase.

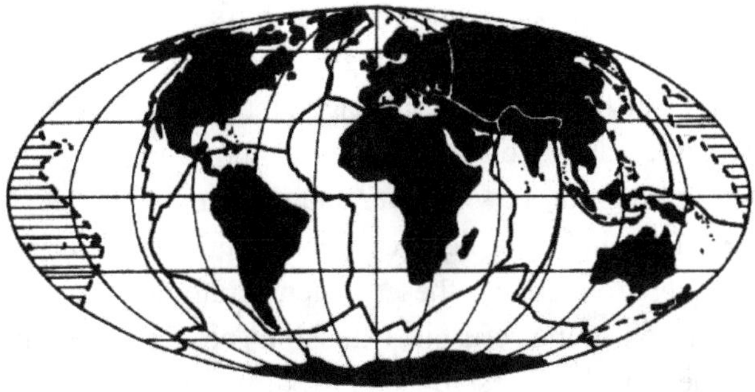

Figure 14.4. Carey's first-order primary crustal plate polygons surrounded by circum-continental spreading "*diapirs*". Each polygon consists of a continent surrounded by its accreted oceanic crust, added during the past 160 million years—now called plates in all tectonic studies. Modified after Carey, 1996.

Carey envisaged that the first adjustment to a decreasing surface curvature would occur at the primary mid-ocean-ridge spreading zones. These adjustments would extend to second rank basins and swells as the crust continued to adjust. Further adjustment would continue down through a hierarchy of fractures and ultimately to ordinary joints as a consequence of a final adjustment to the changing surface curvature. These primary polygonal blocks would then be patterned by second-order polygonal basins and swells, extending through both the continental and developing seafloor crusts.

While each of these historical proposals are feasible, neither Carey nor Rickard were able to appreciate that an increasing radius plate tectonic Earth has been operative for the entire 4,000 million years of known Earth history, not just for the past 160 to 200 million years as implied by these researchers. This lack of appreciation is understandable, because modelling of the continental crusts on small Earth models prior to the Triassic period was not possible before publication of the modern bedrock geological map in 1990. The small Earth modelling studies outlined here are, in effect, the first time this pre-Triassic modelling has been achieved and formerly presented.

Within the context of an increasing radius plate tectonic Earth, when considering the mechanism for changing Earth's surface curvature, it is important to again appreciate that continental crusts are now known to be made up of a broad distribution of cratons, orogens, and sedimentary basins. Each of these continental crustal elements has their own definitional characteristics, which differ fundamentally from the modern seafloor crusts.

In this context, Carey's consideration of primary polygonal crustal blocks must be shifted well back in time to when initial fracturing of the ancient primordial crust first occurred during early-Archaean times. During an increase in early Earth radius, it is envisaged that the mainly brittle crystalline ancient magmatic and volcanic cratonic crust simply fractured and fragmented. Intrusion of new granite magma and volcanic lava then re-stabilized the primordial cratonic crust prior to further fracturing over time.

During Proterozoic times, these ancient cratonic crusts were, in turn, accreted by an on-going crustal extension process operating within a surrounding network of established sedimentary basins and crustal weaknesses. The accumulation of sediments occurred mainly during crustal stretching and extension along the margins of each of the fragmented cratonic crusts. Only during the late-Permian Period did true rupture and breakup of the ancient supercontinental crusts occur to form the modern continents, as well as opening of the modern oceans along each of the newly formed mid-ocean-ridge spreading centres.

14.4. Geosynclines

While the term geosyncline is now rarely used in conventional tectonic studies, it is included here to remain in context with what Rickard and others discussed during the 1960s and 1970s. Additional plate tectonic terms also include subduction, island-arcs, and back arc basins which, strictly speaking, cannot be fully reconciled in their entirety here. On an increasing radius plate tectonic Earth the term geosyncline is, in fact, better suited to the extensional sedimentary basin settings that small Earth modelling studies suggest.

Rickard's 1969 model for geosynclinal development, modified in Figure 14.5, implies that an increase in Earth radius is essentially a radial process. The way the figure is drawn also implies that basin formation, orogenesis, and mountain building are post-continental breakup observations. This model is again a reflection of the lack of appreciation that increasing Earth radius has occurred throughout earlier pre-breakup history, rather than simply being confined to the past 160 to 200 million years.

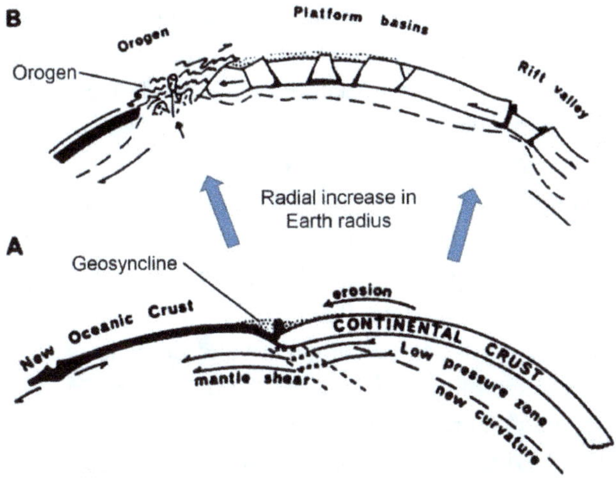

Figure 14.5. Rickard's cross-section of the Earth model for development of a "*geosynclinal trough*" prior to orogenesis. Figure A represents a "*critical stage*" of development where relief of surface curvature is balanced by the strength of the crust and downward acting weight of the sediments. Figure B represents an "*orogenic stage,*" where crustal compression is induced by "*isostatic uplift*" and accompanied by intruded granite magma. Modified after Rickard, 1969.

In this model, Rickard envisaged that a geosynclinal trough formed along the outer margin of a continental crust—left end of the continental crust in Figure 14.5A. Relief of surface curvature to the new Earth radius is then balanced by the strength of the crust and downward acting weight of the sediments deposited within the trough.

During a radial increase in radius of the Earth from A to B, sediments within the geosynclinal trough are then shown to undergo compression and orogenesis during ongoing changes in surface curvature. This may have also been accompanied by intruded granite magma and volcanism.

Rickard further considered that compressional forces leading to orogenesis of the crust during an increase in Earth radius resulted from "*basin inversion.*" This term refers to the uplifting and exposure of basin sediments during change in surface curvature, as distinct from slowly sinking and burial during deposition of the sediments.

Rickard also considered that the orogenic process involved an interaction between the continental and seafloor plates as the Earth proceeded to adjust its surface curvature. He went on to suggest that a lateral outwards movement of the continental plate caused "*...compressional buckling and over-thrusting of the sedimentary fill within a narrow geosynclinal trough.*"

Similarly, he also considered that continental-style geosynclines were developed where continental plates were fractured internally, giving rise to lateral compression as the two adjacent plates moved together during vertical relief of surface curvature.

The delicate state of balance between opposing forces involved in relief of surface curvature was considered by Rickard to account for the complex variety of vertical movements during early geosynclinal activity. This activity eventually led to what he called a critical stage, whereby magmatic activity and rising "*geo-isotherms*"—heat flow—were accompanied by orogenesis.

A model for geosynclinal development and orogenesis was put forward by Carey in the 1970s and 1980s for an Earth undergoing an exponential increase in radius (Figure 14.6). In Carey's model, the development of a simple "*diapiric orogen*"—an upward rising orogenic crust—resulted from initiation of primary stretching in the continental crust due to "*radial expansion.*" Carey further considered this upward rising process would then lead to "*necking*" or thinning of the crust.

The top and bottom surfaces of the continental crust in Figure 14.6 were then said to "*...converge towards zero with time at some 5 kilometres below sea-level*" during which time "*...the mantle rises some 30 kilometres.*" Carey considered that, although the surface of the thinning continental crust subsided steadily, the bottom of the crust, as well as the presence of a "*mantle diaper*"—comprising magmatic rocks such as granite—below it must also rise. This process continued to do so throughout the period of orogenesis. Carey further considered that "*...crustal thinning, caused by the expanding interior, resulting in gravity drive towards isostatic equilibrium, causes all the motions necessary for orogenesis.*"

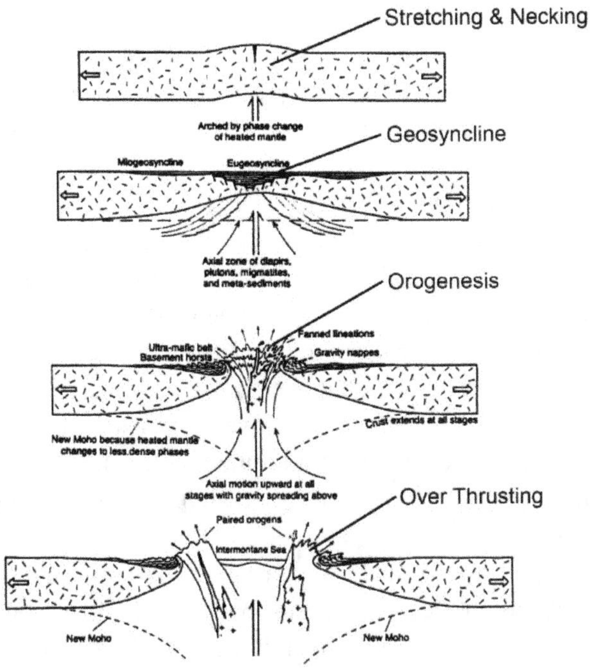

Figure 14.6. Carey's (1996) model for a symmetrical, single phase orogen rising about 100 kilometres. The top figure shows initiation of primary stretching in the continental crust leading to "*necking*" or thinning. The bottom of the crust and mantle diapir below it rises and continues to do so during orogenesis. The middle figures show two contrasting sites of geosynclinal sediment deposition. The lower figures show a continuing and accelerating rise of the diapir signaling the onset of orogenesis and over-thrusting of the geosynclinal sediments to form folded mountains. Carey, 1996.

14.5. Crustal Mechanisms

From an increasing radius plate tectonic Earth perspective, when each of these historical proposals for relief of surface curvature and geosynclines were first suggested, the ancient radius of the Earth was thought to have increased from a Pangaean supercontinent of around 55 percent of the present Earth radius. Increase in radius was then thought to have only occurred during the past 160 to 200 million years. There was very little consideration for changes to the continental crust or for an increase in Earth radius prior to that time.

Because of this lacking, the primitive Earth was considered to have been fully developed by 200 million years ago, implying that it had also been fully developed since early Archaean times.

What the distribution of exposed ancient crusts, as well as the network of ancient continental seas, now show on plate tectonic small Earth models is that the ancient supercontinents, in particular those making up the most ancient cratons and orogens, do indeed show an "*enduring strength in the continental crust*" but not necessarily a prominent "*super-elevation*" of the continents, as suggested by Rickard.

Small Earth modelling studies also suggest that, because of the imperceptibly slow rate of change in Earth radius during the Precambrian times, erosion may have been able to keep pace with changing surface curvature. Because of this erosion, the ancient crusts were never able to truly super-elevate above a relatively flat and featureless landscape.

Some form of super-elevation is, however, considered necessary during these ancient times in order to maintain some form of elevation of the exposed lands which, in turn, was necessary for erosion, transport, and deposition of sediments to occur.

It is envisaged that a subtle enduring strength and elevation contrast between the cratons, orogens, and basins persisted throughout the early Earth history, increasing to include periods of orogenesis and formation of fold mountain belts during Palaeozoic times as the rate of change in Earth surface curvature began to increase.

Super-elevation during these ancient times was, in turn, accompanied by a downward isostatic adjustment for changing surface curvature within each of the continental sedimentary basins, a process that gave rise to formation of geosynclines and ultimately to orogenesis and fold mountain formation.

14.5.1. Crustal Formation

What crustal evidence from the plate tectonic small Earth models shows is that there have been four distinctly different, but inter-related, crustal forming episodes during Earth history. These episodes are distinguished primarily by the differing styles of relief of surface curvature operating and include:

1. A primordial Archaean Earth episode representing formation and stabilization of the most ancient granite, metamorphic, and volcanic cratonic crusts.
2. An extensive Proterozoic episode marked by fragmentation of the ancient Archaean cratonic crust, followed in turn by onset of crustal extension and deposition of sediments within a newly established

network of sedimentary basins. Relief of surface curvature during that time was maintained by on-going erosion of the land surface, reducing it to a relatively flat featureless landscape.

3. A Palaeozoic episode representing a time when supercontinental crustal extension and relief of surface curvature had exceeded the ability to erode the landscape to a flat profile. This episode was then characterized by an increasingly elevated to mountainous landscape prior to eventual supercontinental crustal rupture and breakup during the late-Permian.
4. A Mesozoic and Cenozoic episode represented by rupture and breakup of the supercontinental crust to form the modern continents and opening to form the modern oceans. This episode is characterized by intrusion and preservation of mantle-derived seafloor volcanic lava along each of the mid-ocean-ridge spreading centres.

On an increasing radius plate tectonic Earth, it is envisaged that once the most ancient primordial Archaean Earth crust first cooled and stabilized, the mainly crystalline granite and intruded volcanic rocks remained as a rigid supercontinental cratonic crust covering the entire primitive Earth. It is further envisaged that onset of a very subtle increase in Earth radius was accompanied by simple global-scale fracturing of this brittle supercontinental crust. Fracturing then gave rise to further intrusion of granite magma and volcanic lava along faults and fractures within a newly established network of crustal weakness. This network of crustal weakness also formed the precursor to a primitive network of low-lying sedimentary basins and continental seas.

The extremely small rate of change in both Earth radius and surface curvature occurring during these ancient times was enough to maintain a very limited super-elevation of the central parts of the fragmented crusts and to establish a subtle elevation contrast between the locally elevated lands and low-lying sedimentary basins and seas.

The second, Proterozoic crustal episode was marked by retention of the fragmented primordial cratonic crusts, along with on-going crustal extension within the surrounding network of crustal weakness. This crustal extension process formalized the network of crustal weakness and accompanying sedimentary basins, which are now common to the ancient orogenic belts on Earth today.

It is further envisaged that relief of surface curvature during these Proterozoic times was accompanied by localized super-elevation and further fragmentation of the primitive cratonic crusts, as well as depression of the crusts around the margins within the relatively flexible sedimentary basins.

The extremely long period of time involved during this episode enabled the elevated lands to be eroded into a flat featureless landscape. This time frame allowed the exposed crusts to maintain a limited super-elevation while erosion continued to reduce the elevation contrast between the lands and seas.

This second episode differs from what most geoscientists currently believe. The exposed rocks within the known ancient Precambrian crusts are generally complexly folded and strongly metamorphosed. In conventional tectonic studies this complex folding would imply that the rocks have been subjected to continental collision to form fold mountain belts. Contrary to this belief, as will be discussed later, folding and metamorphism of rocks does not necessarily imply crustal collision, it only requires sustained localized compression.

Because of the extended period of time available during this second crustal episode, it is envisaged that downward relief of surface curvature of the fragmented crusts would have initiated the required prolonged and sustained compression and folding within surrounding sedimentary basins—estimated to be of the order of microns per year—extending over many hundreds of millions of years.

The extensive amount of time available to fold the sedimentary rocks would also mean that erosion of any exposed folded rocks would have continued to maintain an on-going flat landscape. The ancient rocks exposed today do indeed show preservation of the complex folding of these rocks, but not necessarily any evidence of the mature eroded land surface.

During the third Palaeozoic episode Earth radius and surface area began to rapidly increase. Relief of surface curvature during that time was not able to keep pace with the accelerating changes to the land surfaces. These accelerating changes then lead to initiation of orogenesis and formation of elevated fold mountain belts.

As a result of these changes, a feature of this episode was the presence of increasingly active drainage systems where erosion of the landscape and deposition of eroded sediments became prominent. It was during that time that the Earth also began to acquire the varied landscapes that are now more familiar today, albeit devoid of vegetation during much of that time.

Older sediments deposited during the more ancient, smaller Earth radius times were then subject to periods of renewed folding, faulting, erosion, and orogenesis. All of this activity was accompanied by an increasingly elevated and disrupted landscape prior to eventual crustal failure and rupture during the late-Permian to form the modern continents.

This Palaeozoic episode was succeeded by a fourth, post-Permian episode of crustal breakup and opening of the modern oceans extending over the past 250 million years. This fourth episode is characterized by mid-ocean-rifting and on-going intrusion and preservation of modern mantle-derived seafloor volcanic rocks.

During this episode each of the seafloor volcanic crusts, shown as coloured stripes in Figure 14.2, were intruded and preserved at ever-changing Earth curvatures. Relief of surface curvature during this post-Permian episode then shifted from extension-style mechanisms involving sedimentary basin formation and erosion, to completely new continental and seafloor crustal styles.

These new styles mainly involved isostatic crustal uplift and accompanying erosion along the margins of the fragmented modern continents to form mountain plateau, plus on-going mid-ocean-ridge extension and transform faulting within the oceans.

During each of these four fundamental crustal forming episodes, both the sedimentary rocks and seafloor volcanic lava can be considered as episodes of time-dependent deposition or intrusion. In this context, both the sediments and seafloor volcanic rocks were not all deposited at the same time or at the same Earth curvature. Instead, their deposition or intrusion was spread over different periods of time and were formed at the prevailing Earth curvature existing at the time of deposition or intrusion.

14.5.2. Primary Crustal Mechanisms

To fully appreciate the processes involved in relief of surface curvature on an increasing radius plate tectonic Earth, it is necessary to introduce a number of primary crustal mechanisms in order to visualize the way the various crusts may have behaved during an increase in Earth radius.

The basis of these mechanisms is common knowledge in structural and civil engineering studies and are used extensively for designing and testing structural components for engineering projects. These mechanisms are introduced here, at an engineering scale, in the form of a beam (Figure 14.7) and will, in turn, be adapted and modified to the scale of the Earth's crust.

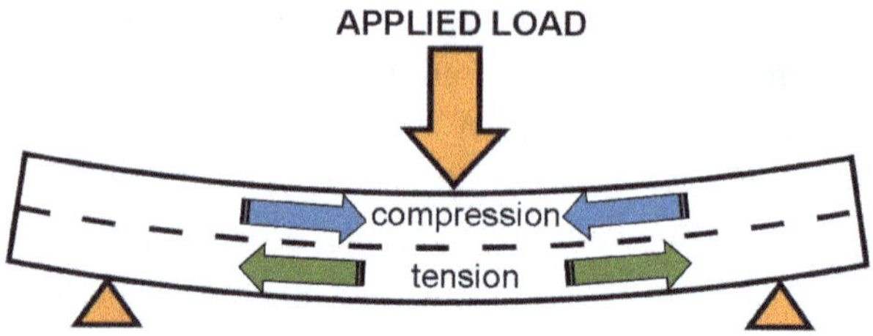

Figure 14.7. A simple horizontal beam supported at both ends with a centrally located applied load.

Consider what happens to a straight horizontal beam supported at both ends, such as a steel beam in a high-rise building. If a heavy load is applied at the mid-point of the beam, it will flex down due to the combined weight of both the load and the beam—shown exaggerated in Figure 14.7. Two things will then happen to the beam. The physical properties of all beams insist that, due to downward flexing caused by the applied load and weight of the beam, the top half of the beam will be in compression, and the bottom half will be in tension, regardless of what material the beam is made of.

Depending on physical strength and applied load acting on the beam, a rigid beam, for example one made of concrete, will be able to resist compression in the top half but will easily break by cracking across the bottom of the beam. In contrast, a flexible beam, for example one made of steel, will be strong in tension but may simply wrinkle or buckle as it sags and compresses along the top surface.

A combination of these two properties is the basis of reinforced concrete, where bars of steel reinforcing stop concrete from cracking under tension and the compressive strength of the concrete stops the steel bars from buckling. Similarly, for a beam made of soft material, such as plastic, where it is weak in compression as well as in tension, the bottom half of the beam will simply stretch, and the top half will wrinkle and compress.

Applying these engineering principles to the scale of the ancient Earth, Figure 14.8A is a schematic cross-section of a small portion of a brittle primordial Earth crust.

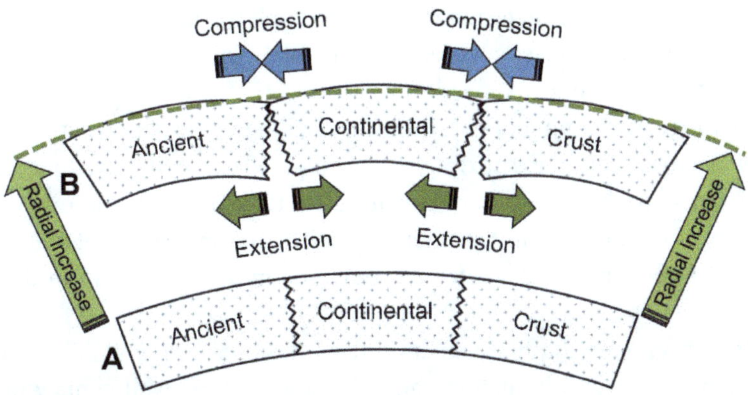

Figure 14.8. A schematic cross-section of a fragment of an ancient supercontinental crust undergoing breakup and fracturing during a change in Earth radius and surface curvature from Figure 14.8A to Figure 14.8B. Note: crustal thicknesses are not to scale.

In this example the crust, which is analogous to the beam in Figure 14.7, is upwardly curved to simulate the curvature of the Earth. The crust is also part of a stabilized ancient craton and comprises a brittle granite craton. During the early-Archaean, this primordial crust covered the entire ancient Earth surface and can be considered analogous physical characteristics to the concrete beam in the previous analogy. During an increase in Earth radius from A to B, the increase in radius generates localized tangential compression at the surface interaction of each fragmented crust, plus tangential extension at the base of each crust.

The Earth's crust in the lower Figure 14.8A can be visualized as a simple, albeit convexly curved beam. In this case, the applied load is the downward force of gravity plus the weight of the crust itself.

Because the crust is brittle, and hence weak in tension, during an increase in Earth radius the crust will initially start to super-elevate and rise in the central part of the crustal fragment, but over time it will eventually fracture and fragment. The smaller, fragmented crusts will then slowly collapse under the influence of gravity and adopt the new surface curvature existing at the time—shown as a dashed green line in Figure 14.8B.

In geology, this process represents a primary mechanism for compression of rocks at the surface, plus extension focused within the lower crust, allowing melting of the lower crust to form new granite magma as well as intrusion of volcanic lava. The tonalite-trondhjemite-granodiorite group of granite rocks common to all Archaean terranes throughout the world may be related to this fundamental compression-extension crustal mechanism and crustal fragmentation process.

Geological mapping of continents around the world show that the ancient cratonic crusts continued to fragment and stabilize until such time as they reached a size where they were able to retain their integrity and resist further large-scale fragmentation. Fragmented ancient cratonic crusts of a number of hundred kilometres in diameter are now commonly preserved in the rock-record.

It is envisaged that the ancient crust would be further stabilized during intrusion of new granite magma and volcanic lava, as well as new material added at the base of the crust sourced and crystalized from the mantle below.

The extremely long timeframe involved in this Precambrian phase—of the order of billions of years—would also see the super-elevated crusts being continually eroded, and the eroded sediments being deposited in surrounding low-lying sedimentary basins. This erosion would have the added effect of thinning and weakening the central parts of the elevated crusts, leading to further crustal fragmentation, extending down to the ordinary faulted and jointed rocks, as proposed by Carey.

As well as a cross-sectional view of this hypothetical crust, it is also important to understand what happens to the fragmented crusts when looking down on the surface of the primitive Earth (Figure 14.9). This consideration is only relevant to an increasing radius plate tectonic Earth where fragmented crusts also undergo crustal jostling.

This crustal jostling process occurs as the fragmented crusts adjust their positions relative to each other during changes in surface area and curvature. Crustal jostling is then seen as a global compensating process of adjustment of each of the crustal fragments.

Figure 14.9 is a hypothetical plan view of the surface of the Archaean Earth showing a number of fragments of ancient supercontinental cratonic crust, each with dimensions of a number of hundred kilometres across. The dashed areas between each of these crusts are a mixture of intruded volcanic lava, as well as sediments deposited in a network of low-lying sedimentary basins.

As each fragment of crust collapses during increasing Earth radius and changing surface curvature, local points of contact between the crustal fragments are shown to be under compression—orange arrows in Figure 14.9. These points of contact act somewhat like hinge points.

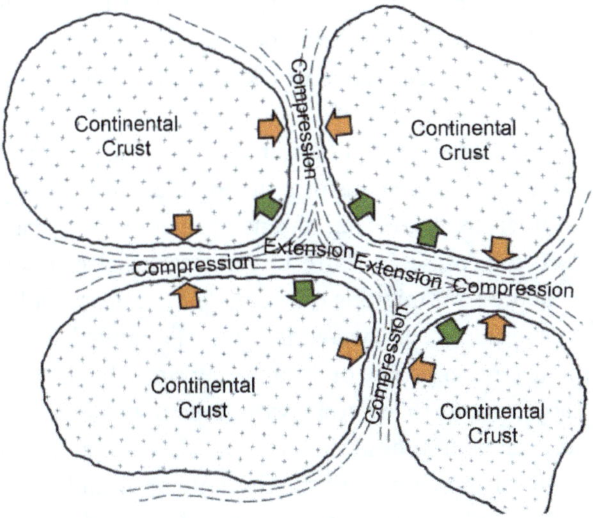

Figure 14.9. Schematic representation of the Earth surface of a brittle fractured primordial supercontinental crust, accompanied by crustal jostling, intrusion of magmatic rocks, and formation of a network of sedimentary basins. Orange arrows show localized compression and green arrows show crustal extension.

In contrast, the intervening areas undergo crustal extension—green arrows in Figure 14.9—and give rise to formation of sedimentary basins. These localized compression-extension observations represent the focal points for both on-going folding of sediments plus further opening and deposition of new sediments and intrusion of volcanic lava within low-lying sedimentary basins.

In addition to the localized crustal jostling in Figure 14.9, on-going changes in surface curvature will also involve regional-scale rotation of the fragmented crusts—which represents more advanced crustal jostling. The outcomes of this rotation and crustal interaction process are now preserved in the rock-record as regional-scale fault and shear movement, as well as folding and metamorphism of the sediments and volcanic rocks as they are heated, compressed, and jostled during periods of cratonization and orogenesis.

14.5.3. Enduring Strength and Super-Elevation

Super-elevation on an increasing radius plate tectonic Earth simply refers to the raised central part of a large fragment of a rigid, or semi-rigid, continental crust. Similarly, enduring strength refers to the ability of the continental crust to remain rigid, and retain its original surface curvature, and hence its ability to super-elevate during increase in Earth radius. Once the enduring strength of the crust is exceeded, gravity-induced relief of surface curvature takes over and, over an extended period of time, the crust collapses and re-equilibrates to the new surface curvature.

From the plate tectonic small Earth model studies, it has been shown that the rate of increase in Earth radius prior to about 2,400 million years ago was of the order of microns, increasing to tens of microns per year. During these ancient times, the increase in Earth circumference also amounted to tens, increasing to hundreds of microns per year. These imperceptibly small Earth circumference measurements were, in turn, distributed around the entire 10,600-kilometre circumference of the ancient primordial Earth, and represent atomic-scale adjustments to the crust per kilometre per year.

Since these ancient times, the rate of increase in Earth radius, and similarly the rate of increase in Earth circumference and surface curvature, has increased markedly. These changes have increased exponentially from microns per year during Precambrian times to 22 millimetres per year increase in radius and 140 millimetres per year in circumference for the present-day Earth. This present-day increase in circumference is now mainly confined to crustal extension within the seafloor crusts, where comparable extension rates of between 10 to 120 millimetres per year are currently recorded from each of the mid-ocean-ridge spreading zones.

The effects of enduring strength and changing super-elevation of a rigid continental crust during changing surface curvature of the Earth can be visualized by way of an example. Consider a fragment of an ancient craton, of say 500 kilometres diameter, that has remained relatively rigid and stable during an increase in Earth radius from an ancient radius to the present-day (Figure 14.10). The calculated difference in mid-point elevation of this craton between the surface curvatures of the ancient Earth of 1,700 kilometres radius—the radius of the Archaean small Earth model—and the present-day Earth is approximately 27 kilometres.

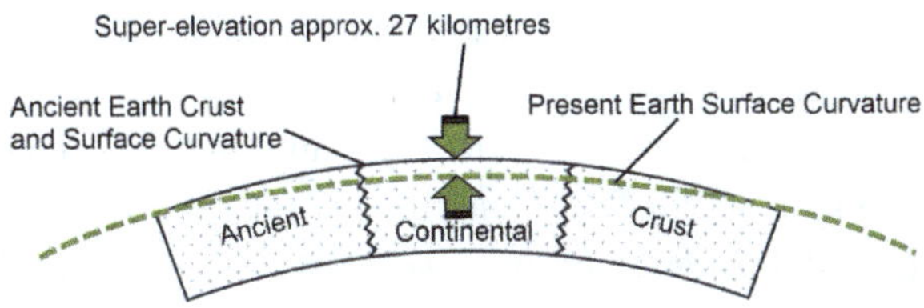

Figure 14.10. Cross section of the Earth showing the magnitude of change in surface curvature during an increase in Earth radius from the primordial Archaean Earth to the present-day. The mid-point super-elevation of the 500-kilometre diameter craton shown amounts to approximately 27 kilometres. Crustal thickness is not to scale.

Over time, the central part of this crust must either fragment and gravitationally collapse by 27 kilometres during changing surface curvature, or it must be eroded by 27 kilometres to maintain the prevailing surface curvature over time. Or, more realistically, a combination of both collapse and erosion must occur.

This order of magnitude in super-elevation is well within the realms of presently observed erosion of these ancient cratons, as currently preserved in the rock-record. Erosion of between 10 to 40 kilometres of vertical crustal thickness is a common feature within each of the ancient cratons present throughout the world today, such as the Pilbara Craton in Western Australia.

The example in Figure 14.10 suggests that fragmented parts of the ancient supercontinental crust may have remained rigid enough to undergo relatively simple super-elevation during a very subtle, but prolonged increase in Earth radius. Fragments of crust up to say 500 kilometres diameter may have had enough enduring strength to remain in an on-going super-elevated state, with enough time available for steady, on-going erosion to maintain the prevailing Earth surface curvature. The Pilbara Craton in Western Australia is again a prime example of this eroded super-elevation state.

Elsewhere, localized fragmentation, folding, crustal extension, volcanic intrusion, and deposition of sediments are also abundantly evident in the rock-record. These processes are particularly evident in the intervening areas between and surrounding the more rigid crustal fragments, which are now often represented by folded orogenic crustal rocks. This suggests that a combination of localized crustal rigidity, fragmentation, and gravity-induced collapse, plus extensive erosion of the crusts was aptly maintained during these ancient times.

In contrast to the most ancient supercontinental crusts, throughout the Proterozoic, and into the following Palaeozoic times, the supercontinental crusts were dominated by crustal extension localized within an established network of geosynclinal sedimentary basins.

Figure 14.11 is schematic only and represents simple fragmentation of an ancient crust. In reality, this fragmentation would also include a considerable amount of plastic deformation of the lower crust as well as faulting and rifting of the brittle upper crust.

In this context, the sediments deposited within the intervening basins shown represent flexible, time-dependent crusts deposited during periods of crustal extension, at the prevailing surface curvature existing at the time of deposition. In this example, there is no requirement for consideration of super-elevation of the sedimentary basins. The relative flexibility of the sediments allows the basins to continue to re-equilibrate to the changing surface curvature through time by simple, gravity-induced down-warping, extension, or folding of the sediments within geosynclinal troughs.

Following breakup of the supercontinents and opening of the modern oceans during the late-Permian Period, crustal extension then shifted from within the continental sedimentary basins to within each of the newly opening mid-ocean-ridge spreading centres. Preservation of this extension and seafloor growth process is now seen as the stripe-like magnetic growth patterns preserved on the seafloors, paralleling each of the mid-ocean-ridges within each of the modern ocean basins. Like the sediments deposited within sedimentary basins, deposition and preservation of this seafloor volcanic lava was at the prevailing ancient Earth surface curvature, present at the time of intrusion.

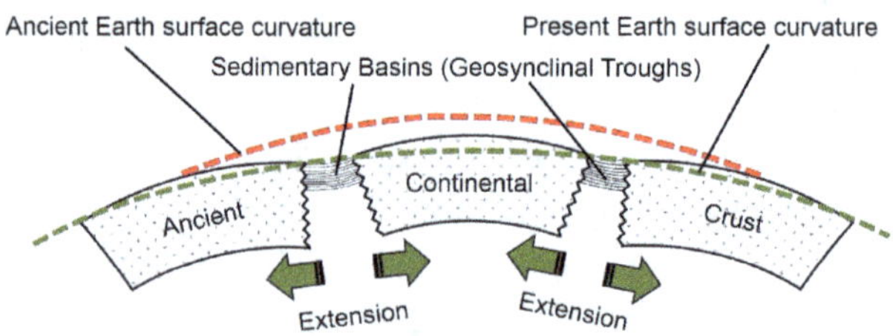

Figure 14.11. A schematic cross-section of an idealized fragment of an ancient supercontinental crust showing geosynclinal sedimentary basin formation during ongoing increase in surface area and Earth radius. Crustal thickness is not to scale.

Modern oceanographic studies now show that intrusion of this seafloor volcanic lava occurs within near-vertical faults and fractures, localized along the full length of the mid-ocean-ridge spreading axes. In effect, as new faults and fractures open, new lava is intruded and is quickly quenched by sea water when it reaches the seafloor. Over time, the lava may accumulate and extend in surface area as a series of stacked vertical dykes.

Because of the physical properties of these stacked vertical dykes, the strength of the seafloor crust is then severely compromised. In this context, relief of surface curvature of the seafloor during changing Earth radius is readily accommodated for by simple vertical movement along the margins of each intruded dyke.

The lack of extensive exposure of seafloor crusts on Earth today suggests that the inherent flexibility of these volcanic rocks is quite high, resulting in a much-subdued enduring strength and super-elevation of these time-dependent seafloor crusts. Larger-scale adjustments to surface curvature can also be seen as fracturing and faulting of the seafloor crusts in all of the modern oceans, and these are shown in the Geological Map of the World as an extensive array of cross-cutting transform faults.

14.5.4. Relief of Surface Curvature (Subduction-Related Features)

In addition to fragmentation, folding, and extension of the crusts, an equally important phenomenon occurs to the crusts during relief of surface curvature. In Figure 14.12 an ancient crust has fragmented and has undergone downward relief of surface curvature from the ancient radius to the present-day Earth radius, as shown by the red and green dashed curved lines respectively.

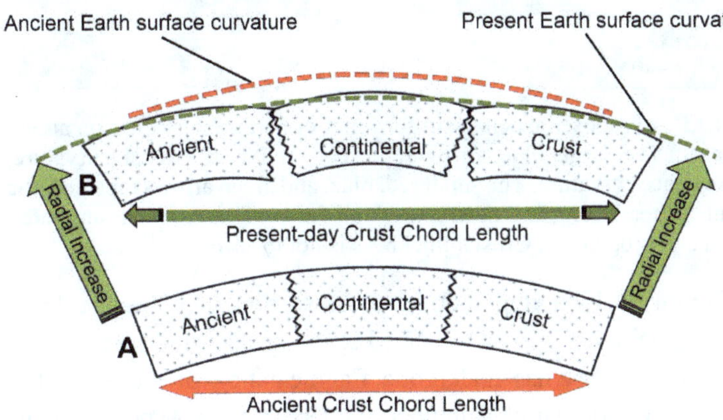

Figure 14.12. A schematic cross-section of an idealized fragment of an ancient supercontinental crust undergoing relief of surface curvature during an increase in Earth radius from Figure A to Figure B. During this relief of surface curvature process the chord length of the fragmented crust is shown to increase over time. Crustal thicknesses are not to scale.

During relief of surface curvature, as the crust flattens, the chord length—the straight-line distance between the outer margins of the crust—as shown by the long horizontal ancient red and present-day green arrows—lengthens. For example, for a crust with a curved surface length of 1,000 kilometres, this increase in chord length is calculated to be approximately 15 kilometres since early Archaean times.

This increase in chord length, in particular where multiple adjacent crustal fragments are involved, provides a means for both ongoing crustal extension within surrounding sedimentary basins, plus periods of localized crustal compression and shortening leading to orogenesis, fold mountain building, and conventional subduction-related observations (Figure 14.13).

The fundamental mechanisms involved in relief of surface curvature of crusts on an increasing radius plate tectonic Earth are highlighted schematically as coloured vector arrows at the left and right ends of Figure 14.13.

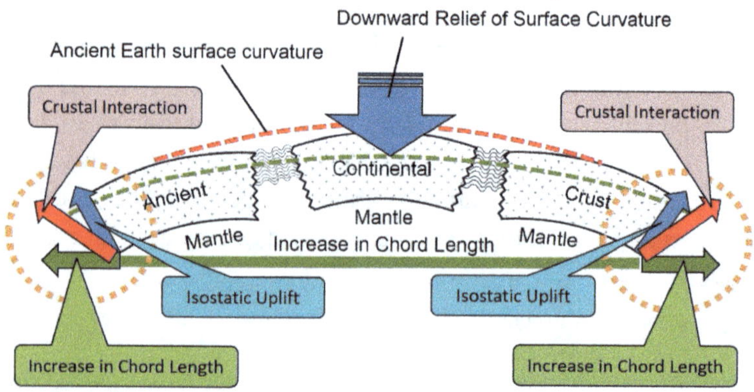

Figure 14.13. A proposed mechanism for relief of surface curvature on an increasing radius Earth. The large blue arrow represents relief of crustal curvature during changing surface curvature. The small red, blue, and green arrows represent the crustal movement vectors along the outer margins of a continent which simulates crustal interaction and orogenesis. Crustal thickness is not to scale.

In Figure 14.13 a hypothetical cross-section of a fragmented ancient supercontinental crust is shown located at the present-day Earth surface. This supercontinental crust has undergone fragmentation of the ancient cratonic crusts, plus extension within a number of intervening geosynclinal sedimentary basins during relief of surface curvature. Folding of the sediments during compressional orogenesis is also evident within the basins. In this scenario, the supercontinental crust now forms part of a modern continent and is separated from other continents by intervening modern oceans and seafloor crusts.

In order for the ancient supercontinental crust to equilibrate to the present-day surface curvature, a downward relief of surface curvature, shown by the large blue arrow, is compensated for by an increase in chord length, as shown by the long green arrow extending between the outer margins of the continent. Isostatic uplift of the continental crust also occurs around the margins of the continent to form an escarpment and associated mountain plateau.

The relative crustal movements associated with this relief of surface curvature are highlighted by the red, blue, and green vector arrows at the left and right sides of the figure. These vector arrows show a green horizontal movement related to an increase in chord length, a blue vertical movement related to isostatic uplift of the outer margin of the crust, and a red net upwards and outwards crustal interaction movement at the interface between the continental crust and the seafloor crust—analogous to conventional subduction-related observations.

This red, net crustal interaction, vector is seen as a prime mechanism for Wadati-Benioff-related seismic observations currently observed along the outer margins of the continental crusts. In addition, chord-related movement along the lower crust-mantle interface is seen as consistent with known crustal movement along the Mohorovičić discontinuity (Moho).

In this context, on an increasing radius plate tectonic Earth, the Moho represents the discontinuity between the crust and mantle during relief of surface curvature and increase in chord length acting over time.

In conventional plate tectonic usage, the Wadati–Benioff zone is considered to be a deep active seismic zone located within a subduction zone. Motion along this zone produces deep earthquakes, the foci of which may be as deep as 700 kilometres.

This same zone and seismic observations on an increasing radius plate tectonic Earth is instead considered to be related to obduction of the continental crust—as distinct from subduction of seafloor crust—as shown by the red crustal interaction vector and outward directed crustal movement shown by the green vector arrows in Figure 14.13.

While schematic, Figure 14.13 shows that there is adequate mechanism for crustal interaction—conventional subduction-related observations—on an increasing radius plate tectonic Earth during downward relief of surface curvature giving rise to the currently observed seismic, Mohorovičić, and Wadati–Benioff zone observations at the margins of continents. The increase in chord length of continental crusts is also again seen as a prime mechanism for generating both Wadati–Benioff-related observations along the outer margins of the continental crusts, as well as Moho-related discontinuity movement along the base of the crusts.

These observations, in turn, then represent a mechanism for the development of long linear mountain escarpments and plateau along the margins of continents, as well as island-arc and back-arc basin observations commonly seen around the western Pacific Ocean rim.

14.5.5. Orogenesis

Gilbert originally defined the term orogeny in 1890 as representing *a period of mountain building*. Burg and Ford in 1997 extended on this definition to define orogeny as a concept referring specifically to the "*...folding of rocks in fold belts.*" This folding was further considered to be "*...characterized by intense internal rock deformation along long linear mountain belts, and, at places, by associated regional synkinematic metamorphism.*" Where, "*regional synkinematic metamorphism*" simply means the alteration of rocks by heat and pressure occurring at the same time as the folding or deformation event.

Burg and Ford further considered that "*...orogenesis encompasses all processes that result in crustal thickening causing uplift and hence high topography*" which may occur over variable intervals of time and may consist of a number of distinct tectonic events.

With changing emphasis to the terminology, in particular with respect to conventional plate tectonic usage, orogeny now refers to forces and events leading to a large structural deformation of the Earth's crust and upper mantle due to the interaction of tectonic plates. Deformation of the crust then results in the formation of long tracts of highly deformed belts of orogenic rocks.

On an increasing radius plate tectonic Earth, orogenesis is seen as resulting from an extended period of crustal compression, as distinct from plate collision, and the definition of orogenesis is here defined as:

> *A sustained compression of rocks between mutually adjacent crusts giving rise to folding of the rocks to form long linear fold mountain belts, and, at places, by associated regional synkinematic metamorphism.*

The fundamental mechanism for orogenesis on an increasing radius plate tectonic Earth involves an extension of the primary relief of surface curvature mechanisms outlined in the previous section. During downward gravity-induced re-equilibration of the ancient crust to changing Earth surface curvature, two fundamental processes are seen as contributing to sustained compression, which may then contribute to orogenesis and formation of fold-mountain belts (Figure 14.14).

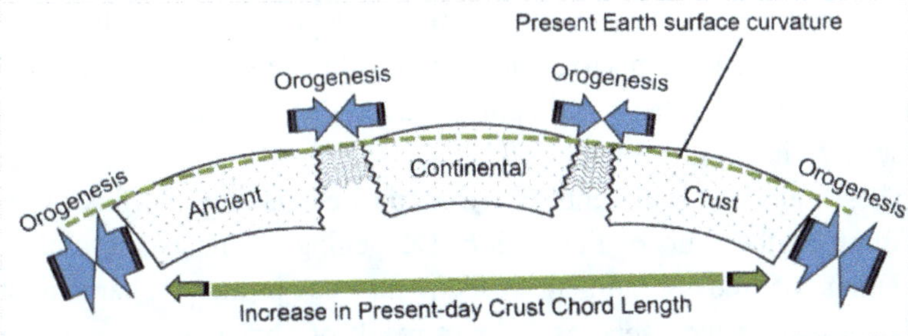

Figure 14.14. A schematic cross-section of fragments of an ancient supercontinental crust undergoing relief of surface curvature and compressional orogenesis within linear geosynclinal sedimentary basins. Crustal thickness is not to scale.

Firstly, as the fragmented ancient crusts gravitationally collapse and re-equilibrate to the changing surface curvature, compression may occur between adjacent cratonic fragments, localized within a surrounding network of geosynclinal sedimentary basins.

Secondly, as the ancient crusts re-equilibrate to the new surface curvature, there is an increase in the overall chord length of the crust. This increase in chord length provides outwardly directed compression along the margins of the crust which may, in turn, be localized within adjoining sedimentary basins.

Both processes are then considered to result in compressional orogenesis and uplift to form long linear fold mountain belts. Changing surface curvature may also include rotation and crustal jostling between the various ancient crustal fragments, which, in turn, may be accompanied by localized folding, faulting, and translational shear movement.

This mechanism for orogenesis on an increasing radius plate tectonic Earth differentiates between pre-breakup of supercontinental crusts and modern post-breakup of continental and seafloor crusts. Orogenesis within the supercontinental crusts is focused primarily within an established global network of crustal weakness and sedimentary basins.

On an increasing radius plate tectonic Earth, there were no large oceans during pre-Triassic times, hence no mechanism, or room, for continental drift to promote continental collision. Similarly, breakup of the supercontinents and opening of the modern oceans is seen as a progressive increase in surface area, involving an accumulation of new seafloor volcanic lava with, again, no provision or need for continental drift or collision.

The mechanism for orogenesis on an increasing radius plate tectonic Earth (Figure 14.14) shows that initially, as Earth radius increases over time, the primitive crustal fragments gradually depress around their margins, allowing eroded sediments to be deposited within newly formed or reactivated sedimentary basins. These sedimentary basins, in effect, are the geosynclines, as discussed in Section 14.4.

These eroded sediments were deposited at the surface curvature prevailing at the time of deposition. Deepening of the basin, and vertical thickening of the accumulated sediments—a primary characteristic of geosynclines—is then seen as a direct result of changing surface curvature over time, with intracratonic compression resulting in orogenesis, or uplift to form fold mountain belts.

14.5.6. Fold-Mountain Belts

A mountain is defined in the Oxford English Dictionary as "a natural elevation of the Earth surface rising more or less abruptly from the surrounding level and attaining an altitude which, relative to the adjacent elevation, is impressive or notable." This impressive mountainous topography can be, and generally is, relatively young, whereas the original formation of the geological structures within the mountains is generally much older and, over time, the mountains may have been eroded and uplifted many times.

It is now known that fold-mountain belts have been formed at various times throughout much of geological history. Over an extended period of time, once formed, these mountains then begin a natural cycle of erosion, decline, and reactivation. What are currently seen as mountains today will eventually be eroded down to simple foothills in the geologically distant future.

In conventional plate tectonic studies, the formation of mountains is portrayed as a random process associated with plate convergence, subduction, or collision of the various supercontinental plates during orogenesis. In this context, the crusts are said to be pushed upwards during collision or lifted upwards during subduction to form mountains.

In this context, the collision or subduction process is said to trigger deformation and thickening of the crust and may be accompanied by magmatic activity. This, in turn, leads to uplift of the crust to form elevated mountains, locally raising the surface of the Earth by over 3,000 metres. Over time, these mountains slowly erode through the action of rivers or glaciers. The eroded material is then washed into surrounding sedimentary basins, or onto the continental shelves. The reduction in weight of the mountains may in turn result in still further uplift of the mountainous terrain.

In 2000, Ollier and Pain noted in their book on mountains that a distinction must be made between orogeny, which deals with the formation of fold belts, and uplift to form mountains. This distinction is not always clear in conventional tectonic usage. In their book, Ollier and Pain showed that, in contrast to folding of rocks to form mountains, modern mountain building instead results from vertical uplift to form "plateau"—large areas of relatively flat uplifted land surfaces—rather than from horizontal compression.

Ollier and Pain provided many examples from major mountain ranges throughout the world to show that most mountains are, in fact, relatively young, eroded plateau, with very little to do with the earlier folding event, or events, as preserved in the rocks below the plateau.

These plateaus, shown schematically in cross section in Figure 14.15, were seen as once being relatively flat, low-lying eroded surfaces, such as now seen along many continental margins. As a result of vertical uplift, the ancient rivers and streams of the time were then reactivated, and new erosion surfaces proceeded to cut across older folded bedrock structures to form new plateau and eroded escarpments.

This observation of Ollier and Pain contrasts strongly with conventional plate tectonic usage, where it is considered that vertical crustal movement to form mountains can only occur from horizontal movement during continental collision or crustal subduction. In considering this, Ollier and Pain pointed out that any conventional tectonic explanation must account for the extended amount of time needed, as well as the stable conditions required, to allow enough time for erosion of the land surfaces to form the observed plateau, in particular where multiple and overlapping plateau events occur. Also,

Ollier and Pain further considered that it must provide an explanation for the rapid vertical uplift that exposed the plateaued land surfaces to renewed erosion, as distinct from the completely separate periods of folding and uplift proposed by conventional tectonics.

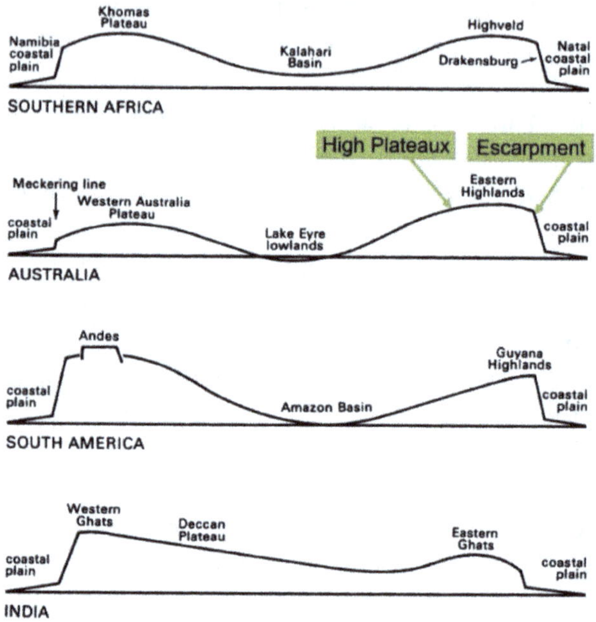

Figure 14.15. Schematic vertical cross-sections of mountains from various continents. In each example great escarpments parallel the coasts and separate a high plateau from the coastal plain. Each continent also has a central depression, bounded by gentle rises to high plateau, beyond which are abrupt outward-facing escarpments and coastal plains. Modified after Ollier, 1985.

In their book, Ollier and Pain presented a model for what they referred to as "*...passive margin continental development and erosion,*" again based on their extensive field-based observations. They concluded from this model that, rather than continental collision, early breakup and rifting to form continents started with a "*rift valley stage*"—a period of crustal extension and faulting. During this stage an ancient coastal plain, located along the edge of the continent, was simply down warped. On-land changes to the existing drainage pattern then occurred and, as seafloor spreading began, part of the coastal plain sank beneath the sea. Erosion was then concentrated on the steeper slopes of the exposed land surface, located between the axis of the warping and the coast, which may have, in time, been uplifted to form a great escarpment. Sediments eroded from the exposed land were then deposited offshore, covering the sunken plain in the process.

On an increasing radius plate tectonic Earth, this coastal down warping of Ollier and Pain occurs naturally around the margins of cratons or continents during changing Earth surface curvature. Relief of surface curvature of the continental crusts then eventually reverses this process, resulting in uplift of the continental margins exposing the plateau and forming an escarpment, followed by further erosion and planation.

14.6. Proposed Crustal Model

The schematic cross-sectional models of the Earth presented below are based on an extension of ideas and mechanisms originally presented by Rickard, Carey, and others during the 1960s to 1980s. From an increasing radius plate tectonic Earth perspective, when these historical proposals were first presented, the ancient radius of the Earth was only thought to have increased from a Pangaean supercontinent of around 55 percent of the present Earth radius. Increase was then thought to have only occurred during the past 160 million years or so.

There was very little consideration in the literature for changes to the continental crust or for any increase in Earth radius prior to that time. The primitive Earth was then considered by these researchers to have been fully developed 200 million years ago, implying that it had also been fully developed since early Archaean times.

Plate tectonic small Earth modelling studies presented here now show that Earth radius has been increasing exponentially throughout the entire geological history of the Earth, extending from early Archaean times through to the present-day. From these small Earth modelling studies, crustal extension, as well as relief of surface curvature within the ancient supercontinental crusts, is shown to have occurred at a very much slower rate, and over a far more extended period of time, than that shown by the modern seafloor crustal spreading rates. The models and ideas of Rickard and Carey are here modified to extend the increasing radius crustal model back to the known beginning of geological time (Figure 14.16).

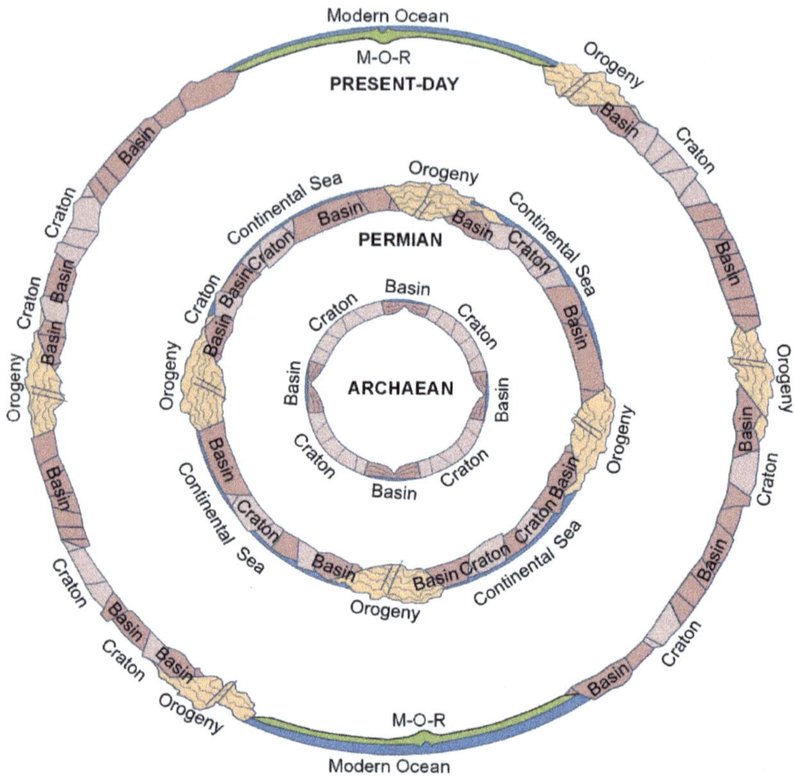

Figure 14.16. Schematic cross-sectional models of an idealized Archaean, Permian, and present-day Earth showing an increasing radius crustal model for development of continental and seafloor crusts. Crustal thicknesses are not to scale.

Note: the crustal thickness depicted in Figure 14.16 is not to scale, and, in reality, would be represented by a very thin skin on the surface of each model.

The schematic model in Figure 14.16 shows three stacked cross-sections of a primitive Archaean and a Permian small Earth model located within a present-day Earth model. The innermost model summarizes the crustal development and tectonic processes operating since initial crustal formation some 4,000 million years ago. The Earth then steadily increased in radius throughout the Proterozoic and Palaeozoic supercontinental stages, to the late-Permian model some 250 million years ago. This model was then followed by a rapidly accelerating increase in Earth radius, commencing with a late-Permian crustal breakup phase to form the modern continents and rifting of the crusts to form the modern oceans.

In Figure 14.16, the inner-most primordial Archaean Earth crust has cooled and stabilized. The mainly crystalline granite and intruded volcanic rocks remain as rigid supercontinental crust, covering the entire Earth. The onset of increasing Earth radius was accompanied by simple, global-scale fracturing of this brittle supercontinental crust.

Ongoing fracturing gave rise to further intrusion of granite magma and volcanic lava along faults and fractures within a newly established network of crustal weakness. This network of crustal weakness also formed the precursor to a primitive network of low-lying sedimentary basins.

The extremely small rate of change in both Earth radius and surface curvature occurring during this ancient time was enough to maintain a very limited super-elevation of the central parts of the fragmented cratonic crust, and to establish a limited elevation contrast between the locally elevated lands and low-lying sedimentary basins.

Between the inner Archaean and the middle Permian small Earth models, Earth radius increased from 27 percent to approximately 50 percent of the present Earth radius. The time interval shown by these two models is represented by the Archaean and Proterozoic Eons as well as the Palaeozoic Era—around 3,750 million years of Earth history.

The Proterozoic Eon is marked by retention of the fragmented primordial crusts, along with on-going crustal extension and deposition of sediments within platform sedimentary basins localized between each of the ancient crustal fragments. This crustal extension process further formalized the global network of crustal weakness and accompanying sedimentary basins.

Relief of surface curvature during these times was accompanied by localized super-elevation and on-going fragmentation of the primitive crusts, as well as depression of the land surface within surrounding low-lying sedimentary basins. The extremely long period of time involved enabled these elevated lands to be continually eroded to a mainly flat and featureless landscape.

Because of the extended period of time available during the ancient Archaean and Proterozoic Eons, downward relief of surface curvature of the fragmented crusts also subjected localized areas to prolonged and sustained compression and folding. This downward relief of surface curvature initiated geosynclinal down-warping within the network of sedimentary basins, followed by orogenesis and mountain building.

The amount of time available to compress these rocks also meant that erosion of the folded rocks exposed at the land surface continued to maintain an on-going, essentially flat landscape while still preserving the complexly folded geology below.

During the Palaeozoic Era—extending from around 540 to 250 million years ago—Earth radius and surface area began to steadily increase. Relief of surface curvature during that time was not able to keep pace with the accelerating changes to the land surfaces, leading to periods of orogenesis and formation of high, elevated mountain belts focused within the established network of geosynclinal sedimentary basins.

As a result of these crustal changes, drainage systems became increasingly active, and erosion of the elevated landscape became prominent. Older sediments deposited during the ancient, smaller Earth radius times were also subject to periods of renewed folding, faulting, erosion, and orogenesis. All of this activity was accompanied by an increasingly elevated and disrupted landscape, prior to eventual crustal rupture and failure to form the modern continents and oceans.

The post-Permian period of crustal breakup and opening of the modern oceans is represented in Figure 14.16 by the change from the middle Permian model to the outer present-day model, representing the remaining 250 million years of Earth history. The unprecedented change in Earth radius, surface area, and surface curvature occurring during this interval of time was characterized by intrusion and preservation of the modern volcanic seafloor crustal rocks.

Relief of surface curvature during this post-Permian time then shifted from extension-style mechanisms involving sedimentary basin formation and erosion, to completely new continental and seafloor crustal styles. These new styles mainly involved crustal uplift and erosion along the margins of the fragmented modern continents to form mountain escarpments and plateau, plus on-going mid-ocean-ridge extension and transform faulting within the oceans.

Throughout geological time, the process of crustal adjustment during gravity-induced change in surface curvature, while simplistic, provides a viable mechanism for both orogenesis and mountain building during an increase in Earth radius. On an increasing radius plate tectonic Earth, this mechanism does not require, or rely on, continental collision, or crustal subduction, to promote horizontal or vertical tectonism.

Multiple phases of orogenic activity are now recorded in the rock-record, and these suggest a crustal extension-orogenesis-extension cyclicity. The radial and tangential vector components of this complex orogenic process (Figures 14.13 and 14.14) result in a continuum of potential orogenic models during on-going relief of surface curvature and crustal interaction.

This continuum of potential orogenic models, along with observations of orogenic development on an increasing radius plate tectonic Earth, are consistent with the fundamental increasing radius definition for orogenesis whereby orogenesis involves a sustained compression of rocks between mutual crusts giving rise to folding of the rocks to form long linear fold mountain belts. This fundamental definition also incorporates the extended definition of Burg and Ford whereby orogenesis is further "*characterized by intense internal rock deformation along long linear mountain belts, and, at places, by associated regional synkinematic metamorphism.*"

Chapter 15

A Paradigm Shift in Palaeogeography

"*Ultimately world reconstructions must be congruent not only with the data from geology and geophysics, but also with palaeobiogeography, palaeoclimatology, and palaeogeography.*" Shields, 1997.

Ancient geography, referred to in the geosciences as palaeogeography, is a study of the Earth's past geography and, in particular, the geography of the ancient land surfaces. It includes studies of the distribution and configuration of ancient continental landmasses, their topographic relief, the distribution of ancient seas and shorelines, and changing climate zones. Palaeogeography is also important in the study and distribution of plant and animal species, where it is referred to as palaeobiogeography.

Evidence for palaeogeography is preserved in many rocks exposed on the Earth today. Sedimentary rocks, for example, are made of rock and mineral particles that have previously been eroded and redeposited from older rocks. These particles often consist of quartz sand, mica, and clay, and may also consist of rock particles or volcanic detritus, weathered from exposed rocks that formed the ancient lands. Once eroded, these particles were then transported and deposited elsewhere by water, wind, or ice, and in the process were further worn and broken down to finer-grained particles. The types of sediments that were deposited and preserved in the rock-record are then dictated by the composition of the original rocks, the prevailing climate, and the environment where they were initially eroded and finally deposited.

As well as showing the inter-relationships between exposed continental lands and intervening ancient seas, geographic information also enables the ancient Pangaea, Gondwana, Laurentia, and Baltica, Laurussia, and Rodinia supercontinents to be quantified on plate tectonic small Earth models. However, because of the limited amount of published geographic information available for Precambrian times, it is not possible to reconstruct the distribution of ancient seas or supercontinents prior to the Cambrian Period.

15.1. Ancient Shorelines

On plate tectonic small Earth models, prior to about 250 million years ago there were no modern oceans, only ancient continental seas. Similarly, before that time, ancient supercontinental crusts existed as a complete continental crustal shell, encompassing the entire ancient Earth. The outlines and configurations of the exposed lands making up the supercontinents were then dictated by the presence of, and changes to, the ancient seas, primarily as a result of changes to the distribution and surface areas of each of the ancient continental sedimentary basins.

The transition from ancient seas to modern oceans only came about when the Pangaean supercontinent first started to rupture and break up during the late-Permian—some 250 million years ago. It was this breakup of Pangaea that initially established the modern continents and intervening modern oceans, as well as initiating draining of waters from the ancient continental seas into the newly opening oceans.

The distributions of ancient shorelines on each of the Cambrian to present-day small Earth models are plotted as blue outlines in Figure 15.1, based on the published evidence of Scotese, 1994, and Smith *et al.*, 1994. The ancient seas and modern oceans are shown as shaded blue areas and the remaining-coloured areas represent exposed lands.

These shorelines, while only showing a generalized simplistic outline, were transferred directly from coastal outlines shown on the published geographical evidence. Because conventional plate tectonic reconstructions need to fragment existing continents to maintain a constant radius Earth, during plotting of this evidence it was found necessary to merge some of the published coastal information, or to simplify it in areas where the information overlapped or interacted on the plate tectonic small Earth geological reconstructions.

The global distribution of ancient shorelines for the Cambrian to present-day small Earth models (Figure 15.1) is shown to be an evolving and progressively changing process. This distribution of coastal shorelines is interpreted to be intimately related to changes in sea-levels over time. On an increasing radius Earth, these changes were, in turn, related to changes in surface curvature and associated tectonism during on-going increase in Earth radius and eventual breakup of the supercontinents.

As well as global-scale sea-level changes, the distribution and variation of ancient sea levels also reflect more regional-scale changes as well as the introduction of new water from along the newly opening mid-ocean-ridge spreading zones.

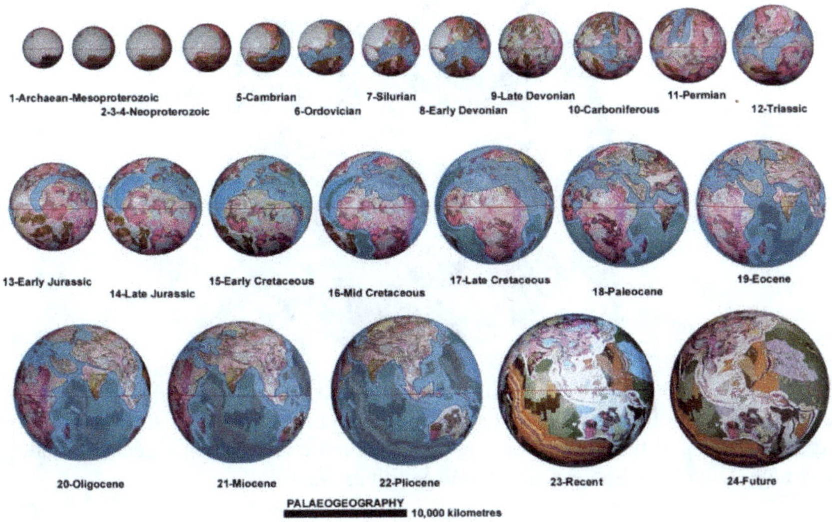

Figure 15.1. Shoreline palaeogeography on Archaean to present-day plate tectonic small Earth models. The ancient shorelines are shown as blue lines, and the ancient seas and modern oceans are shaded blue. Each image advances 15 degrees longitude throughout the sequence to show a broad coverage of palaeogeographic development. Note: there is no published data available for the late-Devonian model or models prior to the Cambrian Period. Data after Scotese, 1994, and Smith *et al.*, 1994.

Precambrian plate tectonic small Earth models—models older than 540 million years ago—show that the distribution of progressively older sedimentary basins continued to form a global network surrounding the most ancient cratonic and orogenic crusts. Without published coastline information to fully confirm this, these sedimentary basins are inferred to represent the distribution of ancient Precambrian continental seas. This inference is substantiated by an abundance of geological evidence from all continents showing that sedimentary basins, and hence continental seas, existed as far back as early-Archaean times.

Figure 15.2 shows in more detail that the exposed supercontinental land surfaces, as highlighted by the surrounding late-Jurassic shorelines and seas, generally coincide with the distribution of the ancient continental cratons and orogens—shown as areas of mainly pink and red coloured rocks. Similarly, the ancient continental seas—highlighted in blue shading—coincide with the distribution of underlying younger sedimentary basins.

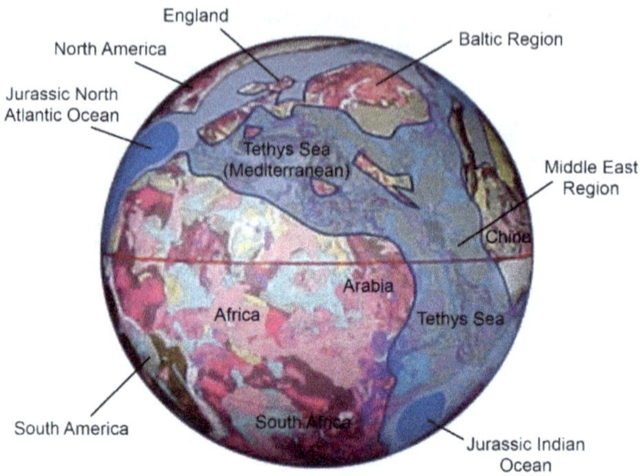

Figure 15.2. Late-Jurassic shorelines, highlighted as heavy blue lines, shown in relation to the shallow continental seas, shaded in blue, as well as the distribution of exposed lands and the newly opening North Atlantic and Indian Oceans—shaded in dark blue. Data after Scotese, 1994, and Smith *et al.*, 1994.

By definition, these sedimentary basins represent low-lying regions, where transported sediments were deposited and preserved. It is to be expected that the margins of these sedimentary basins should coincide with the shorelines of the ancient seas.

The distribution of coastal geography plotted on the plate tectonic small Earth models shows that large conventional Panthalassa, Iapetus, and Tethys Oceans are not required. Instead, these oceans are replaced by continental Panthalassa, Iapetus, and Tethys Seas, which represent precursors to the modern Pacific and Atlantic Oceans, as well as precursors to ancient seas previously covering continental Europe and Asia respectively.

Similarly, configuration of the Rodinia, Gondwana, and Pangaea supercontinents and smaller sub-continents on these small Earth models shows that the aerial distributions and extents of these ancient supercontinents were evolutionary, with no requirement for random conventional dispersion-amalgamation assemblage or breakup cycles. On each plate tectonic small Earth model, the ancient supercontinents are instead defined by progressive changes to the surface areas of sedimentary basins, changes to the distribution of ancient seas and shorelines, and associated changes to sea levels (Figure 15.3).

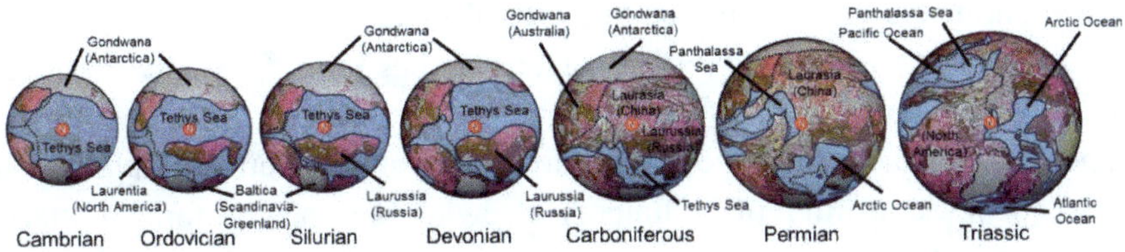

Figure 15.3. The Tethys Sea and Laurentia, Baltica, Laurussia, Laurasia, and Gondwana supercontinental configurations centred over the ancient North Pole, extending from the Cambrian to Triassic Periods. Data after Scotese, 1994, and Smith et al., 1994.

The ability for supercontinental crusts to continue to increase in surface area by crustal extension, localized within each of the sedimentary basins, was finally exceeded during late-Permian times. This process then initiated rupturing and breakup of the supercontinental crusts to form the modern continents and oceans. During late-Permian times, draining of the ancient continental seas was then initiated, which resulted in progressive exposure of the pre-Permian, relatively shallow continental seafloors.

Once exposed, these ancient seafloors were then subjected to the ravages of erosion and vegetative colonization. Eroded seafloor sediments were, in turn, redeposited elsewhere in new low-lying regions, in particular around the newly formed continental margins and marine sedimentary basins.

15.2. Ancient Continental Sedimentary Basins

On an increasing radius plate tectonic Earth, the evolving coastal outlines (Figures 15.1 and 15.3) reflect the prolonged history of changing Earth surface area and surface curvature through time. On each of the small Earth models, sedimentary basin development initially formed a global, inter-connected network, which also reflected the distribution and history of the developing continental seas.

These ancient basins and seas were centred mainly on what are now the European and Asian regions, as well as within Australian and adjoining North American and South American regions. Each of these regions, in turn, show complex sea-level histories, including climate-induced sea-level changes, on-going erosion of the exposed lands, and ultimately to supercontinental crustal breakup, opening of the modern oceans, draining of the ancient seas, and introduction of new waters.

During late-Permian times, the previously long and extended history of supercontinental basin sedimentation began to rapidly decline. As the ancient supercontinental crusts ruptured and crustal breakup was initiated, seawater from the previously existing continental seas began to drain into the newly opening modern oceans. This draining further exposed the ancient supercontinental seafloor sediments to the ravages of erosion, along with subsequent deposition of sediments elsewhere in newly formed marine sedimentary basins.

As the supercontinents continued to break up, the new marine sedimentary basins slowly began to dominate over the older continental basins. The marine sediments were initially deposited in basins located in rift zones between the emerging modern continents, followed later by deposition around the margins of the modern continents.

Supercontinental crustal rupture and breakup continued uninterrupted during Triassic to early-Jurassic times, with new sediments being deposited within the newly opening Pacific Ocean, followed by deposition within the opening South Atlantic and Indian Oceans. The Southern Ocean opened later during Paleocene times—crosscutting and displacing earlier formed ocean basins in these areas.

Each of the marine sedimentary basins formed during these times are shown as white continental shelves and marine plateau areas surrounding each of the modern continents (Figure 15.1). Remnants of these ancient marine sedimentary basins are now located adjacent to the continental shelves, between the lands and the deep abyssal plains of the modern oceans, and as remnant plateaus within older regions of the modern oceans.

15.3. Rise and Fall in Sea Levels

The variation in coastal outlines on plate tectonic small Earth models shows that the total volume of ocean water in the past was very much less than what it is now. This contrasts with conventional studies, where the total volume of ocean waters is considered constant, or nearly constant, over time.

Small Earth modelling also shows that the volume of ocean water has been increasing steadily since Archaean times, and most prominently since the post-Permian crustal breakup and opening of the modern oceans. This post-Permian increase in volume of new water is considered to have occurred in conjunction with intrusion of new volcanic seafloor crust along the global network of mid-ocean-ridge spreading zones. This new water, plus accompanying atmospheric gases, represents escaped volatile elements which occur naturally within the crystal lattices of all molten volcanic rocks.

Petrological studies show that up to fifteen to twenty percent of the weight of a new volcanic rock may comprise entrapped fluid and gaseous elements. Once the volcanic rocks are intruded as lava near the Earth's surface, these volatile elements are then expelled, or boiled off, during formation of new surface lava or seafloor volcanic crust.

On each of the small Earth models, a record of this increase in new water is preserved as changes in the coastal outlines of the supercontinents and newly formed continents. Complicating this process is that major changes in sea levels also occur during merging of two or more ancient continental seas, breakup of the ancient supercontinents, and opening of each of the modern oceans.

Breakup and migration of the supercontinents may have also destabilized and melted ancient polar icecaps existing at the time, which in turn may also affect global climate and sea levels. These major changes in sea levels represent a potential mechanism for periods of mass extinction of ancient plant and animal species.

Depending on the severity of any change in surface area and surface curvature, these sea-level changes may have, in turn, affected regional or even global climate, as well as ocean water circulation patterns, the distributions of marine species, and the type and location of sedimentary deposition. All of these changes ultimately affected the distribution, evolution, or extinction of the various marine and terrestrial animal and plant species as well as disrupting the various established species migration routes.

In conventional plate tectonic studies, these changes to both sea-level and the various rock types, as preserved in sedimentary basins, are interpreted as representing three basic kinds of sedimentary cycle. These sedimentary cycles include:

1. Very long-term cycles: These cycles are said to be caused by the inferred assemblage and subsequent breakup, rifting, and dispersal of supercontinents by seafloor spreading—referred to as Wilson cycles. The cycles take 400 to 500 million years to complete, and the process is thought to have been operating for at least 2,000 million years.
2. Convective overturn in the mantle: This convective overturn is inferred to lead to continental plate collision and rifting events during the assemblage and subsequent dispersal of supercontinents. This cycle is thought to be characterized by rapid global changes to seafloor spreading rates and accompanying changes to sea levels. The

timing of these collisions or rifting events is random but may occur over periods ranging from about 10 to 100 million years.

3. Climate induced global sea-level cycles—referred to as Milankovitch cycles: These cycles are inferred to result from mechanical variations in the Earth's orbit. The cycles are considered to occur over periods of tens to hundreds of thousands of years.

In contrast, plate tectonic small Earth model studies show that there is no requirement for multiple cycles of random continental crustal assemblage, breakup, rifting, or dispersal to form supercontinents. Instead, the ancient supercontinents remained assembled as a complete supercontinental crust throughout the entire 3,750 million years of pre-Triassic Earth's history.

Variation in the distribution and configuration of the supercontinents and seas is then dependent on changes to the surface areas of sedimentary basins during increase in Earth radius. Or, similarly, during periods of orogenesis and post-Permian breakup.

The plate tectonic small Earth model studies show that the pre-Triassic supercontinental crust was subjected to a steady accelerating increase in crustal extension and basin opening, prior to a single phase of crustal rupture followed by supercontinental breakup and opening of the modern oceans.

Climate induced changes to the rock-record, emanating from these crustal extension and sea-level changes, then occurred as a result of disruption to the pre-existing climate zones during subsequent migration of the newly formed continents. These disruptions occurred during apparent migration of the modern continents into or out of the various climate zones during opening of the modern oceans.

The published changes to sea levels for the past 542 million years of Earth history (Figure 15.4) are readily recognizable in the rock-record, in particular by the relative severity of these changes above normal changes to sedimentary deposition or normal species extinction rates. Often, more recent changes to sea-level can be observed along present-day coastlines in the form of raised wave-cut platforms or erosional features seen in rocky ledges and cliffs, as stabilized sand dunes, or as inter-dune swamps and lakes.

The distribution of first-order changes in global sea-level, relative to the present sea-level, is best seen in the smoothed red curve in Figure 15.4. This cycle is referred to in conventional literature as part of the very long-term Wilson cycle.

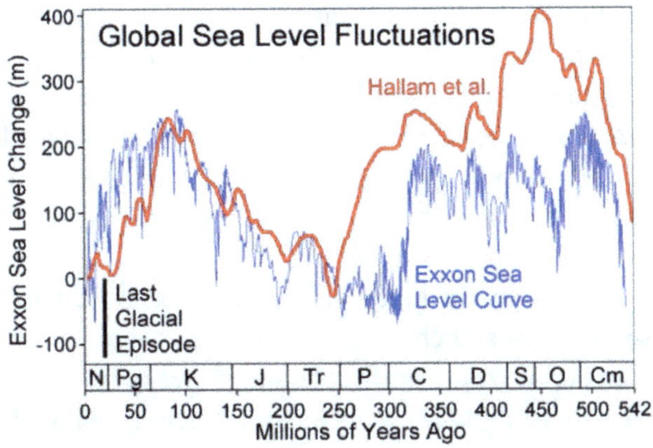

Figure 15.4. Changes in sea-levels during the past 542 million years showing the Hallam et al. and Exxon global sea-level fluctuations for the Phanerozoic Eon. After Hallam et al., and Exxon.

In contrast, on an increasing radius plate tectonic Earth this cycle is the most basic sea-level cycle operating during the past 500 million years of Earth history. Small Earth model studies show that this cycle records the rupturing and breakup of the Pangaea supercontinent during late-Permian times—shown as the rapid decline in sea-levels during the Carboniferous and Permian Periods (C and P)—and formation of the modern continents, including draining of the ancient seas into the newly opening modern oceans.

Sea levels in Figure 15.4 steadily fall from a peak level during Cambrian (Cm) and Ordovician (O) times to a minimum level during the Triassic (Tr). A steady rise during the Triassic to late-Cretaceous (K) times was followed by a steady fall in sea levels to the present-day (N).

Also superimposed on this first-order global cycle are more detailed second-order sea-level cycles, as shown in the blue curve in this figure. These second-order cycles approximate the major first-order changes in sea levels but show greater detail. The effects of these second-order changes were first recognized by naturalists and geologists during the 1700s and 1800s, during the time when geology first became a recognizable science. The recognition of these secondary cycles has traditionally been used in geology to define the major time periods and epochs of Earth history, as well as to correlate sea-level changes and extinction events across each of the modern continents.

On an increasing radius plate tectonic Earth, these first- and second-order sea-level cycles reflect the variation in both coastal outlines and the volume and distribution of available ocean water existing during the past 500 million years. The second-order cycles, in particular, reflect more local variations in sea-level during the merging of previously isolated, smaller continental basins and seas, as shown by the Cambrian to Triassic small Earth models in Figure 15.1. It will be shown later that these local variations, in turn, coincide with local to global-scale extinction events caused by either rapid draining of seas or flooding of land areas.

The sea-level cycles in Figure 15.4, in effect, reflect the progressive changes in surface area and changes in surface curvature operating since the early-Cambrian Period. It is significant that the steady lowering of sea levels in the central part of Figure 15.4 coincides precisely with draining of the ancient continental seas during rupture and breakup of the ancient Pangaea supercontinent. The steady rise in sea levels during the latter part of this cycle coincides with both increases in new water from the newly opening mid-ocean-ridge spreading centres, plus disruption and melting of pre-existing ancient polar icecaps during apparent migration of the continents.

Chapter 16

A Paradigm Shift in Palaeoclimate

"*Climatic features of the Mesozoic and Cenozoic have been shown to depart significantly from latitudinal predictions based on palaeomagnetism.*" Barron, 1983.

Assumptions about the Earth's ancient climate involve the study of three complex systems: ocean circulation, atmosphere, and the past distribution of continents. Ocean circulation, in particular the development of deep ocean currents during opening of modern oceans, is strongly influenced by the presence of land barriers. Similarly, topographic relief on the continents strongly influences the amount of rainfall received from the atmosphere as well as the accumulation of snow and ice in mountainous regions.

Correlation of coal swamps, thick bedded sandstone sequences, and glacial rocks are all considered to be excellent indicators of cold wet climates, while dry climates are indicated by evaporate rocks such as salt deposits, and equatorial regions by the presence of carbonate reef and carbonate-bearing sandstone strata.

Conventional plate tectonic studies suggest that climate during late-Palaeozoic times—around 300 million years ago—was variable and complex. A period of glaciation also occurred at that time which lasted for about 60 million years. This climate was considered by Klein in 1994 to be related to the geography of the ancient Gondwana supercontinent, as well as a subsequent reduction in atmospheric carbon dioxide during assemblage of Pangaea—based on reconstructions on a conventional plate tectonic Earth.

The Permian climate was also considered by Erwin in 1993 to be related to an almost complete Pangaean continental crustal assemblage. The associated patterns of climate and glaciation were further considered to be related to an alternation of greenhouse and icehouse states.

In 1994, Crowley also showed that the Pangaea climate involved a late-Palaeozoic period of glaciation, with a subsequent transition to arid and nearly ice-free conditions during Triassic and Jurassic times. Geological studies indicated to Crowley that aridity during Triassic and Jurassic times was extensive over much of Pangaea, where global evaporite formation and distribution were at a peak.

Other evidence, such as shifts in vegetation patterns, pointed to periods of warming and drying during the late-Jurassic. Some changes were also considered by Crowley to be associated with a rise in sea-level, which was said to have accompanied the breakup of Pangaea.

Cretaceous climates were described by Barron in 1983 as warm and equable. The polar regions in particular were considered to have been much warmer than today, with no evidence for polar ice. During that time, warmth-loving plants and animals were able to spread into high latitudes—as modelled on conventional plate tectonic continental assemblages.

In contrast, Francis and Frakes in 1993 showed there were distinct variations and trends in warming and cooling during the Cretaceous and Barron considered features of the Mesozoic and Cenozoic climates suggested that these ancient climates departed significantly from latitudinal predictions based on palaeomagnetism.

In addition to changes in climate and the associated distribution of plant and animal species, Crowley also showed there were significant variations in the presence of carbon—in the form of coal—during Pangaean times. Organic, carbon-rich layers in the rock-record were found to coincident with peaks in high sea-levels. The largest period of deposition of carbon in the rock-record reflected a time of major coal formation during the Permian and Carboniferous times.

The high carbon present during that time coincided with establishment of land plants on the continents and the high burial rates suggested there was a general absence of organisms capable of consuming decaying organic matter. These carbon changes contributed to the presence of low atmospheric carbon dioxide levels during Carboniferous times, with levels estimated to be comparable to present-day levels, increasing to levels of up to five times the present value during the late-Jurassic.

16.1. Present-day and Ancient Climate Zones

The present-day Earth's surface is divided into five main climatic zones, which are defined by major circles of latitude. The main differences between each of these zones relate to the attitude of the Sun (Figure 16.1). On this figure, a circle of latitude on the Earth's surface is defined as an imaginary east-west circle connecting all locations on the surface with the same north or south latitude. The position of the Equator is fixed at 90 degrees from the Earth's axis of rotation, while latitudes of the remaining circles depend on the tilt of the axis of rotation, relative to the plane of the Earth's orbit around the Sun.

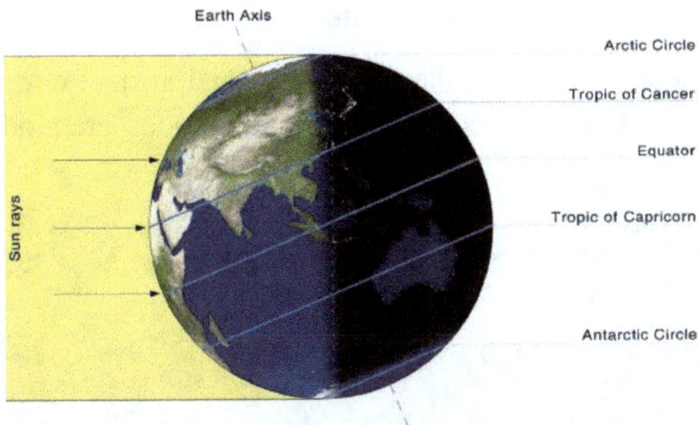

Figure 16.1. Diagram showing the location and derivation of the major circles of latitude on the Earth.

By definition, the positions of the Tropic of Cancer, Tropic of Capricorn, Arctic Circle, and Antarctic Circle all depend on the tilt of the Earth's rotation axis relative to the plane of its orbit around the Sun—called the obliquity of the ecliptic. In year 2000, the mean value of the tilt of the Earth's rotational axis was 23° 26′ 21″. Measurements elsewhere have shown that this axis undergoes small fluctuations over time. This fluctuation has a very small effect on the location of the Tropic and Polar circles and also on the location of the Equator.

The latitudinal values for the established present-day circles of latitude vary slightly over time and each of the circles in Figure 16.1 are presently located at:

1. Arctic Circle: 66° 33′ 44″ north latitude.
2. Tropic of Cancer: 23° 26′ 16″ north latitude.
3. Equator: 0° latitude.
4. Tropic of Capricorn: 23° 26′ 16″ south latitude.
5. Antarctic Circle: 66° 33′ 44″ south latitude.

These circles of latitude also mark the divisions between the five main geographical zones. These zones are defined as follows:

1. The North Polar Region, located north of the Arctic Circle.
2. The North Temperate Zone, located between the Arctic Circle and the Tropic of Cancer.
3. The Tropics, located between the north and south Tropical Circles.
4. The South Temperate Zone, located between the Tropic of Capricorn and the Antarctic Circle.
5. The South Polar Region, located south of the Antarctic Circle.

The locations of each of the present-day and ancient circles of latitude and climate zones on plate tectonic small Earth models are plotted on Figure 16.2.

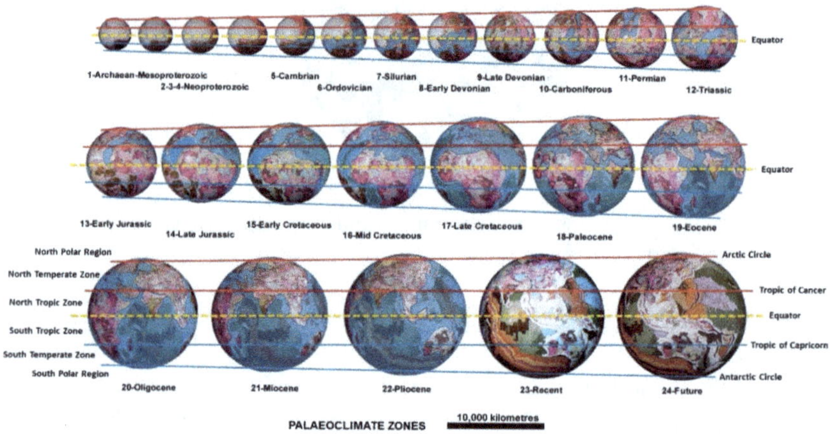

Figure 16.2. The location of climate and geographical zones on plate tectonic small Earth models, based on present-day latitudinal values.

The ancient circles of latitude, while appearing to be non-parallel in this figure, are in fact parallel and their ancient geographical locations have been adopted as being the same as the present-day latitudes. The tapered, non-parallel representation in this figure is meant only to simulate the progressive evolution of these circles and zones between each of the small Earth models.

In reality, this representation may not be strictly correct. If the tilt of the ancient Earth axis varied significantly from what it presently is, then the width and location of these zones would also be affected.

Because of the limited amount of available palaeoclimate data available, absolute confirmation of the Earth's ancient tilt is very difficult to determine. Modern understanding considers that the Earth's interior is molten to semi-molten, hence the angular momentum of the Earth would likely tend to resist any changes to its axis of tilt in a similar way to the motion of a gyroscope. Because of this, for the purpose of this investigation, the Earth's ancient tilt is here considered to be fixed, or near fixed, over time.

16.2. Ancient Coral Reefs

Based on present-day distributions, limestone and coral reefs generally occur within a broad zone located plus and minus 25 degrees of latitude north and south of the equator. The presence of warm water currents may also extend distribution of marine organisms beyond this zone. Within this primary zone, warm ocean waters and currents enable corals and other marine creatures to thrive along the continental shelves of islands and continents. Plotting the distribution of ancient coral reefs on plate tectonic small Earth models will then enable location of the ancient equator, previously established from palaeomagnetic studies, as well as the established climate zones, to be independently verified.

In 1994, Flügel considered that the setting and location of Pangaean reefs on conventional plate tectonic reconstructions is unusual. Compared with the distribution of modern reefs, most Pangaean reefs were shown to be formed in and around continental basins rather than along exposed continental margins flanking large oceans. This observation of Flügel contradicts conventional plate tectonic reconstructions of Pangaea where the growth of Permian and Triassic reefs supposedly took place along the margins of, or in basins adjacent to, a large, inferred east-west trending Tethys Ocean.

This inferred Tethys Ocean was, in turn, located at the western end of an even larger inferred Panthalassa Ocean. Additionally, the Pangaean reefs were also interpreted as being located on, and associated with, small continental fragments—such as small continents or islands. Evolutionary crises affecting the marine species from each of these sites were then considered by Flügel to correlate with a major end-Triassic reef extinction event.

In contrast, on an increasing radius plate tectonic Earth, the presence of a large, inferred, Tethys Ocean, and similarly an even larger Panthalassa Ocean, is not required. Instead, prior to the late-Permian supercontinental breakup, these hypothetical oceans are shown to form a network of ancient continental seas, coincident with a global network of sedimentary basins.

On small Earth models, the ancient Tethys Sea is also shown to have, at times, extended from equatorial regions through to the North Polar Region, in particular during early-Palaeozoic times. This distribution of seas enabled warm equatorial sea currents to circulate freely into high northern Polar Regions, accompanied by warm-water marine organisms.

The distribution of published occurrences of mid- to late-Palaeozoic carbonate reef deposits—around 300 to 250 million years ago—is shown on the Permian plate tectonic small Earth model in Figure 16.3 (data from Flügel, 1994), along with the location of established tropical climate zones and circles of latitude.

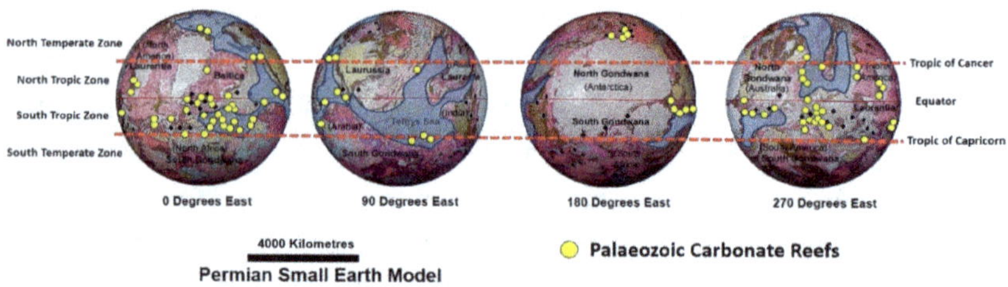

Figure 16.3. Distribution of Palaeozoic carbonate reefs (yellow dots) as well as ancient Permian coastlines and continental seas, plotted on a Permian plate tectonic small Earth model. Reef data after Flügel, 1994.

Also shown is the distribution of ancient continental seas, previously established from the distribution of published coastal outlines on the small Earth models. This distribution of carbonate reefs is based on published records and does not necessarily represent the entire global distribution of reefs in lesser studied regions.

In this figure, there is some minor disparity between the plotted reef locations and the published ancient shorelines. This is only apparent because the reef data extend back in time a further 100 to 200 million years to times when these coastlines were different to those shown on this Permian Pangaean small Earth model.

In this figure, the distribution of carbonate reefs is, in general, shown to neatly straddle the equator and shows a good correlation with the adopted equatorial Tropic zones. Outlying reefs shown within the northern Temperate Zone may reflect the presence of warm tropical Tethys Sea currents extending into these regions.

On an increasing radius plate tectonic Earth, the ancient climate zones remained as fixed zones of latitude. Similarly, the presence of living marine organisms was dependent on the presence and location of preferred climate zones and environmental conditions existing at the time. During supercontinental breakup and migration of the modern continents, the marine organisms were then forced, over tens to hundreds of millions of years, to continuously relocate and re-establish new reefs or colonies in order to remain within their preferred climate zones.

It should be noted that migration of the reefs and marine species is, however, only apparent—it is the continents and established ancient reefs that move, not the climate zones or marine species.

In this context, if an ancient continent and its attached reef migrate out of a preferred climate zone, the marine organisms living on this reef would likely gradually die out. This migration would then leave behind the abandoned carbonate sediments or ancient coral reefs.

In contrast, if a continent migrated into a tropical climate zone, the marine organisms would likely begin to establish a new reef or enlarge on an existing reef. As the continents continue to migrate, any new reef colonies established in these regions would then be subject to further on-going relocation or extinction over time.

The distributions of Palaeozoic through to Jurassic carbonate reefs are shown as various coloured dots plotted on the early-Jurassic plate tectonic small Earth model in Figure 16.4. These reefs are again shown in conjunction with the anticipated climate zones. Also shown on this figure is the early opening of the modern Pacific and Atlantic Oceans, shaded in blue, as well as the locations of the early-Jurassic coastlines shown as heavy blue lines.

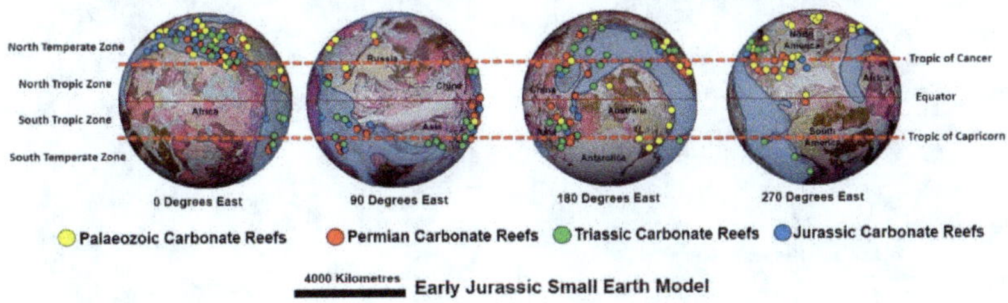

Figure 16.4. Distribution of Palaeozoic to Jurassic carbonate reefs as well as ancient Jurassic coastlines and opening modern oceans plotted on an early-Jurassic plate tectonic small Earth model. Reef data after Flügel, 1994.

On an increasing radius plate tectonic Earth, the late-Palaeozoic to Jurassic times represent a transitional period of Earth history, documenting the opening and rapid development of the early North and South Pacific, North Atlantic, and Arctic Oceans. During opening and dispersal of the modern continents, earlier formed carbonate reef communities were then subject to fragmentation, disruption, and migration away from their previously established tropical zones, into new and often different climatic zones.

The various reef communities in Figure 16.4 highlight this period of disruption, where the continents, and most of the abandoned ancient reefs, are shown to have migrated well north into temperate climate zones. This is highlighted by the distribution of ancient Palaeozoic reefs, shown as yellow dots, which can be compared to their original distributions in Figure 16.3.

The distribution of the youngest Jurassic reefs—blue dots in Figure 16.4—is reasonably well constrained within the Jurassic Tropic Zone, but they also extend north into the newly opening Mediterranean Sea. This may either represent a holdover of marine species in this region, or the presence of warm water currents that have extended reef development into these areas.

In addition, during the late-Triassic Period—the time interval between Figures 16.3 and 16.4—a pre-existing land connection between California in North America and Queensland in Australia was breached (Figure 16.5). This breaching resulted in the ancient Panthalassa Sea, located to the north of Australia, merging with the newly opening South Pacific Ocean to form a single modern Pacific Ocean during the Jurassic.

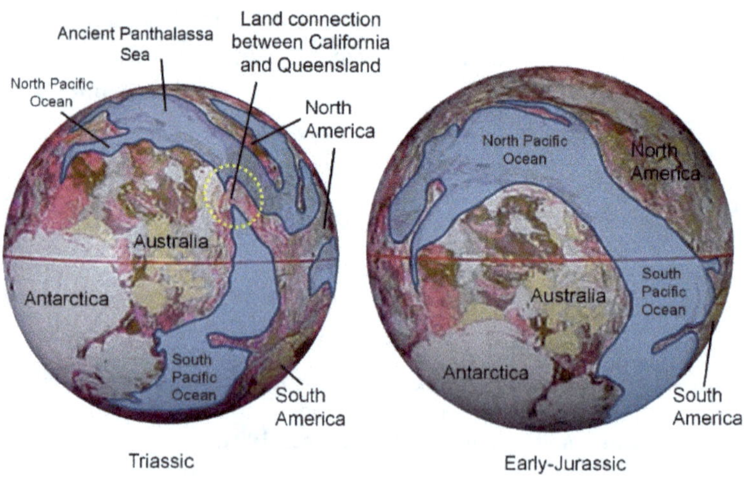

Figure 16.5. Triassic to early-Jurassic breaching of a land connection between the ancient Panthalassa Sea and the opening South Pacific Ocean to form the modern Pacific Ocean.

This breaching globally disrupted previously separate sea levels existing at the time, along with disrupting the Western Interior Seaway in North

America. Breaching also coincided with the end-Triassic reef extinction event as noted by Flügel in 1994.

This Pacific Ocean breaching event, plus similar breaching events in the Atlantic Ocean, then opened up new avenues for species migration and development into other oceans. Marine organisms remaining after breaching were then able to migrate into, or away from, areas previously not favourable or accessible to colonization.

By the early-Jurassic Period, the ancient continental seas were continuing to drain from the ancient Pangaea supercontinent. This draining of the seas exposed the older Palaeozoic reef communities, forcing the marine organisms responsible for reef development to migrate into new locations, to evolve, or perish over an extended period of time lasting many tens of millions of years.

The continued presence of Triassic and Jurassic reefs suggests that these marine organisms had indeed migrated and were able to keep pace with the migrating continents and lowering sea-levels.

16.3. Ice Ages

An ice age is defined as a period of long-term reduction in the temperature of Earth's surface and atmosphere, resulting in the presence or expansion of continental and polar ice sheets and alpine glaciers. Within a long-term ice age, individual pulses of cold climate are termed glacial periods, and intermittent warm periods are referred to as interglacials. On Earth, an ice age implies the presence of extensive ice sheets in either the northern or southern Polar Regions, or both.

There have been at least five major ice ages in the Earth's geological past, referred to as the Huronian, Cryogenian, Andean-Saharan, and Karoo ice ages, and the current Quaternary glaciation. Outside these ages, the Earth was ice-free even at high latitudes.

Glacial related rocks from the earliest, well established, Huronian ice age formed around 2.4 to 2.1 billion years ago during the early-Proterozoic Eon. The next well-documented Cryogenian ice age, suggested as being the most severe, occurred from 850 to 630 million years ago during the late-Proterozoic. It has been suggested by others that the end of this ice age was responsible for the subsequent late-Proterozoic and Cambrian explosion of life forms, although this suggestion is controversial.

The Andean-Saharan ice age occurred from 460 to 420 million years ago during the late-Ordovician and Silurian Periods. The Karoo ice age occurred at intervals from 360 to 260 million years ago in South Africa during the Carboniferous and early-Permian Periods. Correlatives are also known from Argentina within the ancient supercontinent Gondwana.

The current ice age, the Quaternary glaciation, started about 2.58 million years ago during the late-Pliocene when the spread of ice sheets in the Northern Hemisphere began, although an ice cap began forming much earlier in Antarctica some 33 million years ago. Since then, the Earth has seen cycles of glaciation with ice sheets advancing and retreating on 40,000-to-100,000-year time scales.

The Earth is currently in an interglacial period and the last glacial period ended about 10,000 years ago. All that remains of the present glaciation is the Greenland and Antarctic ice sheets and smaller glaciers such as on Baffin Island, Canada, and various mountain glaciers.

The proffered cause of ice ages is poorly understood for both the large-scale ice ages and the smaller ebb and flow of glacial and interglacial periods within an ice age. The consensus in the literature is that several factors are important, including atmospheric composition, the motion of tectonic plates resulting in changes in the relative location and amount of continental and oceanic crust on the Earth's surface, variations in solar output, the orbital dynamics of the Earth–Moon system, the impact of relatively large meteorites, or eruptions of super-volcanoes.

On an increasing radius plate tectonic Earth, the cause of ice ages is also poorly understood. Speculation is, however, made here on the cyclicity of ice ages and their potential relationship with Galactic cycles, along with the possible gross scale seasonal effect on Earth's climate.

A Galactic cycle, also known as a cosmic year, is the duration of time required for the Solar System to orbit once around the centre of the Milky Way Galaxy. Estimates of the length of one orbit vary from 225 to 250 million terrestrial years, with Harvard-Smithsonian astronomers recently measuring a precise value of 226 million years, accurate to within 6 percent.

It is speculated that, just like the Earth's yearly Solar Cycle, a seasonal effect may be present as the Solar System orbits around the Milky Way Galaxy, giving rise to subtle temperature variations similar to Earth's winter-summer seasonal cycle.

In Figure 16.6, the simulated locations of Galactic winter cycles are highlighted in green shading. These are shown in relation to the five major ice ages on Earth, highlighted in magenta shading. The Galactic winters are plotted as an average 226-million-year cycle, with a 50 million year spread replicating the duration of a hypothetical ice age on Earth.

While speculative, the locations of the known ice ages in relation to the Galactic winters in Figure 16.6 show a relatively close approximation, which is considered to warrant further consideration.

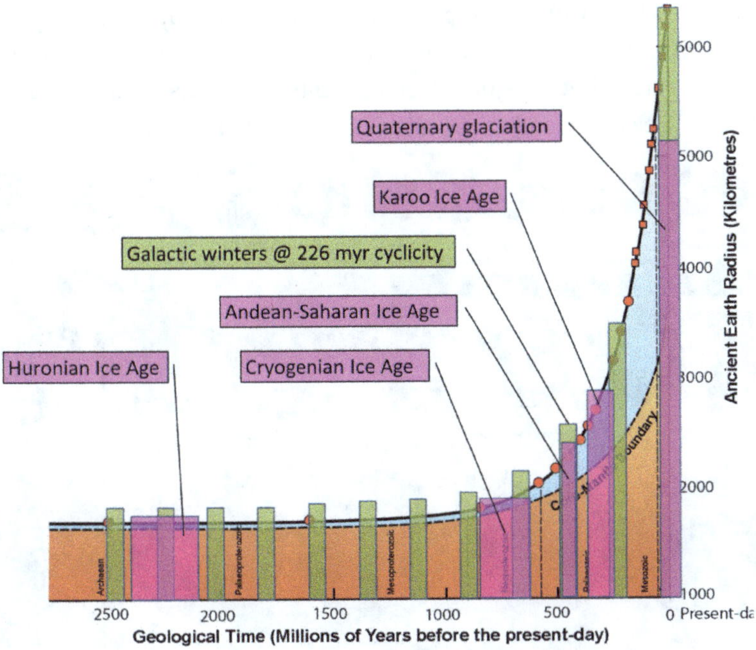

Figure 16.6. Galactic winter cycles, highlighted in green shading, are shown in relation to the five major ice ages on Earth highlighted in magenta shading. The Galactic winters are plotted with an average 226-million-year cycle and a 50 million year spread, shown plotted in relation to the small Earth radius curve.

16.4. Distribution of Ancient Polar Regions

The locations of ancient North and South Poles were previously established from plate tectonic small Earth modelling of available palaeomagnetic pole data in Chapter 12. From these pole locations, it is then possible to show the locations of each of the north and south polar regions.

These polar-regions can be established by locating the polar circles of latitude—areas north and south of 66° 33′ 44″ latitude. For the purpose of this investigation, it is assumed that the tilt of the Earth's ancient rotation axis was the same, or very similar to what it is now, and hence circles of latitude were similar to those on the present-day Earth.

Figure 16.7 shows locations of the North Polar Region, centred over the ancient North Pole on each of the small Earth models. Similarly, Figure 16.8 shows locations of the South Polar Region, centred over the ancient South Pole.

Also shown on each of these figures is the published distribution of known glacial rocks and formations—shown as red dots, after Hambry and Harland, 1981, as well as the presence of known ice sheets shaded in white. Both the glacial rocks and icesheets are shown to coincide with locations of the highlighted five major glacial events. The distribution of ancient continental seas and modern oceans is shown as pale blue.

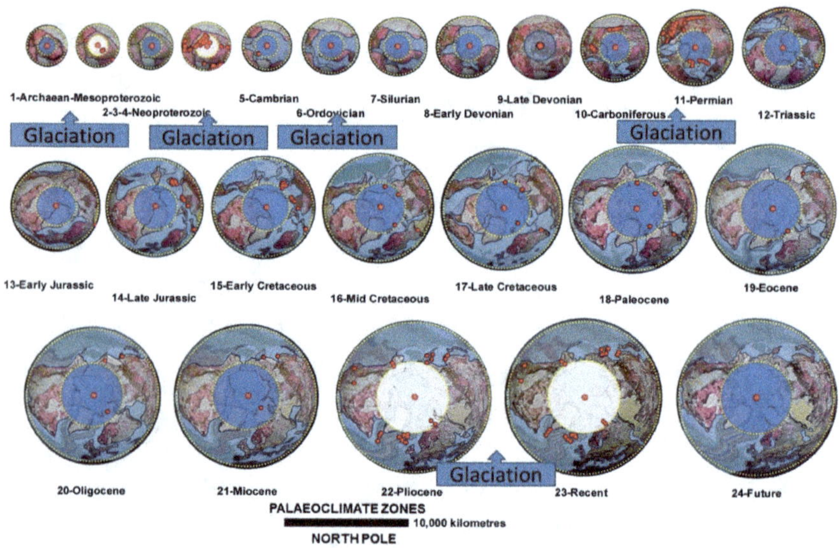

Figure 16.7. The locations of ancient North Polar Regions are shaded blue on the plate tectonic small Earth models. Glacial events are highlighted, the presence of known icesheets is shaded white, and ancient seas and modern oceans are shaded pale blue. Glacial data after Hambry and Harland, 1981.

The opening of the modern oceans beneath an established icesheet may expose any existing ice to the influence of circulating ocean currents. This opening may then change the icesheet from a permanent continental sheet—such as the modern Antarctic icecap—to a seasonal marine sheet—such as the modern Arctic icesheet. This change may, in turn, affect the presence, size, and extent of ice cover within these climate zones and directly affect global climate, sea levels, and the distribution or decline of various plant and animal species.

For example, migration of the South Pole and opening of the South

Atlantic Ocean is highlighted in Figure 16.8 on the Triassic to Oligocene small Earth models. Here, next to no glacial rocks were present, which suggests that any ice sheet present during that time may have been small to absent.

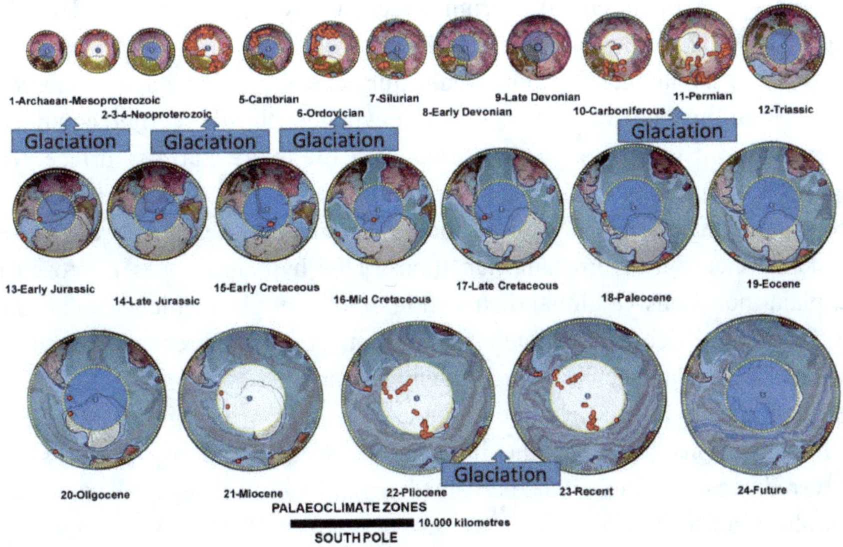

Figure 16.8. The locations of the ancient South Polar Regions are shaded in blue on the plate tectonic small Earth models. Glacial events are highlighted, the presence of known icesheets is shaded white, and ancient seas and modern oceans are shaded pale blue. Glacial data after Hambry and Harland, 1981.

16.5. Ancient Glacial Record

Evidence for ancient ice sheets and glacial occurrences in the rock-record comes from various sources. These include the presence of striated rock surfaces and rock formations created by the passing of a glacier, large foreign rocks embedded within sedimentary strata, and the accumulation of characteristic rock debris. This rock debris accumulates where a glacier or ice-sheet melts and includes deposits that occur adjacent to streams flowing from a melting glacier, through to debris rafted well out to sea by icebergs, where melting later deposits the debris on the seafloor.

The generic term glacigenic will be used here to include all glacial-related rocks and formations. In this context, glacigenic simply means rock formations or rock debris that have a glacial origin and includes debris that was deposited both on the land and on the sea floors.

Evidence from glacial studies suggested to Eyles and Young in 1994 that, during the late-Archaean to early-Proterozoic, Earth surface temperatures may have been 30 degrees Celsius less than now and the Sun's luminosity was about 70 percent of the present value. These low Archaean temperatures were considered by Eyles and Young to have been the result of an insulating effect caused by development of a primitive crust in conjunction with declining geothermal temperatures in the mantle and crust.

Conventional plate tectonic studies further suggest that glacial conditions during that time may have existed as a snowball Earth, where frigid conditions were inferred to have extended throughout the entire Earth's surface. The Snowball Earth hypothesis was first presented by Hoffman *et al.* in 1998, who considered that the Earth's surface became entirely, or nearly entirely, frozen at least once during Precambrian times. This hypothesis was proposed to explain the extensive global distribution of ancient glacigenic rock deposits, in particular deposits located in ancient equatorial regions based on conventional plate tectonic Earth assemblage models.

Opponents of this hypothesis contest the presence of global glaciation, with objections based on the implications surrounding an ice- or slush-covered ocean, as well as the perceived difficulty of the Earth to escape from an all-frozen condition. There are also a number of unanswered questions proposed by these opponents, including whether the Earth was a full snowball, or a slush ball with a thin equatorial band of open, or seasonally open water.

The snowball Earth hypothesis is refuted in this study. The abundance of normal, water-lain sedimentary rocks preserved within Archaean and Proterozoic terranes throughout the world today, including warm-water carbonate rocks, is testament to the presence of ancient rivers, streams, and seas, as well as variable climate conditions during these Precambrian times. The presence of sedimentary and warm-temperature rocks demonstrates that some form of latitudinal distribution of climate zones must have been present during Precambrian times, which may have also been separated into distinct zones, as they are today.

16.5.1. Early-Proterozoic Huronian Glaciation

Apart from small glacigenic occurrences from South Africa and western North America, evidence for glaciation during the Archaean Eon is scarce. This evidence suggests that the Archaean record, in contrast to what Eyles and Young inferred from low surface temperature evidence, may have been essentially barren of glacial deposits.

On a primordial plate tectonic Archaean Earth, this lack of glacigenic occurrences may have also simply been due to a shortage of surface water required to form icesheets. Similarly, a low elevation contrast between the lands and seas would have limited the ability to form mountain glaciers, and hence glacial erosion to form glacigenic debris. Small, known glacial occurrences in South Africa and western North America may also simply represent all that is left after subsequent erosion of the ancient Archaean terranes.

In contrast to Archaean times, plate tectonic small Earth models show the Proterozoic to be characterized by an extensive network of stable sedimentary basins and shallow seas. There is abundant evidence in the rock-record for widespread glaciation occurring during early-Proterozoic times from several regions of North America, Finland, South Africa, Australia, India, and Scandinavia. Other Proterozoic glacigenic deposits occur elsewhere in the Baltic region as well as the east-central and northern Russian regions.

The glacigenic deposits from each of these region's range in age from about 2,560 to 1,620 million years. In some areas there is evidence for multiple glaciations separated by warmer interglacial periods, with associated sedimentation and chemical weathering of the exposed land surfaces. This chemical weathering is indicative of warm surface temperature conditions, in particular within the equatorial region.

In contrast, the mid-Proterozoic Era is shown from the rock-record to be a non-glacial period. The only reported occurrence of glacigenic deposits is from Russia, but this occurrence has an uncertain age. This lack of mid-Proterozoic glacial activity may therefore indicate the presence of warmer climates persisting during that time.

16.5.2. Late-Proterozoic Cryogenian Glaciation

During the late-Proterozoic Era, there is again abundant evidence from the rock-record for widespread and long-ranging periods of glaciation existing during that time. These periods of glaciation lasted from around 1,000 million years ago into the early-Cambrian Period—to around 500 million years ago.

Most regions where late-Proterozoic glacigenic rocks have been found show evidence of several periods of glaciation during this major event. These glacial periods occurred on a scale of tens of millions of years and are similar to glaciations previously observed for the early-Proterozoic Era. Each of these individual glacial events were also preceded, separated, and followed by periods of relatively warm inter-glacial climates.

In 1989, Chumakov and Elston noted that late-Proterozoic glaciers recorded from Russia descended to sea-level in low latitudes close to the ancient equator. In 1994, Eyles and Young also showed that most late-Proterozoic glacial deposits accumulated as marine sediments along the ancient continental margins. In contrast, carbonate rocks, interbedded with many of the preserved glacigenic deposits, show that many of these glacigenic rocks were deposited in restricted warm-water seas prior to onset of continental uplift along the basin margins.

16.5.3. Precambrian Glaciation

The known Archaean and Proterozoic glacigenic occurrences are plotted on the late-Proterozoic plate tectonic small Earth model in Figure 16.9. These occurrences include all deposits from each of the early- and late-Proterozoic glacial events as well as the few Archaean glacigenic occurrences.

The Precambrian glacigenic deposits are plotted together because palaeomagnetic evidence shows that, on an increasing radius plate tectonic Earth, the North and South Poles were relatively stationary throughout these ancient times. During these times, a south-polar icesheet was located in what is now west Central Africa and a north-polar icesheet was located in what is now Mongolia-China—shown shaded white in Figure 16.9. The ancient poles remained at these locations throughout the Precambrian Eras, and their continued presence and locations extended well into the early-Palaeozoic Era.

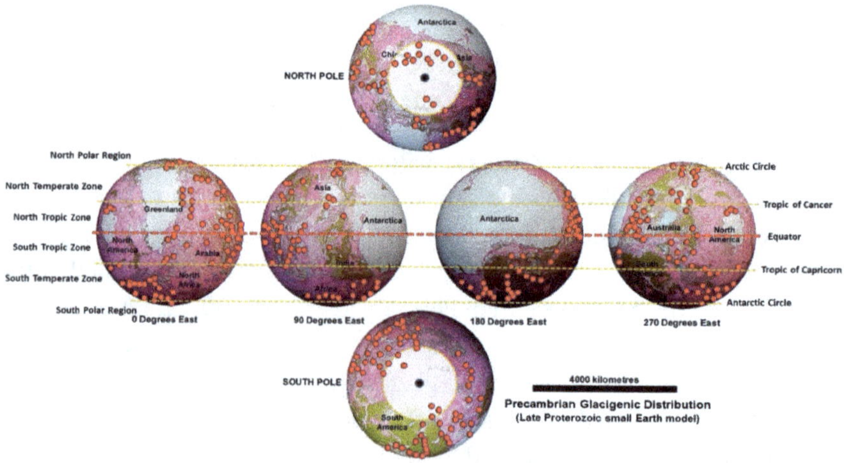

Figure 16.9. Precambrian glacigenic deposits plotted on a late-Proterozoic plate tectonic small Earth model. Known Precambrian glacigenic occurrences are shown as red dots, and the locations of polar icecaps are shaded white. Glacigenic occurrences after Hambry and Harland, 1981.

Figure 16.9 shows that the global distribution of Precambrian glacigenic deposits extended throughout each of the ancient Archaean and Proterozoic terranes. As noted by Eyles and Young in 1994, most of these Precambrian glacigenic occurrences were deposited as marine sediments. In most cases this observation is confirmed in Figure 16.9 by their close association with an identified network of Proterozoic sedimentary basins and shallow continental seas—shown as khaki-coloured terranes.

The very long period of time available during this Precambrian history, along with the extremely small rate of increase in Earth radius, also suggests that the elevation contrast between the lands and seas may have been quite low during these times. This low elevation contrast then limits the potential for elevated topography suitable for forming mountain glaciers.

To explain the distribution of Precambrian marine glacigenic rocks on an increasing radius plate tectonic Earth, the dimensions of ancient Earth must be taken into consideration. During the Archaean to late-Proterozoic Eras, the surface distances extending from the ancient poles to the ancient equator varied from about 2,600 increasing to 3,200 kilometres respectively, compared to about 10,000 kilometres now. When the radius of the ancient polar region is subtracted from these distances, say 1,000 kilometres—this reduces the maximum distances to between 1,600 to 2,200 kilometres.

These dimensions suggest that transport and dispersal of debris by icebergs fragmented from a polar or floating icesheet may be a viable means of dispersing marine glacigenic debris during Precambrian times. The network of continental seas and circulation currents on the ancient Earth would then enable icebergs to readily disperse and migrate into warmer equatorial regions, prior to complete melting.

This melting would have also been accompanied by progressive dropping of large amounts of rock debris onto the seafloor. The presence of interbedded warm-water carbonate rocks, including iron-silica-rich banded iron and chert sediments, is then justifiable during warm interglacial times. Similarly, the erosion and transport of glacigenic debris by melt-water streams on the lands may explain the presence of glacigenic rocks deposited on and around the margins of many of the ancient lands.

16.5.4. Early-Palaeozoic Andean-Saharan Glaciation

A significant early-Palaeozoic glacial episode, peaking at about 440 million years ago, is recorded from late-Ordovician rocks in West Africa,

Morocco, and Saudi Arabia. Less well-defined deposits occur in Scotland, Ireland, Normandy, South Africa, Spain, Ethiopia, and Portugal, and small mountainous ice-centres also occur in eastern North America. In some areas, the late-Ordovician glacial episode lasted into the following early-Silurian Period, in particular those recorded from South America and possibly Europe.

Plate tectonic small Earth modelling studies show that, during the Palaeozoic Era, the ancient Earth underwent a steady to rapid increase in both radius and surface area, along with changes to surface curvature. The Palaeozoic Era is noted for its onset of complex geosynclinal and orogenic activity within sedimentary basins, along with an increase in deposition of eroded sediments.

This activity implies that, on small Earth models, there was an increasing elevation contrast between the lands and sedimentary basins during the Palaeozoic Era along with an increasing potential for elevated topography suitable for forming mountain glaciers.

The distribution of early-Palaeozoic glacigenic occurrences is shown plotted as red dots on the Ordovician plate tectonic small Earth model in Figure 16.10, along with published coastal outlines and ancient seas shaded blue.

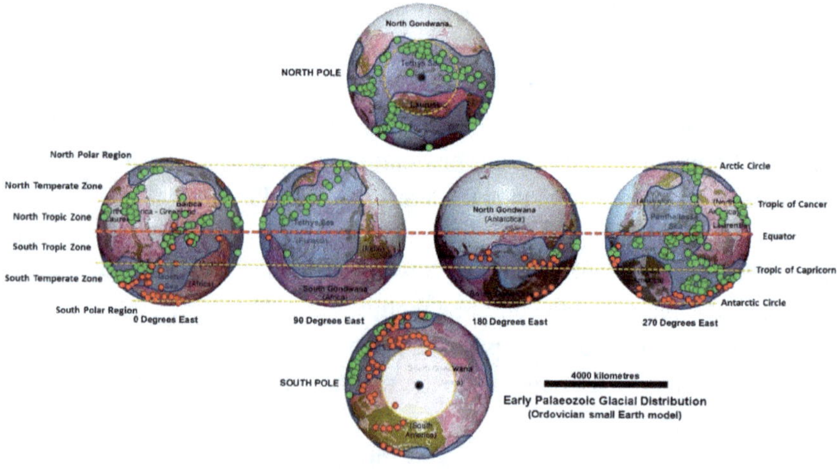

Figure 16.10. Early-Palaeozoic glacigenic deposits on an Ordovician plate tectonic small Earth model, shown in relation to climate zones and distribution of anthozoan corals. Known glacigenic occurrences are shown as red dots, anthozoan corals are shown as green dots, the South Pole icecap is shaded white, and continental seas are shaded blue. Glacigenic occurrences after Hambry and Harland, 1981, anthozoan corals after Paleobiology Database, 2015.

On an increasing radius plate tectonic Earth, the distribution of early Palaeozoic glacigenic rocks is restricted to the southern hemisphere during this glacial period. No significant glacigenic deposits are recorded from the northern hemisphere. This distribution coincides with an exposed South Gondwana supercontinent, which was centred over the South Pole, and an extensive Tethys Sea centred over the North Pole. The presence of glacigenic debris within the southern hemisphere suggests there was a continental icesheet located over the South Pole—shaded white in Figure 16.10—whereas the North Pole was effectively ice free during Ordovician times.

The distribution of glacigenic rocks in Figure 16.10 also highlights the preferential deposition of these rocks within the South Temperate climate zone. This distribution represents glacigenic material eroded from the South Polar Region and transported and deposited by movement of ice emanating from the icesheet. Similarly, there is also some evidence shown in this figure for on-going deposition of marine glacigenic rocks within the equatorial regions via melting icebergs.

In Figure 16.10, the ancient Tethys Sea is shown to extend from equatorial to North Polar latitudes, which suggests that both the northern sea temperatures and climates were moderated by warm equatorial sea currents. The presence of anthozoan corals—green dots in Figure 16.10, along with additional marine species such as molluscs and brachiopods, would imply that sea water conditions throughout the North Polar Region were then warm enough to support tropical marine species.

On an increasing radius plate tectonic Earth, a slow southern apparent migration of the ancient South Pole commenced during mid- to late Palaeozoic times, prior to eventual rupture and breakup of Pangaea and opening of the modern Atlantic Ocean. The presence of early Silurian glacigenic rocks persisting into South America is testament to a slow waning of the African icesheet during its migration into Brazil. Low-latitude glacigenic deposits in Europe, persisting to the Early Silurian, may also suggest the presence of small highland ice-centres in these regions.

16.5.5. Late-Palaeozoic Karoo Glaciation

A late-Devonian period of glaciation is recorded from Brazil and Bolivia, and by the following early-Carboniferous Period this glaciation had extended from Bolivia into Argentina and Paraguay. By the mid-Carboniferous Period glaciation had spread even further and extended into parts of Antarctica, Australia, central and southern Africa, the Indian sub-continent, northern Asia, and the Arabian Peninsula. By the late-Carboniferous Period a very large area of the ancient Gondwana supercontinent was then experiencing glacial conditions which persisted into the following late-Permian times.

The distribution of Carboniferous and Permian glacigenic debris is shown as red dots on the Permian plate tectonic small Earth model in Figure 16.11. These deposits are shown in relation to the predicted climate zones plus established continental seas and shorelines. To highlight the ongoing influence of warm Tethys Sea currents into the North Polar Region, the distribution of Ordovician anthozoan corals is again shown as green coloured dots (data after the Paleobiology Database, 2015).

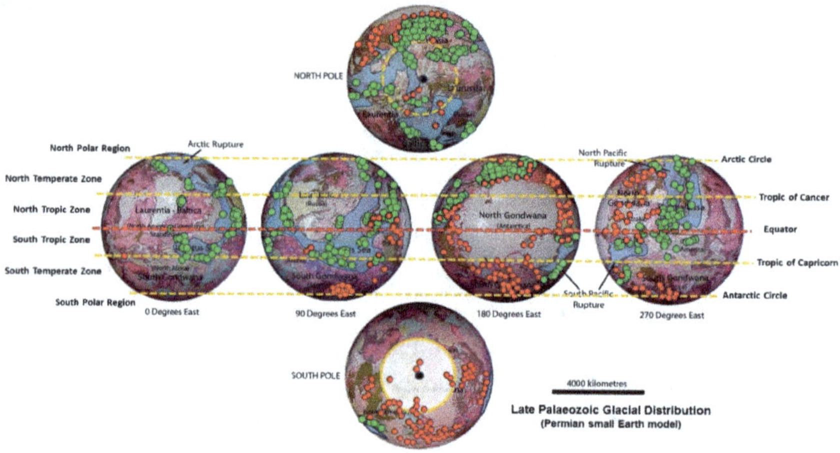

Figure 16.11. Late-Palaeozoic glacigenic deposits on a Permian plate tectonic small Earth model shown in relation to climate zones and continental seas. Known glacigenic occurrences are shown as red dots, anthozoan corals are shown as green dots, the South Pole icecap is shaded white, and continental seas are shaded blue. Glacigenic occurrences after Hambry and Harland, 1981, anthozoan corals after Paleobiology Database, 2015.

The distribution of late-Palaeozoic glacigenic deposits on the Permian plate tectonic small Earth model extends from high-southern to high-northern latitudes. This distribution, in particular the extensive distributions within the temperate and tropical zones, suggests the presence of mountain glaciation with lesser iceberg activity within the continental seas. This is consistent with geographical studies where the Palaeozoic Era is noted for its increasing elevation contrast between the lands and sedimentary basins, along with an increasing elevated topography suitable for forming mountain glaciers.

The distribution of southern hemisphere glacigenic deposits coincides with an exposed South Gondwana supercontinent, which included a prominent continental ice-sheet—shaded white in Figure 16.11. Glacigenic deposits in the northern hemisphere coincide with an exposed North Gondwana supercontinent, as well as lesser deposits in northern extensions of the ancient Tethys continental sea.

It should be noted that the distribution of continental seas in Figure 16.11 represents distributions for the early-Permian Period only. In contrast, the distribution of both glacigenic rocks and corals represents distributions for the entire Carboniferous and Permian Periods, hence the disparity in some coral distributions.

In Figure 16.11, since the end-Ordovician, the ancient Tethys Sea has been shown to have reduced in surface area to form a number of separate and discrete continental seas. The distribution of these seas continued to extend from equatorial to high northern latitudes. Their reduced surface areas, and connectivity, suggests that the northern sea temperatures and climates were only partially moderated by warm equatorial currents.

However, the presence of anthozoan corals—green dots in Figure 16.11—along with additional marine species such as molluscs and brachiopods, would imply that conditions throughout the North Polar Region still remained warm enough to support tropical marine species.

The presence of marine glacigenic debris within high northern latitudes may also suggest the presence of mountain glaciers, plus the distribution of glacigenic debris by melting icebergs.

No direct geologic evidence for glaciation exists for the following Triassic, Jurassic, or Cretaceous Periods, which, on an increasing radius plate tectonic Earth, coincides with breakup and opening of the modern oceans. During that time the South Pole migrated along the west coast of Africa to South Africa, then across the opening South Atlantic Ocean.

This apparent migration resulted in destabilization and melting of the ancient south-polar icesheet, along with destabilization of the climate and sea levels. Eyles and Young, as well as Frakes *et al.* in 1992 and 1993, also considered that ice-rafted material from Jurassic and Cretaceous rocks in Siberia and Cretaceous rocks in Australia may have indicated seasonally cold mountainous conditions in these areas.

16.5.6. Late-Cenozoic Quaternary Glaciation

The late-Cenozoic glacial era covers the time interval from around 2.58 million years ago to the present-day. Glacial events during this time are geologically relatively recent, with well-documented observations. Glaciation within high mountainous regions is extensive, and the Earth is at present experiencing a relatively warm interglacial period.

In 1994, Eyles and Young considered the late-Cenozoic glacial events to be the result of cyclical climatic changes which began about 60 million years ago. During that time, it was noted that a north-polar icesheet periodically extended south into North America and Europe, and a south-polar icesheet may have extended well into the Southern Ocean.

On an increasing radius plate tectonic Earth, the Cenozoic Era—commencing around 65 million years ago—coincided with the establishment of symmetrical seafloor spreading and on-going opening of each of the modern oceans. More importantly, the climatic changes occurring during that time coincided with the Antarctica continent migrating from an equatorial position into the south-polar region (Figure 16.12). A permanent continental icesheet was then established over Antarctica by about 33 million years ago.

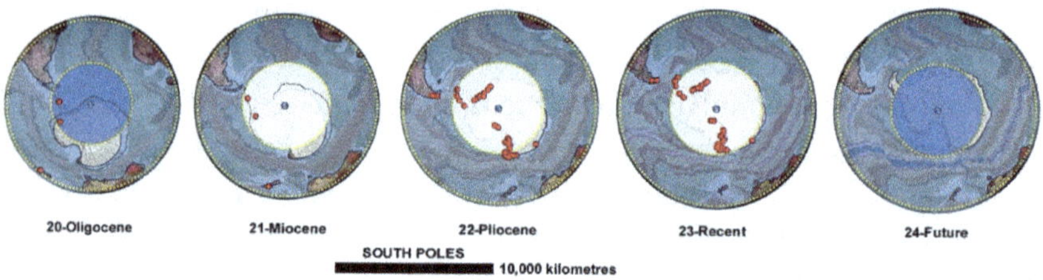

Figure 16.12. Cenozoic plate tectonic small Earth models centred over the South Polar Region showing migration of the Antarctican continent onto the South Pole. The presence of glacigenic rocks is shown as red dots, and the Antarctic ice sheet is shaded white. The presence of an ice sheet into the future is speculative and not shown. Glacigenic rocks after Hambry and Harland, 1981.

Migration of Antarctica into the South Polar Region (Figure 16.12) commenced during Oligocene times—around 60 million years ago. Since Oligocene times the migration of Antarctica, accompanied by the opening of the surrounding Southern Ocean, continued steadily to its present location centred over the South Pole.

Antarctica is currently continuing to migrate across the South Polar Region beneath the existing polar icesheet, and, on an increasing radius plate tectonic Earth, Antarctica will continue to migrate across this region into the future.

As a result of this migration, the Antarctic icesheet will be subject to melting and refreezing along the outer edges as the continent attempts to displace the icesheet away from the South Polar Region. This migration is continuing today, which may have a direct influence on the presently observed melting of the South Polar icesheet, as well as its long-term influence on climate.

Chapter 17

A Paradigm Shift in Understanding Extinction

"*A eustatic drop in sea level of 100 to 150 metres is a major physical consideration for the Maastrichtian [end-Cretaceous]. At the very least, this event caused a major loss of stratigraphic resolution of extinction events; likely, it was a contributor to extinctions both in the sea and on land.*" Dodson & Tatarinov, 1990.

An extinction event, also known as mass extinction, is defined as a *widespread and geologically rapid decrease in the amount of life on Earth*—where geologically rapid refers to events lasting millions to tens of millions of years. Such an event is identified by a change in the variety and diversity of life forms occurring when the rate of extinction increases with respect to the rate of speciation—the evolutionary process by which new biological species arise.

As the majority of organisms on Earth are microbial, and therefore difficult to detect in the rock record, extinction events are best observed in the biologically more complex organisms, rather than from the total diversity or abundance of life on Earth.

Published literature suggests that over 98 percent of all recorded species of life forms are now extinct. Based on the fossil record, the background rate of extinction on Earth has also been estimated by others to be about two to five families of marine invertebrates and vertebrates every million years. This background rate of extinction represents the natural rate of dying out of species as they either evolve to replace pre-existing species, or new species emerge to take their place.

Since life began on Earth, several major mass extinction events have significantly exceeded this background extinction rate. In the past 540 million years there have been five major events, where over 50 percent of animal species existing at the time have died out (Figure 17.1).

It is acknowledged in literature that mass extinctions seem to be associated with the past 540 million years or so of Earth history, occurring after the large complex organisms came into existence. Before that time, extinction rates were considered much lower and hence less detectable.

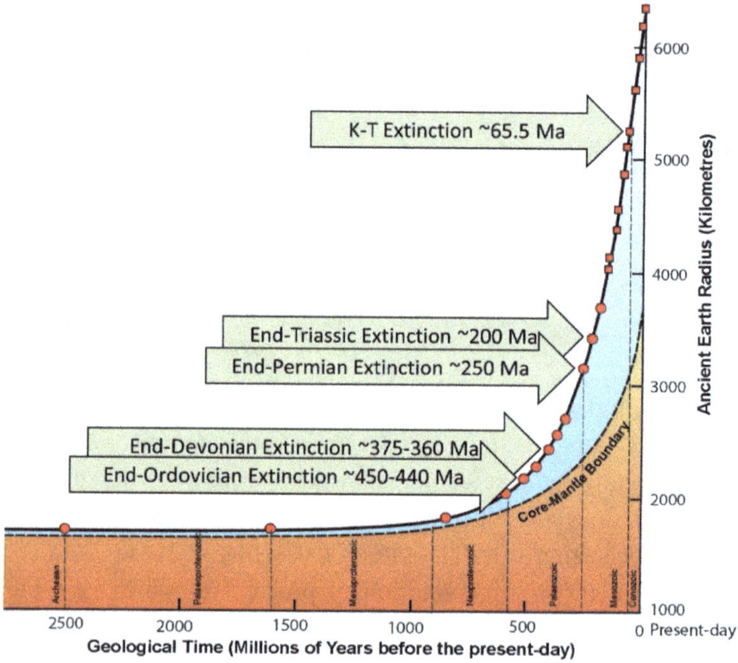

Figure 17.1. The location of key extinction events over the past 540 million years highlighted in green, shown in relation to the plate tectonic small Earth radius curve.

The most common causes of mass extinction currently postulated by others in the literature include.

1. **Flood basalt events**: Where formation of large volcanic eruptions may have produced dust, sulphureous oxides, and carbon dioxide, which may have inhibited photosynthesis and caused food chains to collapse on land and at sea.
2. **Sea-level falls**: Falls in sea-level reduce the continental shelf area, the most productive part of the ocean sufficient to cause a marine mass extinction, as well as disrupting weather patterns to also cause extinctions on land. Variations in sea-level are known to be associated with most of the known mass extinction events.
3. **Impact events**: The impact of a sufficiently large asteroid or comet is thought to cause food chains to collapse, both on land and at sea,

by producing dust and particulate aerosols which may inhibit photosynthesis.

4. **Sustained and significant global cooling**: This cooling is thought to make the Earth's climate more arid by locking up much of the planet's water in ice and snow. This aridity is further considered to potentially kill many polar and temperate species and force others to migrate towards the equator, thus increasing competition for tropical species.
5. **Sustained and significant global warming**: This global warming is thought to make the Earth's climate wetter by melting ice and snow and thus increasing the activity of the water cycle. It may also expand the area available for tropical species, thus killing temperate species or forcing them to migrate towards the poles, possibly causing severe extinctions of polar species.
6. **Anoxic events**: These events are said to occur when the middle and upper layers of the ocean become deficient, or totally lacking, in oxygen. The potential causes are complex and controversial, but all known instances are associated with severe and sustained global warming, mostly caused by massive volcanism.
7. **Oceanic overturn**: Overturn is a disruption of ocean circulation which lets surface saline-rich water sink to the bottom. This disruption brings anoxic deep water to the surface and kills most of the oxygen-breathing organisms that inhabit the surface and middle depths. It is thought to occur either at the beginning or the end of a period of glaciation.
8. **A nearby nova, supernova, or gamma ray burst**: A nearby gamma ray burst is thought to be potentially powerful enough to destroy the Earth's ozone layer, leaving organisms vulnerable to ultraviolet radiation from the Sun. These gamma ray bursts are fairly rare, occurring only a few times in a given galaxy per million years.
9. **Global Tectonics**: Movement of the continents contributes to extinction by initiating or ending ice ages, by changing ocean and wind currents and thus altering climate, or by opening seaways or land bridges which expose previously isolated and poorly adapted species to competition.
10. **Other hypotheses**: Including the spread of new diseases, super-volcanic events, the eventual warming and expanding of the Sun, and

human activities—including changes in climate, overhunting, overfishing, invasive species, and habitat loss.

While each of these proposals is feasible, it is not the intent of this chapter to discuss the individual merits of these proposals. On an increasing radius plate tectonic Earth, the small Earth modelling studies show that, during early-Palaeozoic to present-day times, there have been a number of drastic and prolonged changes to sea-levels, which coincide precisely with the known extinction events.

On these models, major changes in sea-levels are shown to occur as a result of separation, or merging of previous ancient continental seas, as well as onset of geosynclinal activity and orogenesis, breakup of the ancient supercontinents, opening of modern oceans, and draining of the ancient continental seas. Depending on the severity of these events, it is considered that sea-level changes may also adversely affect regional to global-scale climate, as well as ocean-water circulation patterns, species habitats, and the type and location of sedimentary deposition.

17.1. End-Ordovician Extinction Event

An end-Ordovician extinction event occurred over a period of 10 million years at the Ordovician to Silurian transition, extending from 450 to 440 million years ago. Within this time frame, two events have been documented, which show that 27 percent of all families, 57 percent of all genera, and 60 to 70 percent of all species were killed off. Together, these events are ranked by many scientists as the second largest of the five major extinctions in Earth's history in terms of percentage of genera that became extinct.

Plate tectonic small Earth modelling shows that, during the early-Palaeozoic times, there was a single, on-going and well-established network of interconnected continental seas, which collectively defined each of the surrounding supercontinents. Modelling (Figure 17.2) demonstrates that, while the overall configuration of this ancient continental sea remained the same, there was a reduction in surface area between the Ordovician and Silurian small Earth models. This reduction in surface area gave rise to separation of the original network into at least two discrete seas, which severely disrupted established coastal communities.

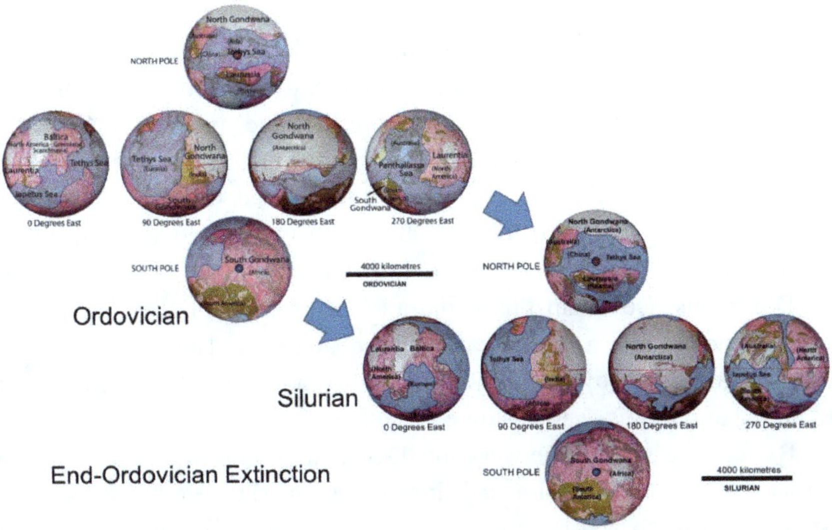

Figure 17.2. Ordovician and Silurian plate tectonic small Earth assemblages showing the distribution of ancient coastlines and seas during the end-Ordovician extinction event. The ancient Tethys, Iapetus, and Panthalassa Seas form part of a global network of continental seas surrounding each of the exposed supercontinents. Coastlines after Scotese, 1994, and Smith *et al.*, 1994.

Extinction during the end-Ordovician, in particular over the 10-million-year timeframe involved, is considered to have occurred as a result of increasing changes to surface curvature and elevation of the lands during initiation of Palaeozoic geosynclinal and orogenic activity. This activity deepened the seas, and progressively drained waters from the shallow coastal shelf areas. These changes then disrupted the configuration of previously existing coastlines and the overall distribution of continental seas. Crustal changes in one area during changing surface curvature, globally affected coastlines elsewhere by changing the overall sea-levels. This change in sea-levels then disrupted favoured continental shelf habitats, as well as established migration routes.

17.2. Late-Devonian Extinction Event

A late-Devonian extinction event is known to have occurred over a period of 15 million years at the Devonian to Carboniferous transition, extending from 375 to 360 million years ago. During that time, a prolonged series of extinctions eliminated about 19 percent of all families, 50 percent of all genera, and 70 percent of all species. This extinction event is thought by others to have lasted longer, for as much as 15 to 20 million years, and there is also some evidence for a series of extinction pulses during that time.

On an increasing radius plate tectonic Earth, the timing of the late-Devonian extinction event places it within the latter part of the Gondwana supercontinental configuration. During that time, the previously well-established network of interconnected continental seas was being subjected to ever-increasing adjustments to Earth surface curvature, geosynclinal activity, and orogenesis.

During the Devonian times, the ancient continental seas continued to further separate into a number of deep isolated seas. The geosynclinal and orogenic activity also provides substance for the series of extinction pulses occurring during this event, as noted by others.

By comparing the network of Devonian continental seas with the Carboniferous seas in Figure 17.3, it can be seen that there is a considerable difference in both coastal outlines and surface areas of these seas, when compared to the older ancient seas in Figure 17.2. This difference highlights the geographical observation that changing surface curvature and tectonic activity during that time was severely disrupting the land surface, giving rise to an increasing elevation contrast between the lands and seas.

This contrast of elevation, in turn, suggests there was an increase in orogenesis and mountain building, which in turn increased the exposure and rate of erosion of lands. These changes also influenced changes to the distribution and depths of the seas, as well as corresponding coastal outlines of the exposed supercontinents.

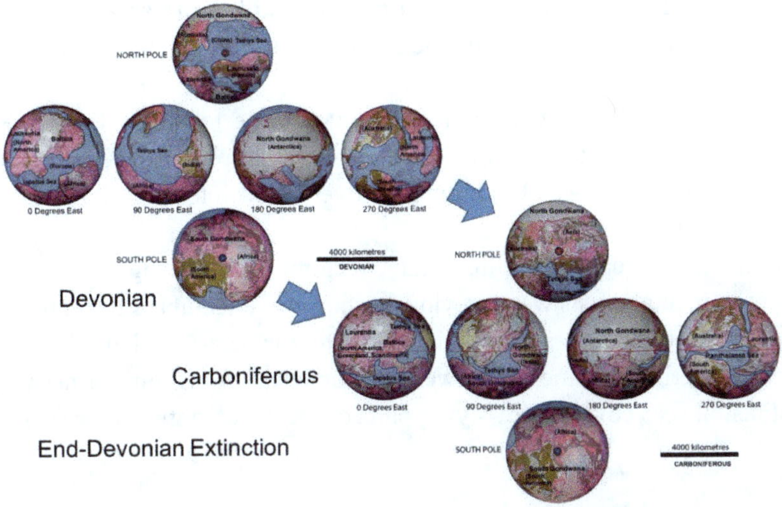

Figure 17.3. Devonian and Carboniferous plate tectonic small Earth assemblages showing the distribution of ancient coastlines and seas during the end-Devonian extinction event. The ancient Tethys, Iapetus, and Panthalassa Seas form part of a disrupted global network of continental seaways surrounding each of the exposed supercontinents. Coastlines after Scotese, 1994, and Smith et al., 1994.

On an increasing radius plate tectonic Earth, it is considered that the late-Devonian extinction event primarily occurred during these Gondwanan times as a result of increasing changes to surface curvature, severe disruption to the configuration of the coastlines, and isolation of continental seas into a number of relatively deep and restricted seaways.

17.3. End-Permian Extinction Event

An end-Permian extinction event occurred around 250 million years ago, at the Permian to Triassic transition. This event is considered by most researchers to have been the largest of the five major extinctions in Earth's history, devastating life on Earth and estimated to have killed off 57 percent of all families, 83 percent of all genera, and 90 to 96 percent of all marine and land species.

Uncertainty exists in literature regarding the overall duration of the end-Permian extinction event. Recent research shows that different groups of species became extinct at different times. An older theory, still supported in the literature, shows that there were two to three major extinction pulses, 5 million years apart, separated by periods of extinctions well above the background level. The final extinction pulse killed off about 80% of all remaining marine species, while the other losses occurred during the first and intervening pulses.

Research indicates that the long recovery time for this extinction event was due to successive waves of extinction as well as prolonged environmental stress to organisms, which inhibited recovery well into the following early-Triassic Period.

On an increasing radius plate tectonic Earth (Figure 17.4), the supercontinental crusts were previously shown to encompass the entire Earth as a single Pangaean supercontinent up to the late-Permian. During this extended period of time, there were no modern oceans, only a network of continental seas covering low-lying parts of the supercontinental crusts.

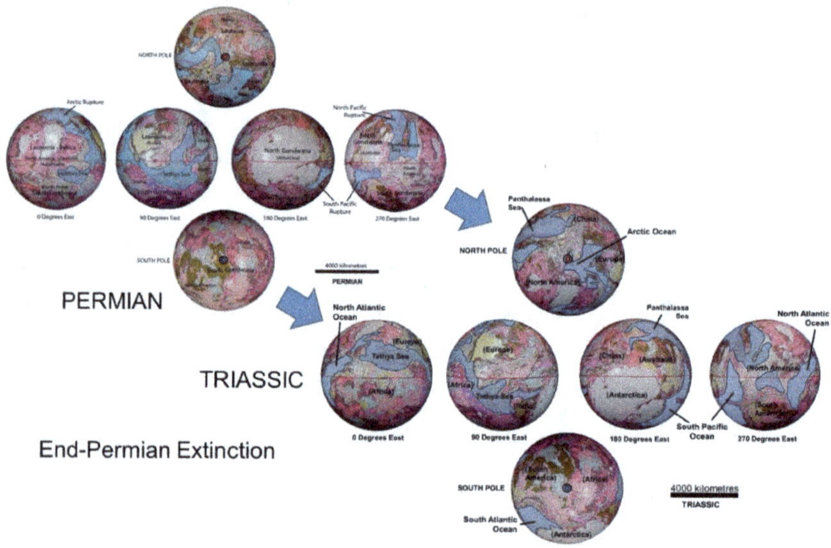

Figure 17.4. Permian and Triassic plate tectonic small Earth crustal assemblages showing the ancient coastline distribution as well as remnants of the ancient Pangaea supercontinent during the end-Permian extinction event. The figure also shows the locations of Permian continental rupture commencing in the north and south Pacific and Arctic Ocean regions to form the modern oceans. Coastlines after Scotese, 1994, and Smith et al., 1994.

The transition from ancient seas to modern oceans came about when the ancient Pangaea supercontinental crust first started to rupture and break up some 250 million years ago. The end-Permian extinction event is therefore significant, as it coincides with breakup of the Pangaean supercontinent to form the modern continents and opening of the intervening modern oceans.

Breakup of the Pangaea supercontinent during the late-Permian initiated draining of the waters of the ancient continental seas into these newly opening modern oceans. In addition, new waters were continually being expelled from newly emerging mid-ocean-ridge spreading zones, in conjunction with extrusion of new seafloor basaltic crusts.

This crustal breakup also destabilized previously existing polar icecaps and the subsequent migration of the modern continents, in turn, destabilized established climate zones. Draining of the continental seas then exposed the ancient sea-beds as new lands, which in turn gave rise to rapid and unprecedented changes in sea-levels, erosion of the lands, and disruption of species habitats.

On an increasing radius plate tectonic Earth, the end-Permian extinction event is considered to have resulted from a prolonged period of breakup and disruption to the pre-existing supercontinents and seas. This extinction was a direct result of breakup of the Pangaean supercontinent to form the modern continents and draining of the continental seas to form the precursors to the modern oceans.

The post-Permian times are then seen as representing a shift in the style of extinction events occurring on Earth. These events shifted from simple coastline and sea-level changes, occurring prior to the Permian Period, to more disruptive sea-level changes during catastrophic merging of previously isolated seas and oceans during post-Permian times.

17.4. End-Triassic Extinction Event

An end-Triassic extinction event occurred some 200 million years ago at the Triassic to Jurassic transition. About 23 percent of all families, 48 percent of all genera—20 percent of marine families and 55 percent of marine genera—and 70 to 75 percent of all species became extinct.

This event is noted for the extinction of most of the non-dinosaurian archosaurs, as well as most therapsids—a group of the most advanced terrestrial vertebrates, including the ancestors of mammals—and most of the large amphibians, leaving the dinosaurs with little competition to contend with on land.

On an increasing radius plate tectonic Earth, by the end of the Triassic Period, breakup of the Pangaean supercontinent and opening of the modern oceans was well established (Figure 17.5). Draining of the ancient continental seas into the modern ocean basins was also well advanced, with only remnant seas remaining within the Mediterranean to Middle Eastern Tethys Sea regions. As well as these remnant seas, small Earth modelling shows that a number of modern oceans also remained isolated.

It is significant to note that the end-Triassic represents the period of time when the previously separated North and South Pacific Oceans merged to form a single modern Pacific Ocean. This mergence is highlighted in Figure 17.5 by comparing the Triassic and Jurassic small Earth models, where a land connection between North America and Australia is shown to have been breached.

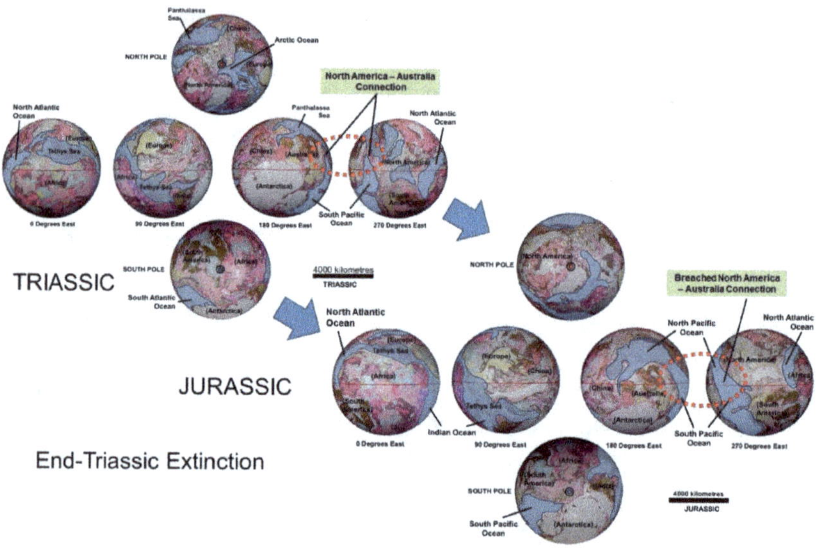

Figure 17.5. Triassic and Jurassic plate tectonic small Earth crustal assemblages showing the ancient coastline distribution as well as breakup of the ancient supercontinents to form the modern continents and oceans during the end-Triassic extinction event. The figure highlights the location of continental breaching in the north and south Pacific to form the modern Pacific Ocean. Coastlines after Scotese, 1994, and Smith et al., 1994.

This breaching allowed the ancient North and South Pacific Oceans to merge as a single ocean, along with draining of the North American Western Interior Seaway. In doing so, the previously separate sea-levels within each of these early oceans and seas then equilibrated—where sea-level in one ocean rose and sea-level in the other fell.

On an increasing radius plate tectonic Earth, the end-Triassic extinction event is considered to be the direct result of continental breaching between North America and Australia and merging of the previously separate North and South Pacific Oceans to form a single Pacific Ocean.

Breaching and merging of these two separate oceans then resulted in drastic changes to global sea-levels, in particular the Western Interior Seaway, as the previously separate sea-levels equilibrated, disrupting the various habitats and established species migration routes. Merging of these oceans also allowed previously separate marine and terrestrial species to migrate and colonize new habitats, potentially displacing or eliminating less adaptive species.

17.5. End-Cretaceous Extinction Event

The end-Cretaceous extinction event—commonly referred to as the K-T event—is the most publicized and well known of all the extinction events, occurring around 65.5 million years ago at the Cretaceous to Paleogene transition. This extinction event was a large-scale mass extinction of animal and plant species, most notably the dinosaurs, in a geologically short period of time. During that time, about 17 percent of all families, 50 percent of all genera, and 75 percent of all species became extinct. In the oceans, the number of marine species was also reduced to about 33 percent.

It is currently hypothesized by others that the end-Cretaceous extinction event was caused by one or more catastrophic events, including either an asteroid impact or an increase in terrestrial volcanic activity.

In 1980, a team of researchers lead by Luis Alvarez recognized that concentrations of the element iridium, located at the Cretaceous–Paleogene time boundary, were higher than in surrounding sedimentary layers. This anomaly gave support for consideration of a collision of the Earth by a large celestial object—similar to the Shoemaker-Levy impact observed on Jupiter in 1994. This potential celestial object was later identified as the Chicxulub crater located in Central America.

Other researchers though, consider the extinction event was more gradual, resulting instead from sea-level and climate changes already occurring during the late-Cretaceous, which may or may not have been aggravated by impact events or increased volcanic activity.

Based on fossil evidence, the end-Cretaceous dinosaur extinction, for instance, was considered by Dodson and Tatarinov in 1990 to have been a highly selective world-wide phenomenon, which lasted for approximately 8 million years. They went further to consider that although disappearance of the dinosaurs was a spectacular aspect of the Cretaceous extinction, it actually represented the loss of rather few species and was only a small aspect of the entire extinction of species.

On plate tectonic increasing radius small Earth models, the end-Cretaceous extinction event coincides with breaching and opening of the Southern Ocean, located between Australia and Antarctica (Figure 17.6). During the late-Cretaceous, the Australia to Antarctica assemblage had exposed land connections adjacent to Western and Eastern Australia, with an isolated shallow sea located adjacent to Southern Australia—known as the Eucla Basin. These land connections were then partly breached along the Western Australian coastline, causing flooding of the Eucla Basin, prior to complete breaching and opening of the Southern Ocean during Paleocene to Eocene times—around 65 million years ago.

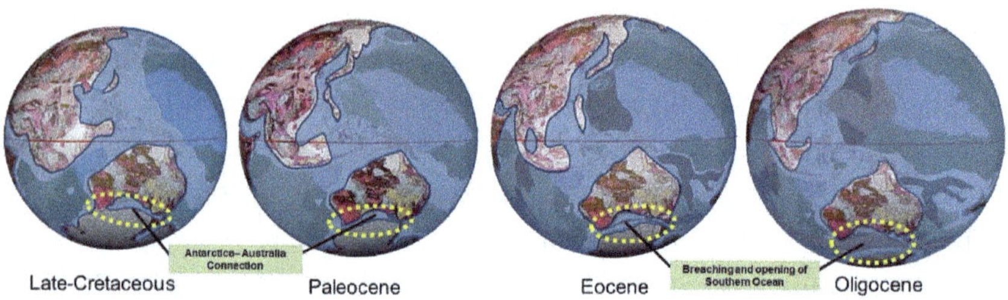

Figure 17.6. Late Cretaceous to Oligocene plate tectonic small Earth models showing land connections between Australia and Antarctica, prior to breaching during the Paleocene to form the Southern Ocean. Coastlines after Scotese, 1994, and Smith et al., 1994.

This breaching event is highlighted in Figure 17.7, showing the continued decline in surficial areas of the Western Interior Seaway of North America during the late-Cretaceous and Paleocene times. The Western Interior Seaway is an important area for dinosaur study and has provided a large amount of data for interpretation of the end-Cretaceous extinction event.

It is significant to note that this regression of the Western Interior Seaway into the Gulf of Mexico coincides with the extended duration of the end-Cretaceous extinction event, as noted by Dodson and Tatarinov.

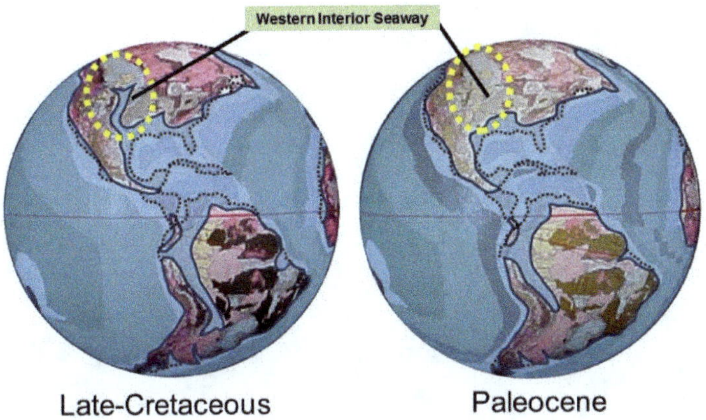

Figure 17.7. Late Cretaceous and Paleocene plate tectonic small Earth models showing regression of seas within the Western Interior Seaway of North America). Coastlines after Scotese, 1994, and Smith et al., 1994.

On an increasing radius plate tectonic Earth, breaching between Australia and Antarctica during initiation and opening of the Southern Ocean resulted in a period of disruptive sea-level change. This continental breaching and sea-level change coincided precisely with the end-Cretaceous extinction event. These observations support the conclusions of others where the extinction event was more gradual, resulting from drastic sea-level changes already occurring during the late-Cretaceous, extending over a period of approximately 8 million years.

Chapter 18

A Paradigm Shift in Palaeobiogeography

"*Plate Tectonic reconstructions and proposed movements of continents do not correspond to the known or necessary migration routes and directions of biogeographical boundaries.*" Meyerhoff et al., 1996.

Biogeography is defined as *a study of the distribution of plant and animal species and ecosystems in geographic space and through geological time*. These distributions are known to vary in a regular manner depending on their latitude, elevation, isolation, and habitat area. Biogeography is also seen as an integrative field of inquiry that unites concepts and information from the various fields of ecology, evolutionary biology, geology, and physical geography. Ancient biogeography—known as palaeobiogeography—goes one step further to include ancient geographic information, in particular in relation to tectonic studies.

This information is used in conventional studies to help constrain hypotheses on the timing of biogeographic events, such as extinction, and provides insights into the regional development and evolution of all life forms.

Published literature suggests there are significant inconsistencies in both biogeographic and geographic reconstructions in conventional plate tectonic studies. In 1992, Smiley reported that the distributions of land-based plants and animals existing during late-Palaeozoic, Mesozoic, and Cenozoic times, as well as many marine creatures, just do not support continental drift-based plate tectonic reconstructions.

The evidence instead suggested to Smiley that the positions of both the ancient continents and ancient poles have been relatively stable since late-Palaeozoic times. This continental and polar stability contrasts strongly with the complex, random, amalgamation-dispersal-amalgamation plate motions required for conventional plate tectonic reconstructions. From this, Smiley concluded that "*Plate Tectonic theories create more biogeographic problems than those already existing.*"

In 1989, Noël recognized from his extensive studies into the global distribution of fruit and nut trees that all related species, as well as many animal species, must have a common ancestry. The only way for the various species to have spread naturally is for the areas of population to have been in close proximity in the past and to have since moved apart during "*Continental Drift*". From this, Noël concluded that "*...distributions are all readily explicable on the assumption that the current land areas of the Earth were once all physically linked...* [because]*...the Earth has since expanded.*"

Just like Noël, in 1990 Cox showed that the distribution of various plant and animal species on the present Earth form a number of distinct east-west and north-south distributional patterns. He considered that these patterns "*...better substantiate the alternative theories, such as Earth expansion, which do not require any form of complex plate motions.*"

This observation was further supported by the work of McCarthy in 2003 where he showed that faunal and floral species matches across the Pacific Ocean were good. He also showed that species that matched on opposite sides were absent in between, disproving any ad-hoc stepping-stone or land-bridge explanations for species migration and development.

Similarly, in 1990 Stevens, in discussing the various marine faunal links across the Pacific Ocean, showed that a number of distinctive Jurassic marine fossil species from New Zealand have very marked similarities to species from Mexico. On conventional plate tectonic reconstructions, a direct link between these now remote continents is impossible because of the need to include the presence of vast Panthalassa and Tethys Oceans.

To maintain any form of faunal link between New Zealand and Mexico on conventional plate tectonic reconstructions, additional links would be required through polar regions between New Zealand, West Antarctica, and South America. Stevens showed that these faunal species links just do not occur on any of these now remote continents.

Stevens suggested instead that tighter palaeobiogeographical assemblages could be achieved by assuming an increasing radius Earth thesis. By reducing the Earth radius back in time, the need for a theoretical Panthalassa or Tethys Ocean is then drastically reduced. This reduction then makes direct links between New Zealand and both Central America and South America a physical possibility and, in turn, simplifies the required migration routes between New Zealand and Mexico.

In 1996, Meyerhoff *et al.* carried out an extensive study of the Cambrian to present-day plant and animal regions of the Earth. They showed that the "*...assemblage and movements of the continents on Plate Tectonic reconstructions do not correspond to the known, or necessary, migration routes shown by these established biogeographical regions.*" In most cases the discrepancies between the actual physical distributions of species and the conventional plate tectonic continental assemblages were impossibly large, so large that not even an approximate match could be made.

This observation of Meyerhoff was also substantiated by the continental reconstructions of Shields. In 1996, Shields considered that none of the existing plate tectonic or partial Earth expansion assemblages of continents—referring to the partial expanding Earth reconstructions of Owen in 1976—surrounding the Indian and Pacific Oceans appear to completely satisfy all of the constraints imposed by palaeobiogeography.

18.1. Biogeographic Data Modelling

The quality and quantity of faunal and floral data available for biogeographical studies is now extensive. The brief selection of species data presented in this chapter is used as examples only in order to emphasize distributions and inter-relationships on increasing radius plate tectonic small Earth models. Modelled data are based on the published Paleobiology Database (PaleoBioDB, 2015).

The navigator used to access this data can be opened from *https://paleobiodb.org/navigator/*. The reader is encouraged to visit this navigator in order to compare and contrast data plotted on increasing radius small Earth models with the same data plotted on conventional plate tectonic assemblages.

The selection of data in this chapter, plotted from the PaleoBioDB, represents known fossil sites present throughout the world. Distributions of the various fossil species shown on each of the small Earth models represent the distribution of known fossils, and these should be visualized along with their relationships with adjoining continents. The preservation and discovery of these fossil sites is, however, governed by a number of factors, including the:

- Presence of strata suitable for preservation of fossils.
- Remaining outcrop of suitable fossiliferous strata.
- Degree of erosion or weathering of fossiliferous strata.
- Burial of fossiliferous strata by younger cover rocks.

- Remoteness of the site.
- Research effort and available research funding.
- Sampling bias from better studied sites.
- Political stability of the various countries.

In addition to the fossil data, it should be noted that the coastal outlines shown on each of the small Earth plate tectonic models presented here represent a snapshot in time depicting coastal outlines at the beginning of each period or epoch shown. In reality, these coastlines were constantly shifting and changing during changes to Earth surface curvature through time.

In contrast, the various fossil species plotted on the small Earth models represent known distributions for the full extent of each time period represented. As such, some of the plotted data may be at variance with the coastlines and distribution of ancient seas shown on each of the small Earth models.

A brief evolutionary history of life on an increasing radius plate tectonic Earth is shown schematically in relation to known extinction events and changing Earth radius in Figure 18.1. In this figure, the presence of ancient microbial life forms coincides with the extremely long Precambrian times.

Plate tectonic small Earth modelling has previously suggested that throughout this extended period of time the ancient Earth had a relatively flat featureless landscape, devoid of mountains, and included an extensive network of relatively shallow continental platform seas. Deposition of sediments within these shallow seas included chemically precipitated siliceous chert, carbonate, and banded iron formation rocks.

In contrast to the Precambrian, Figure 18.1 shows that the rapidly changing Palaeozoic times coincided with the development and evolution of all complex life forms on Earth. Environmental change during these times went from stable featureless land conditions to increasingly elevated topography, to increased disruption and geographic isolation of the lands and continental seas.

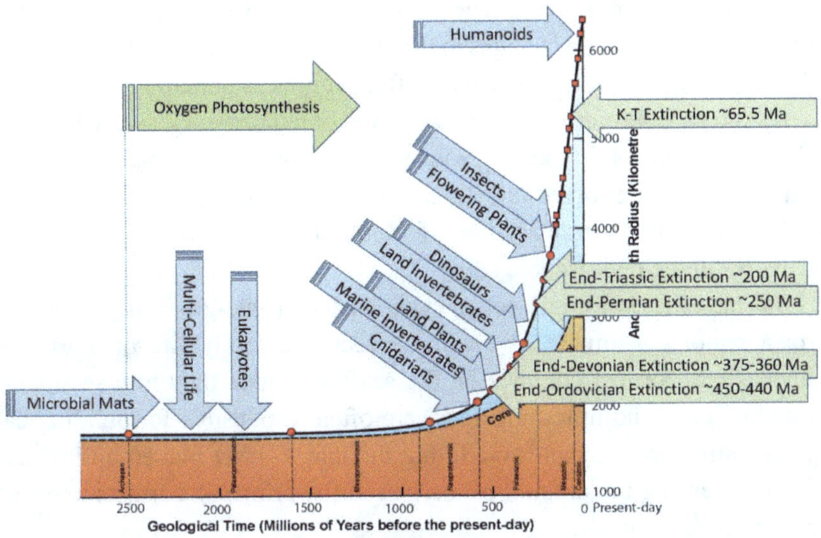

Figure 18.1 Summarized evolutionary history of life on an increasing radius plate tectonic Earth in relation to known extinction events. Key evolutionary events are highlighted in blue in relation to the exponential increase in Earth radius curve.

On an increasing radius plate tectonic Earth, the degree of crustal and environmental change occurring during these Palaeozoic times had the capacity to markedly drive evolutionary change in all life forms.

Similarly, extinction of species has been shown to be a by-product of sea-level change, disruption to the distribution of established seas, and the inability of species to keep pace with these environmental changes. The evolution of species during these times was then subsequently driven by the need to survive amongst these ever-increasing changes to the Earth and environment.

18.2. Precambrian Life

Evidence from palaeontology suggests that life began on Earth approximately 3 to 3.5 billion years ago. The mechanism by which life emerged is uncertain and various hypotheses are still being formulated. Since then, life has evolved into a wide variety of forms which biologists have classified into a hierarchy of taxa—a group of one or more populations of organisms.

The evolution of oxygenic photosynthesis during the early- to mid-Proterozoic then enabled the various primitive microbes existing at that time to play an important role in oxygenation of the atmosphere from about 2,400 million years ago. These changes in the atmosphere were considered by others to have then increased the microbe's effectiveness as nurseries of evolution.

By the early Proterozoic, the Earth was populated exclusively by single-celled prokaryotic life forms. From these simple beginnings all advanced forms of life later evolved and spread throughout the rest of the Earth. The dominance of early prokaryotic life took the form of stromatolites and relatively simple microscopic cells.

Stromatolites are single-celled cyanobacteria (blue-green algae), and they form a major constituent of the fossil record of early life on Earth. The cyanobacteria formed colonies on the seafloor where they trap sediment to form algal mats and limestone and were often responsible for building early reef-like structures. Cyanobacteria are thought to have played an important role in increasing the amount of oxygen in the primeval Earth's atmosphere through photosynthesis.

After the Cambrian diversity of marine animals, grazing stromatolite mats by newly emerging herbivores greatly reduced the occurrence and distribution of stromatolites in most environments.

Multi-cellular life during Precambrian times was initially composed only of eukaryotic cells and the earliest evidence for this is from around 2,100 million years ago. The earliest evidence of more complex eukaryotes with organelles—specialized cells—dates from around 1,850 million years ago. Further specialization of cells for different functions first appeared between 1,430 million years ago—thought to be a possible fungus—to 1,200 million years ago—a possible red alga.

At that time, before the advent of predatory life forms, the cyanobacteria and eukaryotic algae were relatively free from predation. They were able to multiply rapidly in marine aquatic environments, limited only by the supply of nutrients essential to their growth.

The distribution of known late-Proterozoic eukaryote fossil sites (Figure 18.2) highlights a global distribution coinciding with the network of ancient seas—shown as khaki-coloured strata. This network will become more evident after mapped ancient seas are shown for the following Palaeozoic Era.

In this figure, the distribution of eukaryote fossil sites is shown to favour ancient equatorial to temperate regions, extending within shallow seas into high northern latitudes. The presence of eukaryotes in high northern latitudes suggests that either the algae were tolerant of cooler waters, or warm equatorial waters extended into the North Polar Region effectively moderating the extent or presence of a North Polar icesheet.

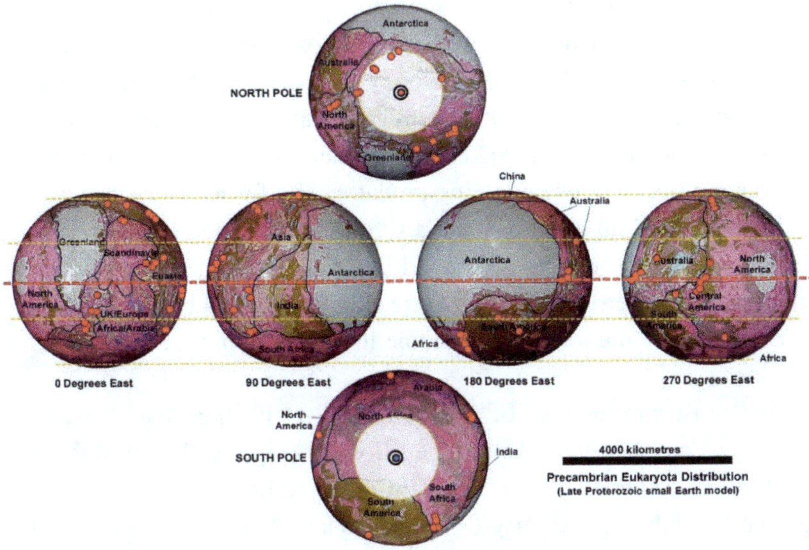

Figure 18.2. Distribution of Precambrian eukaryota plotted on a late-Palaeozoic plate tectonic small Earth model. Eukaryota data are shown as red dots in relation to climate zones and late-Proterozoic north and south polar ice sheets, shaded white. Black lines represent outlines of named continental cratonic crusts. Palaeontological data after PaleoBioDB, 2015.

By the end of the Precambrian, life in the ancient seas had begun to diversify rapidly. Nearly all new life forms emerging during that time were soft-bodied, mainly jellyfish-like animals and worms. These, along with early-Cambrian life forms, were mainly herbivores which fed on the abundant algae that blanketed the ancient sea floors and included the cnidarians, distinguished by the presence of specialized cells used mainly for capturing prey.

18.3. Early-Palaeozoic Life

On an increasing radius plate tectonic Earth, there is not a lot of difference between the late-Proterozoic and the early-Palaeozoic—comprising the Cambrian and Ordovician Periods. Exposed lands defined the distribution of ancient supercontinents and surrounding sedimentary basins defined the various continental seas. The slowly increasing rate of change in Earth radius had initiated an increasing elevation contrast between the lands and the seas which, in turn, initiated an erosional cycle and the beginnings of an elevated topography. Increasing changes in Earth surface curvature also resulted in subtle changes to the distribution of continental seas and shorelines.

The early-Palaeozoic is renowned for an explosion of new life forms that left behind external shells and skeletons composed of durable minerals. It has been suggested by others that a chemical change within the seas may have triggered production of these skeletons. Other suggestions included the need for protection from newly evolving predatory life forms.

Both of these suggestions mark a significant change in the basic structure of ecosystems throughout the world, which remained confined to the continental seas. Terrestrial realms were still barren of all but simple spore-like forms and before the middle-Palaeozoic there were no insects, vertebrates, or plant forms.

The most conspicuous of marine animals with hard parts that evolved during the Cambrian Period were the trilobites (Figure 18.3). Trilobites were hard-shelled, segmented creatures with multiple body segments and jointed legs that lived during the early-Cambrian to late Permian Periods (570 to 230 million years ago). They are considered to be one of the earliest complex life-forms and are one of the key signature creatures of the Palaeozoic Era.

The smallest known trilobite species was under a millimetre long while the largest included species from 30 to over 70 centimetres in length. With such a diversity of species and sizes speculation on the ecology of trilobites includes planktonic, swimming, and crawling forms where they filled a varied set of feeding niches, most notably as detritivores, predators, or scavengers.

In Figure 18.3 the distribution of trilobites is shown on the Ordovician plate tectonic small Earth model to extend throughout all of the interconnected continental Tethys, Iapetus, and Panthalassa Seas, extending from mid southern to high northern latitudes. The presence of warm equatorial Tethys Sea waters extending into the North Polar Region restricted the presence of northern sea ice during the Ordovician glacial period, enabling the trilobites, along with many other species, to thrive in these areas.

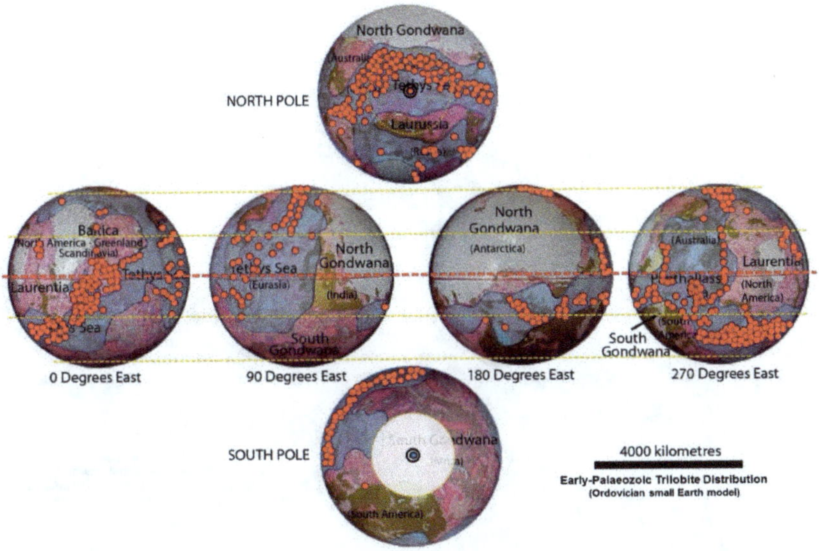

Figure 18.3. Distribution of early-Palaeozoic trilobite species plotted on an Ordovician plate tectonic small Earth model. Trilobite data are shown as red dots in relation to ancient climate zones, an early-Palaeozoic South Polar icesheet, shaded white, and the distribution of ancient continental seas. Palaeontological data after PaleoBioDB, 2015.

An important group of predators that arose during the late-Cambrian were the nautiloids which, along with the molluscs, belong to the class Cephalopoda. These exclusively marine animals are characterized by bilateral body symmetry, a prominent head, and a set of arms or tentacles modified from the primitive molluscan foot. Cephalopods prevailed throughout the seas, extending from the abyssal plain to the sea surface. Early cephalopods underwent pulses of diversification during the Ordovician Period to become widely dispersed and increasingly dominant in the Paleozoic and later Mesozoic seas.

On the Ordovician plate tectonic small Earth model in Figure 18.4, the cephalopods show an extensive distribution coinciding with distribution of the shallow continental Tethys, Iapetus, and Panthalassa Seas. Again, warm sea waters extended from the equator to the North Polar Region effectively moderating polar climates and allowing the cephalopods to thrive within the northern reaches of the Tethys Sea.

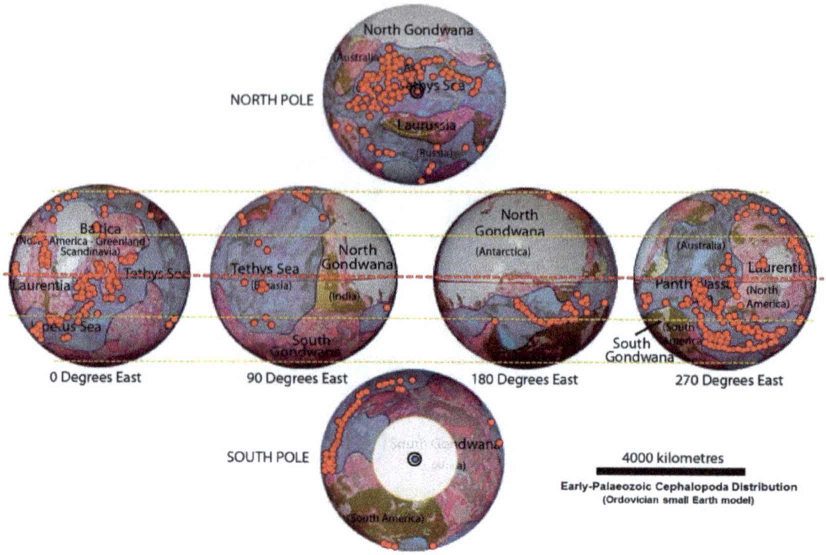

Figure 18.4. Distribution of early-Palaeozoic cephalopoda plotted on an Ordovician plate tectonic small Earth model. Cephalopod data are shown as red dots in relation to climate zones, an early-Palaeozoic South Polar icesheet, shaded white, and the distribution of ancient continental seas. Palaeontological data after PaleoBioDB, 2015.

The adaptive radiation of marine animals with skeletons during this time was disrupted during the latter part of the Cambrian by several minor mass extinctions. The last of the Cambrian mass extinctions eliminated large numbers of nautiloids and trilobite genera. Three groups of animals present, but not highly diversified, during the Cambrian times emerged to become important species during the following Ordovician and included the articulate brachiopods, the graptolites, and the conodonts. These species are all important index fossils for Ordovician times.

Brachiopods are marine animals with hard shells on the upper and lower surfaces, unlike the left and right arrangement in bivalve molluscs. The earliest confirmed brachiopods have been found in early-Cambrian strata, inarticulate forms appearing first, followed soon after by articulate forms. At their peak during the Palaeozoic the brachiopods were among the most abundant filter-feeders and reef-builders. Brachiopod fossils have also been useful indicators of climate changes during the Paleozoic era.

When global temperatures were low, as in much of the Ordovician, the difference in temperatures between the equator and poles created different fossils at different latitudes. During warmer periods, such as during much of the Silurian, there were smaller differences in sea temperatures and all seas at the low to middle latitudes were colonized by the same few brachiopod species.

The distribution of brachiopods is shown on the Ordovician plate tectonic small Earth model in Figure 18.5 to have an extensive distribution coinciding with the distribution of trilobites and cephalopods, extending throughout the Tethys, Iapetus, and Panthalassa Seas from mid southern to high northern latitudes.

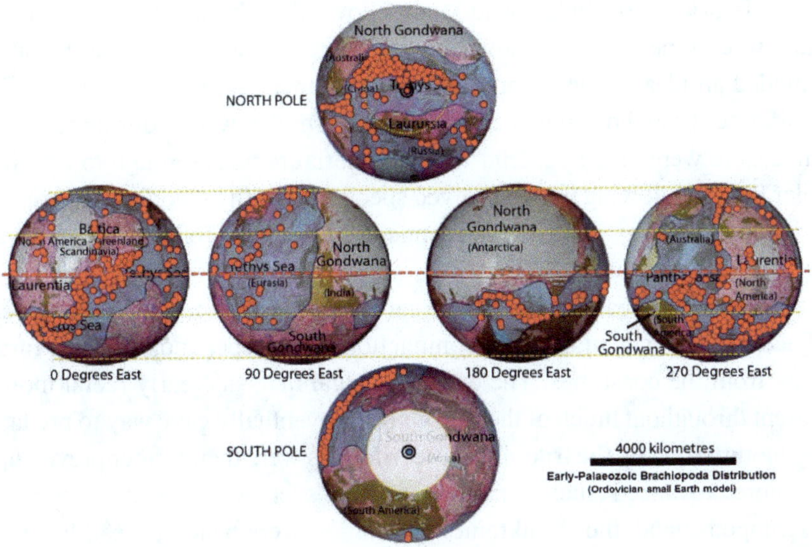

Figure 18.5. Distribution of early-Palaeozoic brachiopods plotted on an Ordovician plate tectonic small Earth model. Brachiopod data are shown as red dots in relation to climate zones, an early-Palaeozoic South Polar icesheet, shaded white, and the distribution of ancient continental seas. Palaeontological data after PaleoBioDB, 2015.

Graptolites are an extinct group of animals whose fossils include benthonic species but were more often planktonic colonial animals known chiefly from the late-Cambrian times to the early-Carboniferous Periods. Fossils are often seen as fragile forms preserved as carbonized impressions in dark shales with a worldwide distribution. The preservation, quantity, and gradual change in graptolites over a relatively short geologic timescale allow the fossils to be used as important index fossils for dating Palaeozoic rocks.

Most fossil graptolites look like small saw blades, with the teeth of the saw formed by short open branches from a main tube. Careful study of the microscopic structure of the tubes of graptolites shows that they were very similar to the tubes of small worm-shaped animals.

Graptolite distribution is shown on the Ordovician plate tectonic small Earth model in Figure 18.6. Graptolite distribution again coincides with the distribution of the ancient Tethys Iapetus and Panthalassa Seas, extending from mid-southern to high-northern latitudes. The preservation of graptolites within black shales suggests that the deeper parts of these ancient seas were devoid of oxygen and conditions were anoxic.

This brief early-Palaeozoic study shows that the network of relatively shallow continental seas shown on the plate tectonic small Earth models provided an ideal environment for Precambrian algal mats and early reefs to proliferate throughout the ancient Earth. Warm sea waters during the early-Palaeozoic were able to extend from equatorial regions through to the North Polar Region allowing newly evolved species to readily colonize and populate throughout much of the interconnected ancient Tethys, Iapetus, and Panthalassa Seas.

These warm seas enabled stable carbonate shelf environments to develop adjacent to the ancient lands, in conjunction with deeper anoxic basins further away from the coastlines. The extensive algal mats and early reefal mounds present throughout much of the Precambrian eventually gave way to predation and bioturbation by the trilobites. The trilobites were then further preyed upon by more advanced shelly marine species, such as the cephalopods and brachiopods, and the planktonic graptolites were able to readily spread throughout the deeper seas.

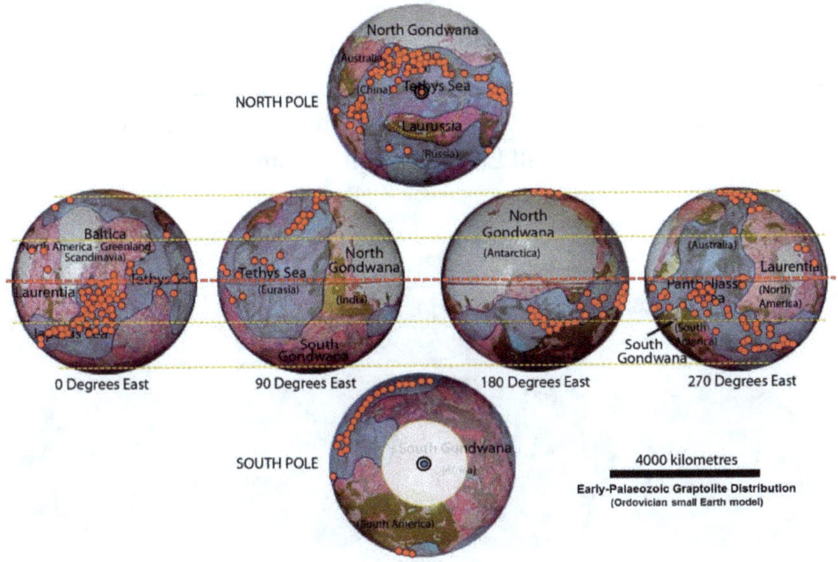

Figure 18.6. Distribution of early-Palaeozoic graptolites plotted on an Ordovician plate tectonic small Earth model. Graptolite data are shown as red dots in relation to climate zones, an early-Palaeozoic South Polar icesheet, shaded white, and distribution of ancient continental seas. Palaeontological data after PaleoBioDB, 2015.

18.4. Mid-Palaeozoic Life

The middle-Palaeozoic comprises the Silurian and Devonian Periods. On an increasing radius plate tectonic Earth, the mid-Palaeozoic was a time of increasing changes to Earth surface area and curvature, along with significant changes and disruptions to the distribution and extent of continental seas. These changes were manifested in a number of prominent extinction events occurring at the end-Ordovician and end-Devonian Periods. All of these changes gave rise to extensive disruptions to pre-existing marine faunal habitats, forcing existing species to adapt, migrate, or perish.

After the end-Ordovician extinction event broad, relatively shallow Silurian and Devonian continental seas were slowly repopulated and once again teamed with life. Within the tropical zones, a diverse community of organisms built large, prominent reefs. More advanced predators emerged to include the early vertebrates in the form of jawed fishes.

The Devonian is also noted for the progressive colonization of lands by new forms of plants and insects. Towards the end of the Devonian Period the first vertebrate animals then crawled onto the land before a wave of mass extinction again swept away large numbers of aquatic species.

The anthozoa are a group of animals that include the sea anemones and corals. The distribution of mid-Palaeozoic anthozoan corals is shown on the Devonian plate tectonic small Earth model in Figure 18.7.

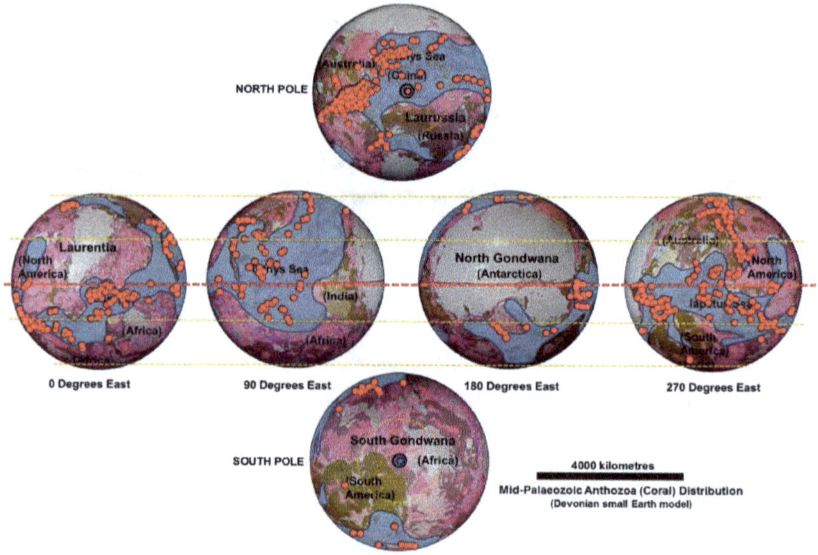

Figure 18.7. Distribution of mid-Palaeozoic anthozoan corals plotted on a Devonian plate tectonic small Earth model. Anthozoa data are shown as red dots in relation to climate zones, mid-Palaeozoic Polar Regions, and distribution of ancient continental seas. Palaeontological data after PaleoBioDB, 2015.

Typically, anthozoans were attached to a substrate creating entirely new and rich coral reef ecosystems. The reefs provided a rich source of food and provided physical protection for the shores. The anthozoa include corals that built great reefs in tropical waters, as well as sea anemones, sea fans, and sea pens.

They also have a long and diverse fossil record, extending back at least 550 million years. The oldest anthozoans are thought to be descendants of polyp-like and sea pen-like fossils known from the late-Precambrian, while the first mineralized coral-like organisms appeared during the Cambrian period.

In this figure the distribution of the Tethys, Iapetus, and Panthalassa Seas mirror the effects of the end-Ordovician extinction event with disruptions to the coastal outlines and surficial extents of the seas. During this time, the circulation of warm sea currents around the ancient Laurentian supercontinent was disrupted but very little disruption to circulation occurred within the North Polar Region. Distribution of the anthozoan corals, and other species, throughout these mid-Palaeozoic seas may then represent a hold-over effect from the configuration of earlier seas.

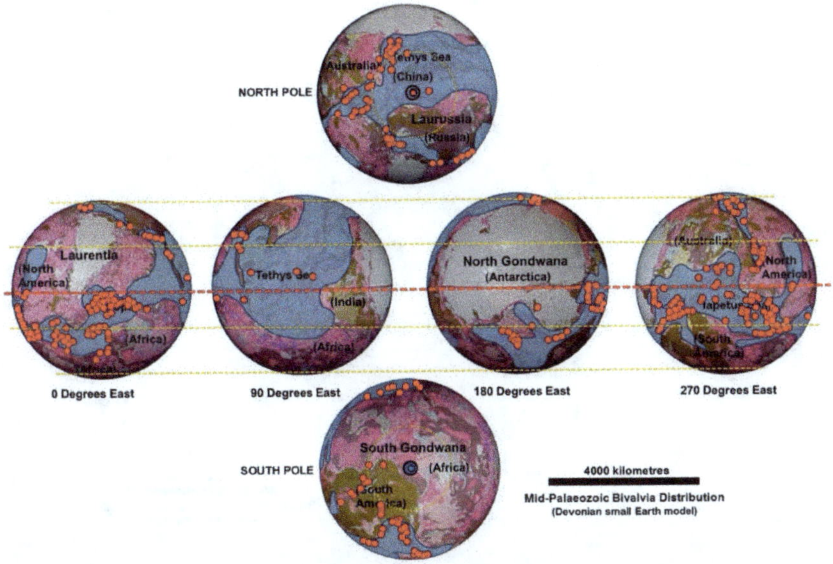

Figure 18.8. Distribution of mid-Palaeozoic bivalves plotted on a Devonian plate tectonic small Earth model. Bivalve data are shown as red dots in relation to climate zones, mid-Palaeozoic Polar Regions, and distribution of ancient continental seas. Palaeontological data after PaleoBioDB, 2015.

The bivalves comprise a class of marine and freshwater molluscs that have laterally compressed bodies enclosed by a shell consisting of two hinged parts. The distribution of bivalves is shown on the Devonian plate tectonic small Earth model in Figure 18.8. This distribution suggests a preference within the ancient Iapetus and parts of the Tethys Seas surrounding Laurentia, with lesser colonization within the more distal seas.

The first occurrences of bivalves are found in early-Cambrian strata, but it was not until the early-Ordovician that bivalve diversification, both taxonomic and ecological, exploded in the fossil record. During the early-Ordovician, there was a great increase in the diversity of bivalve species and by the early-Silurian gills were becoming adapted for filter feeding. During the Devonian and Carboniferous Periods siphons first appeared which, with a newly developed muscular foot, allowed the animals to bury themselves deep in the seafloor sediment.

The gastropods are shelly animals belonging to the phylum Mollusca. The first gastropods were exclusively marine, with the earliest representatives of the group appearing in the late-Cambrian.

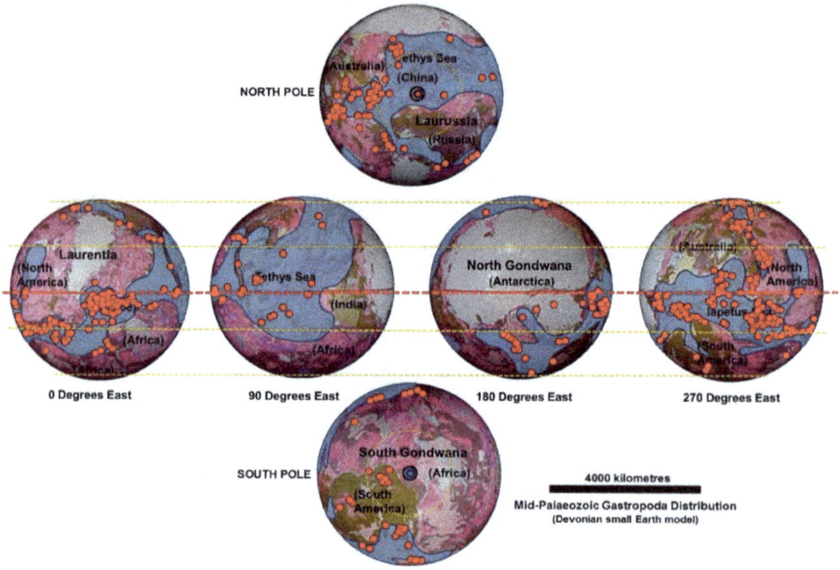

Figure 18.9. Distribution of mid-Palaeozoic gastropods plotted on a Devonian plate tectonic small Earth model. Gastropoda data are shown as red dots in relation to climate zones, mid-Palaeozoic Polar Regions, and distribution of ancient continental seas. Palaeontological data after PaleoBioDB, 2015.

The distribution of gastropods is shown on the Devonian plate tectonic small Earth model in Figure 18.9. Like the bivalves, the figure again highlights a preference of distribution within the ancient Iapetus and parts of the Tethys Seas surrounding Laurentia, with lesser colonization within more distal seas.

By the Ordovician period, gastropods were a varied group present in a range of aquatic habitats. Most of the gastropods of the Palaeozoic Era belong to primitive groups, a few of which still survive. By the Carboniferous period many of the shapes seen in living gastropods can be matched in the fossil record. It was mainly during the latter Mesozoic Era that the ancestors of many of the living gastropods then evolved.

Vertebrates include the jawless fishes and the jawed vertebrates, which include the cartilaginous fish (sharks and rays) and the bony fishes. The defining characteristic of a vertebrate is the presence of a vertebral column, in which the notochord has been replaced by a segmented series of stiffer vertebrae separated by mobile joints.

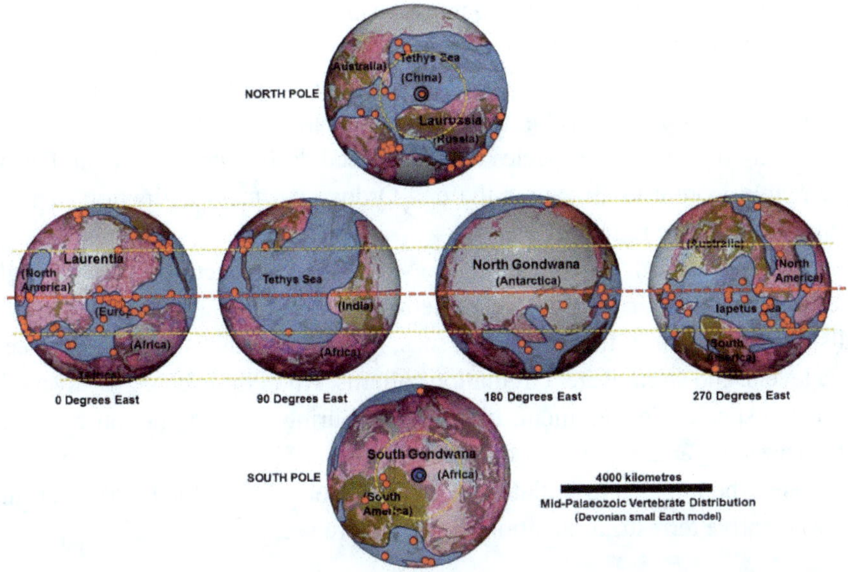

Figure 18.10. Distribution of mid-Palaeozoic vertebrates plotted on a Devonian plate tectonic small Earth model. Vertebrate data are shown as red dots in relation to climate zones, mid-Palaeozoic Polar Regions, and distribution of ancient seas. Palaeontological data after PaleoBioDB, 2015.

The distribution of mid-Palaeozoic vertebrates is shown on the Devonian plate tectonic small Earth model in Figure 18.10. The distribution of the vertebrates shown on these models suggests a preference for equatorial and temperate regions, with lesser colonization within the North or South Polar Regions.

These vertebrates originated during the Cambrian Period and all of the early vertebrates lacked jaws and relied on filter feeding close to the seabed. The first jawed vertebrates appeared in the late-Ordovician and became common during the Devonian Period. The Devonian then saw the demise of virtually all jawless fishes, except for lampreys and hagfish, as well as a group of armored fish that dominated much of the late-Silurian. The Devonian also saw the rise of the first labyrinthodonts, which were a transitional form between fishes and amphibians.

This brief introduction to the distribution of mid-Palaeozoic life forms on an increasing radius plate tectonic Earth shows that the ongoing network of relatively shallow continental seas provided an ideal environment for early reefs and marine creatures to continue to thrive and proliferate. Warm sea waters during the mid-Palaeozoic continued to extend from equatorial regions through to the North Polar Region allowing evolved species to readily colonize and populate throughout much of the ancient seas.

Many of the marine species that flourished during the Ordovician Period were able to re-establish after the end-Ordovician mass extinction to form renewed adaptive radiation.

Changing Earth surface curvature was, however, becoming increasingly active, in particular later in the Palaeozoic. These changes fragmented and disrupted the distribution of ancient seas, including changes to surficial areas, sea levels, and warm water circulation patterns. All of these changes adversely affected species development, in particular during mass extinction events at the beginning and end of this time interval. Changes to the environment also provided the impetus to initiate and drive colonization of the lands, sparking new adaptive terrestrial radiations never before seen.

18.5. Late-Palaeozoic Life

The late-Palaeozoic interval of time includes the Carboniferous and Permian Periods. During that time, new groups of animals and plants exerted major influences over the type and accumulation of sediments up to the end-Permian mass extinction event. During Carboniferous, skeletal debris from marine organisms accumulated to form widespread limestone reefs, and spore-bearing trees stood in broad swamps contributing wood and foliage to the formation of vast coal beds. The late-Palaeozoic was also marked by major changes to the ancient climate, leading to a glacial event spreading over the South Polar Region during the Carboniferous and then receding again during the Permian.

On an increasing radius plate tectonic Earth, the late-Palaeozoic was a time when the ability of continental crusts to continue extending within established sedimentary basins was reaching its limit, eventually leading to onset of crustal rupture and breakup during the late-Permian. Breakup of the continental crusts during the late-Permian initiated opening and formation of the modern continents as well as rifting to form modern, relatively deep, oceans. All of which gave rise to the major end-Permian extinction event and the demise of many plant and animal species on both the lands and in the seas. Draining of the ancient continental seas into the newly formed oceans then led to contraction of established coal swamps, plus exposure and drying of the lands along with the accumulation of evaporites.

Marine life during the late-Palaeozoic was similar to that of the late-Devonian times, with new adaptive radiations occurring after the end-Devonian extinction event. The distribution of ammonites, for example, is shown on the Permian plate tectonic small Earth model in Figure 18.11.

The ammonites were predatory, squid-like creatures that lived in coil-shaped shells, with some growing to more than a metre across. Like other cephalopods, ammonites had sharp, beak-like jaws inside a ring of tentacles that extended from their shells to snare prey such as small fish and crustaceans. Ammonites first descended from straight-shelled cephalopods dating back to the Devonian. They were prolific breeders, lived in schools, and are among the most abundant fossils found today. The ammonites became extinct 65 million years ago during the end-Cretaceous extinction event.

In Figure 18.11, the aerial distribution and surficial areas of the ancient seas had greatly diminished after the end-Devonian extinction event. This diminished surface area reflects the advanced stages of geosynclinal activity and orogenesis prevalent during the mid- to late-Palaeozoic times. The disparity of some plotted ammonite data in this figure also reflects the relatively rapidly changing coastal outlines and distribution of seas during these times.

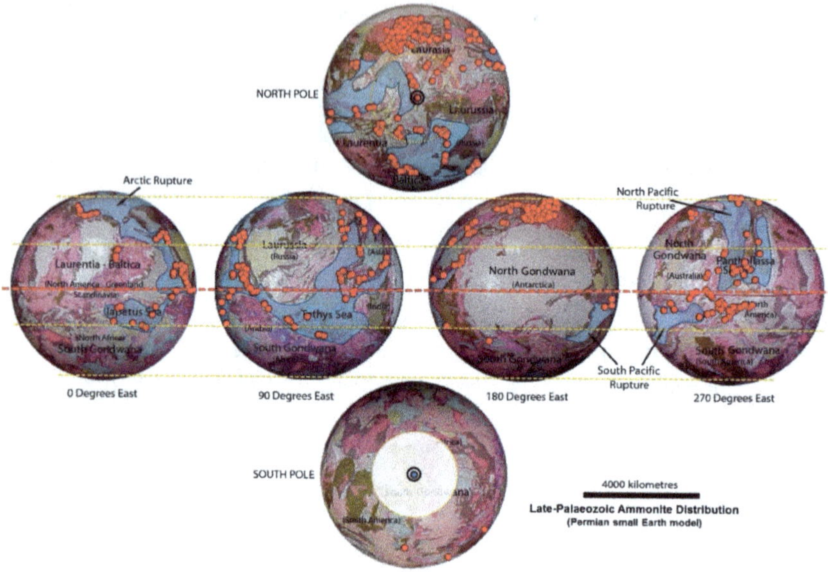

Figure 18.11. Distribution of late-Palaeozoic ammonites plotted on a Permian plate tectonic small Earth model. Ammonite data are shown as red dots in relation to climate zones, a late-Palaeozoic South Polar icesheet, shaded white, and distribution of ancient continental seas. Palaeontological data after PaleoBioDB, 2015.

Crinoids are marine animals that once lived both in shallow water and at depths as great as 6,000 metres. The Crinoids generally had a stem used to attach themselves to a substrate, but many lived attached only as juveniles and became free-swimming as adults.

The distribution of crinoids is shown on the Permian plate tectonic small Earth model in Figure 18.12. The disparity in distribution again reflects the relatively rapidly changing coastlines and distribution of seas during these times. The distribution of crinoids also highlights the preference for coastal habitats associated with fringing reefs within the equatorial and warmer temperate regions, with lesser distribution within high northern latitudes.

The crinoids underwent two periods of abrupt adaptive radiation. The first occurred during the Ordovician, the other after the end-Permian extinction event. Some thick limestone beds dating to the mid- to late-Palaeozoic are almost entirely made up of disarticulated crinoid fragments.

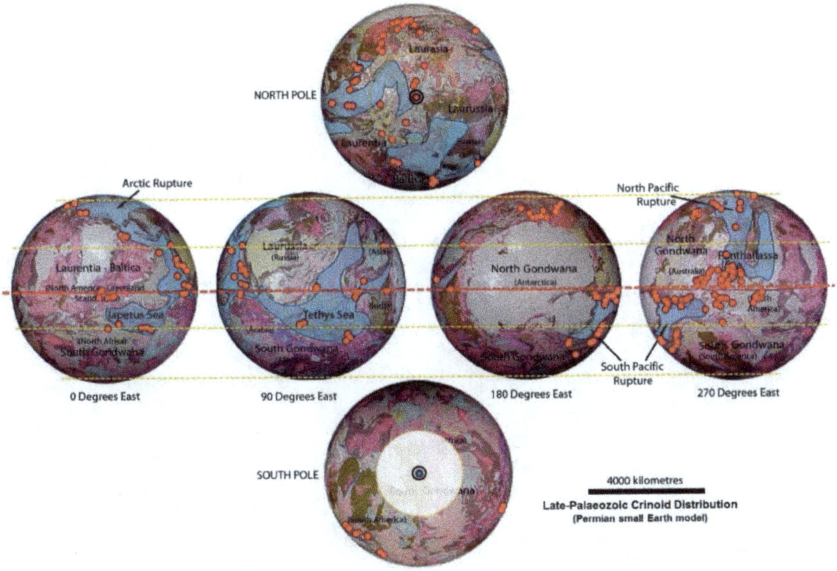

Figure 18.12. Distribution of late-Palaeozoic crinoids plotted on a Permian plate tectonic small Earth model. Crinoid data are shown as red dots in relation to climate zones, a late-Palaeozoic South Polar icesheet, shaded white, and distribution of ancient continental seas. Palaeontological data after PaleoBioDB, 2015.

The late-Palaeozoic was also characterized by profound changes to life on the land, including evolution of the insects and colonization of broad lowland swamps by spore-bearing plants. Evidence by others suggests that an algal scum first colonized the land some 1,200 million years ago, but it was not until the Ordovician Period, around 450 million years ago, that land plants began to appear. These plants diversified during the late-Silurian Period, around 420 million years ago, and by the middle of the Devonian Period most of the features recognized in plants today were present, including roots, leaves and secondary wood. By late-Devonian times seeds had evolved and plants had reached a degree of sophistication that allowed them to form forests of tall trees (Figure 18.13).

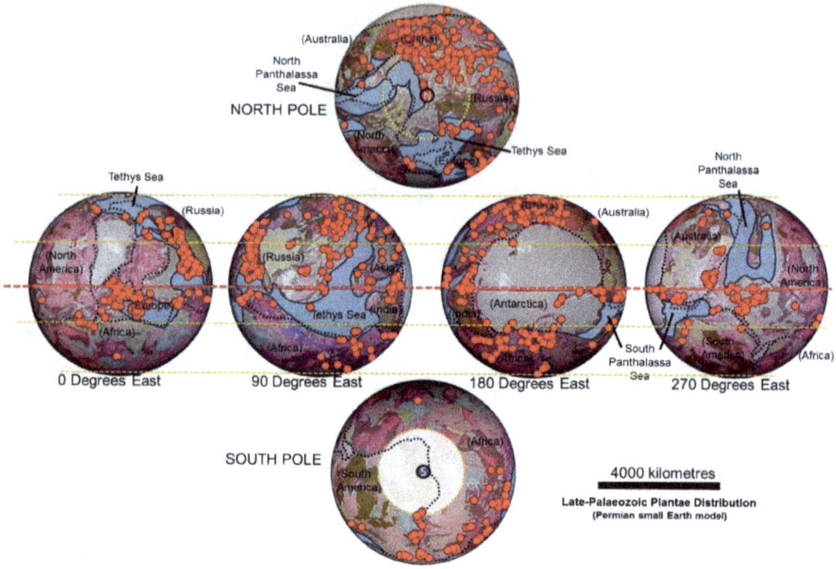

Figure 18.13. Distribution of late-Palaeozoic plant species plotted as red dots on the Permian plate tectonic small Earth model. Plant data are shown as red dots in relation to climate zones, a late-Palaeozoic South Polar icesheet, shaded white, and distribution of ancient continental seas. Palaeontological data after PaleoBioDB, 2015.

The distribution of plant fossils is shown on the Permian plate tectonic small Earth model in Figure 18.13. During the late-Palaeozoic, a southern ice sheet was present over Southern Gondwana and the presence of plants in the high northern latitudes suggests that temperatures were again moderated in these regions.

Evolutionary innovation continued after the Devonian period. Most plant groups were relatively unscathed by the end-Permian extinction event, although the structures of many communities changed. This change may have then set the scene for the evolution of flowering plants during the Triassic Period, commencing 250 million years ago, which exploded during the Cretaceous and Paleogene.

The latest major group of plants to evolve was the grasses which became important during the mid-Paleogene, from around 40 million years ago. The grasses, as well as many other groups, evolved new mechanisms of metabolism to survive the low carbon dioxide levels and warm, dry conditions of the tropics existing over the past 10 million years.

Reptiles belong to the class Reptilia and comprise today's turtles, crocodiles, snakes, lizards and tuatara, along with some of the extinct ancestors to mammals. The earliest known reptiles originated around 315 million years ago during the Carboniferous Period, having evolved from advanced reptile-like amphibians that had become increasingly adapted to life on dry land.

The distribution of reptiles is shown on the Permian plate tectonic small Earth model in Figure 18.14.

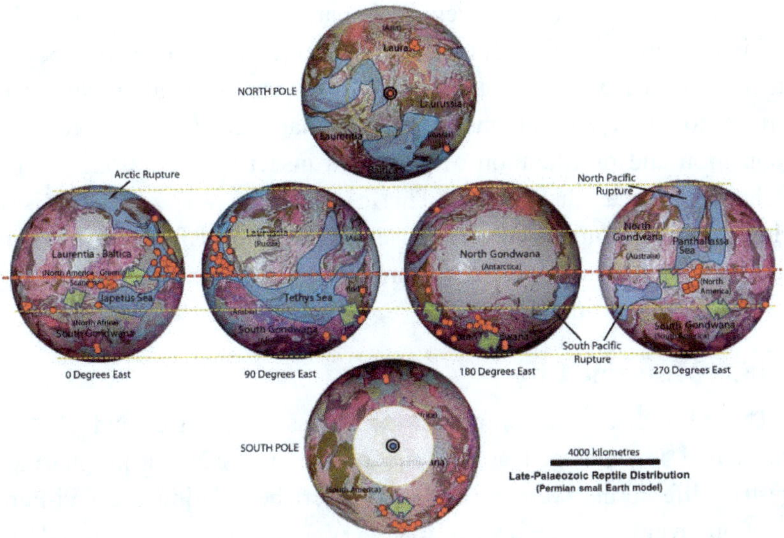

Figure 18.14. Distribution of late-Palaeozoic reptiles plotted on a Permian plate tectonic small Earth model. Reptile data are shown as red dots in relation to climate zones, a late-Palaeozoic South Polar icesheet, shaded white, and distribution of ancient continental seas. Suggested migration routes are highlighted by green arrows. Palaeontological data after PaleoBioDB, 2015.

Also highlighted on this figure are suggested migration links between the various ancient continents. These links, highlighted by green arrows, represent possible migration routes for both the reptiles and succeeding dinosaur life forms prior to breakup of the Pangaean supercontinent during the late-Permian. Some of these links remained within the early-Triassic allowing the early dinosaurs to migrate and populate many of the modern continents prior to isolation during the late-Triassic. Important provincial centres for the reptiles included western North America, western and central Europe, and South Africa.

The distribution of late-Palaeozoic life forms on an increasing radius plate tectonic Earth, briefly outlined here, continues to highlight the importance of an ongoing network of relatively shallow continental Tethys, Iapetus, and Panthalassa Seas in providing an ideal environment for reefs and proliferation of marine creatures. During that time increasing geosynclinal activity and orogenesis severely disrupted the distribution of continental seas, resulting in isolated, often relatively deep seaways. Warm sea waters during the late-Palaeozoic were mainly confined to the equatorial and northern hemisphere regions. A southern polar icecap was well established, with possible seasonal icing within the northern polar region, in particular in mountainous regions.

Disruptions to the continental seas, along with elevated topography, further increased erosion of the lands with extensive coal-bearing swamps confined to low-lying regions. The late-Palaeozoic is well noted for the colonization and proliferation of plant and insect species, along with early vertebrates such as the reptiles. This colonization was severely disrupted during the late-Permian with the onset of breakup of Pangaea to form the modern continents and opening to form the modern oceans, coincident with the end-Permian extinction event.

18.6. Mesozoic Life

The Mesozoic Era comprises the Triassic, Jurassic, and Cretaceous Periods and is renowned as the age of the dinosaurs. Prior marine and terrestrial life forms were severely impoverished during the end-Permian extinction event. In the oceans, recovery included modern reef-building corals, the ascendancy of molluscs, swimming reptiles, and new kinds of fishes. Non-avian dinosaurs appeared in the late-Triassic and became the dominant terrestrial vertebrates early in the Jurassic, occupying this position for about 135 million years.

Mammals remained small and unobtrusive. Flying reptiles and primitive birds first appeared, and near the end of the era flowering plants replaced conifers as the dominant forms of terrestrial vegetation. The rapid rise of flowering plants is thought to have been assisted by co-evolution of pollinating insects. The climate of the Mesozoic is said to have varied, alternating between warming and cooling periods.

On an increasing radius plate tectonic Earth, this era coincides with breakup of the ancient Pangaean supercontinent to form the modern continents and opening of the modern oceans. This era also coincides with apparent migration of the South Pole across the opening South Atlantic Ocean, migrating from South Africa as it moved north, across to Antarctica as it moved into the South Polar Region. During that time, both the north and south poles were centred over oceans, and there were no polar ice caps. The era was also punctuated by the end-Triassic extinction event during merging of the North and South Pacific Oceans and the era ended with the end-Cretaceous extinction event.

Current knowledge suggests that dinosaur evolution followed changes in vegetation, along with changes in the physical locations of the continents. Gymnosperms, a group of seed-producing plants represented mainly by the conifers, became well-established during the late-Triassic. The dominance of gymnosperms continued into middle- and late-Jurassic times and these plants provided an important food source for the dinosaurs. A major change in the early-Cretaceous was also marked by the evolution of flowering plants, along with the first grasses, which continued into the late-Cretaceous. This change in vegetation enabled several groups of herbivorous dinosaurs to adapt and evolve more sophisticated ways to process food.

The following figures will focus on the various dinosaur-related lineages to highlight their global distributions in context with opening of the modern oceans, and dispersal of the modern continents during the Mesozoic Era.

During the slow recovery from the end-Permian extinction event, a previously obscure group of animals called the archosaurs—a group that includes the extinct dinosaurs and whose living representatives consist of birds and crocodiles—became the most abundant and diverse terrestrial vertebrates during the Triassic. The distribution of archosaurs is shown on the early-Triassic plate tectonic small Earth model in Figure 18.15.

Triassic witnessed the appearance of several new groups of archosaurs, some of which have living descendants today. From their provincial ancestral reptile distributions (Figure 18.14) the archosaurs dispersed widely during the Triassic Period, extending from south polar to high northern latitudes. After the end-Triassic extinction event, most archosaurs did not survive into the Jurassic, except for the crocodilian lineage, as well as the dinosaurs and the champsosaurs, which all appeared at about the same time during the late-Triassic.

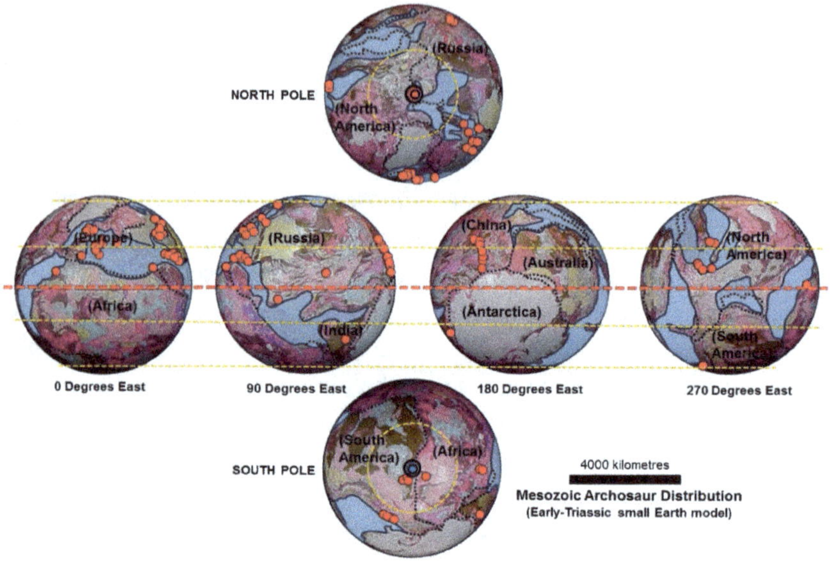

Figure 18.15. Distribution of Mesozoic archosaurs plotted on an early-Triassic plate tectonic small Earth model. Archosaur data are shown as red dots in relation to climate zones and distribution of ancient continental seas and modern oceans. Palaeontological data after PaleoBioDB, 2015.

The dinosaurs were a diverse group of animals that first appeared during the Triassic period (Figure 18.16). They were the dominant terrestrial vertebrates until the end of the Cretaceous. The dinosaurs are divided into two orders, the Ornithischia and Saurischia. This division is based on the evolution of the pelvis into more bird-hip and lizard-hip like structures, as well as details in the vertebrae, the development of armor, and the possession of a predentary bone in the front of the lower jaw used to clip off plant material.

The early distribution of dinosaurs is shown on the late-Triassic plate tectonic small Earth model in Figure 18.16. Important provincial centres coincide with their ancestral reptile centres and include locations in Europe—located in the northern Temperate Zone, North America—located in the Equatorial Zone, and South Africa—located in the South Polar Region.

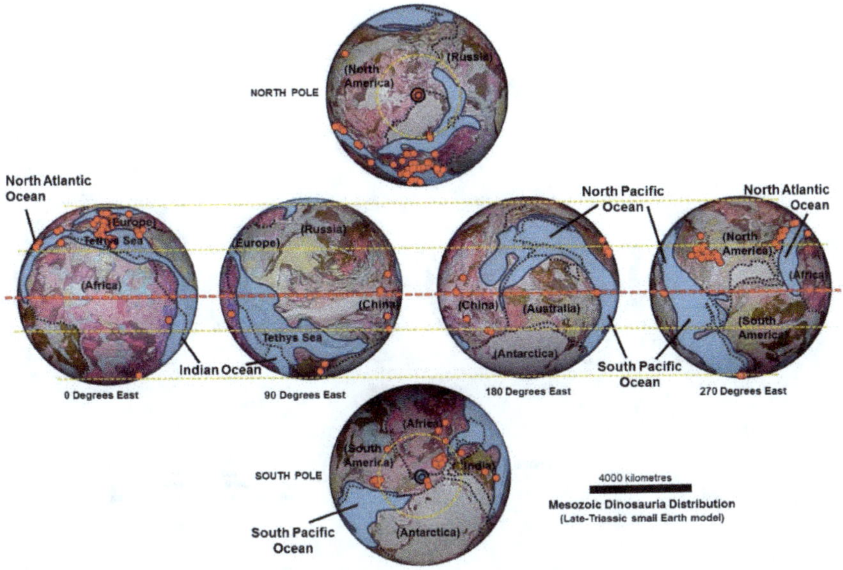

Figure 18.16. Distribution of Mesozoic dinosauria plotted on a late-Triassic plate tectonic small Earth model. Dinosauria data are shown as red dots in relation to climate zones and distribution of ancient continental seas and modern oceans. Palaeontological data after PaleoBioDB, 2015.

The first dinosaurs appearing in the late-Triassic period were not major components of the fauna. However, the early-Jurassic dinosaurs were diversifying rapidly to become dominant occupants of many major terrestrial adaptive environments. It was also during the late-Jurassic that the bird lineage diverged from its flightless theropod ancestors. These first birds enjoyed an explosion in diversity during the Cretaceous period and beyond to the present-day. The end-Cretaceous extinction event led to the extinction of most dinosaur groups at the close of the Mesozoic Era.

The Ornithischia are a group of medium-to-large, herbivorous dinosaurs, shown on the late-Jurassic plate tectonic small Earth model in Figure 18.17. The ornithischia includes one of the earliest discovered dinosaurs, iguanodon, as well as the crested hadrosaurs. The ornithischia had a pelvis that superficially resembled a bird's pelvis in which the pubis pointed backwards. Being herbivores, they were numerous and sometimes lived in herds, while many were also prey animals.

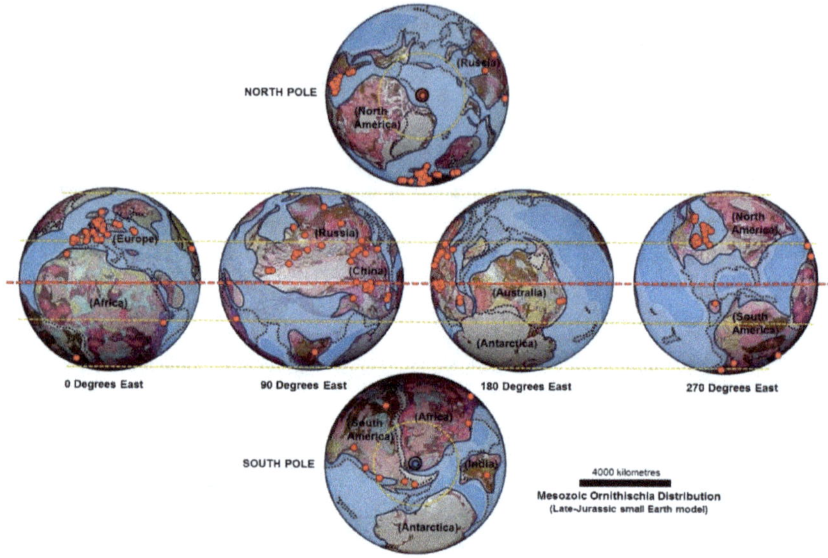

Figure 18.17. Distribution of Mesozoic ornithischia plotted on a late-Jurassic plate tectonic small Earth model. Ornithischia data are shown as red dots in relation to climate zones and distribution of ancient continental seas and modern oceans. Palaeontological data after PaleoBioDB, 2015.

In this figure, as with the dinosauria in Figure 18.16, there is still an element of provincial distribution centred on locations in North America and Europe, plus a new centre in China. By the late-Jurassic, indications are that the South African ornithischia had either died out or had migrated to adjoining southern South America.

During that time the South Pole had migrated into South Africa and breaching between Africa and South America and opening of the Atlantic and Indian Oceans had commenced. This opening effectively isolated South Africa from adjoining southern South America, India, and Antarctica. Elsewhere, other dinosaur species were becoming increasingly isolated in Australia, India, and the Asian and Russian continents.

The saurischian dinosaurs are traditionally distinguished from ornithischian dinosaurs by their three-pronged pelvic structure, with the pubis pointing forward. The oldest known dinosaurs, from the middle-Triassic of South America, were saurischians. The saurischians form two major sub-groups: the Sauropoda and the Theropoda. The sauropoda were large herbivores, such as apatosaurus and diplodocus. The theropoda were bipedal carnivores. Living birds have common ancestors to the theropod lineage.

The distribution of saurischian dinosaurs is shown on the early-Cretaceous plate tectonic small Earth model in Figure 18.18. By the early-Cretaceous these dinosaurs were rapidly diversifying and populating most continents on Earth. From small provincial centres the saurischians were also evolving independently as continental breakup and changing sea levels formed a number of isolated island continents.

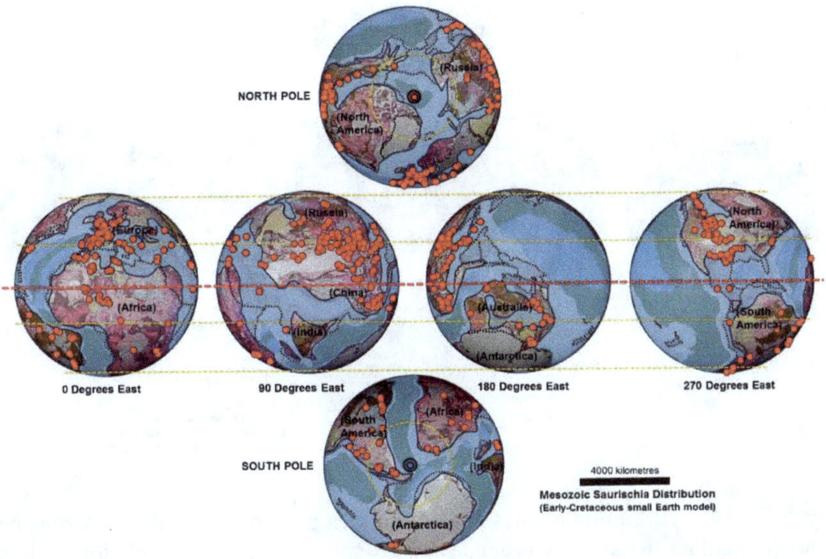

Figure 18.18. Distribution of Mesozoic saurischia plotted on an early-Cretaceous plate tectonic small Earth model. Saurischia data are shown as red dots in relation to climate zones and distribution of ancient continental seas and modern oceans. Palaeontological data after PaleoBioDB, 2015.

True sauropods, such as diplodocus, first appeared in the late-Triassic and began to diversify during the mid-Jurassic, about 180 million years ago. They had very long necks and tails, relatively small skulls and brains, and erect limbs. The nostrils of these animals were located high up on the skulls, and another unusual feature which appeared in some of the later sauropods was rudimentary body armor. In addition to their wide geographic distribution, sauropods are one of the most long-lived groups of dinosaurs, spanning some 100 million years, extending from the early-Jurassic to the late-Cretaceous.

Sauropods were a group of four-legged, herbivorous animals with a relatively simple body plan that varied only slightly throughout the group. Early relatives of the sauropods, the late-Triassic plateosaurs or prosauropods, may have occasionally stood on their hind legs.

The distribution of sauropod dinosaurs is shown on the mid-Cretaceous plate tectonic small Earth model in Figure 18.19. From the mid-Cretaceous through to the late-Cretaceous their distributions remained much the same and fossil evidence shows they were more prolific during late-Cretaceous times.

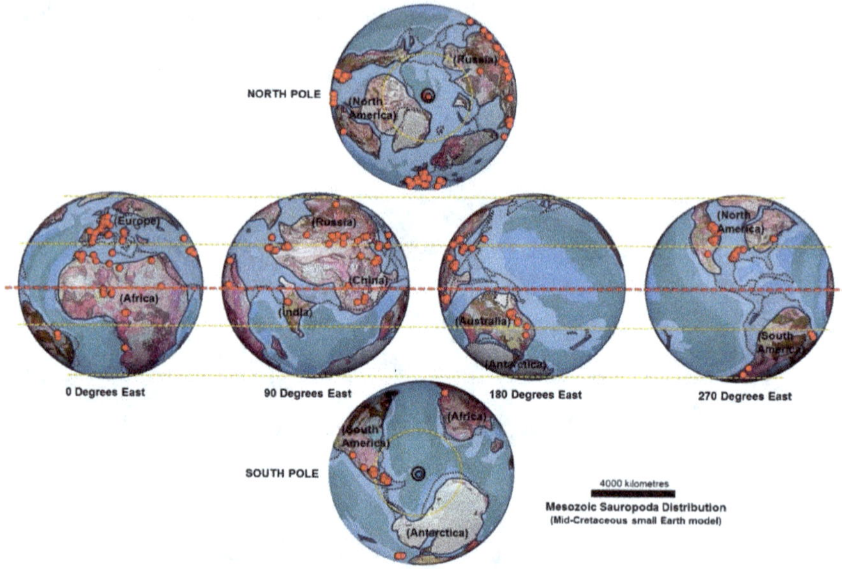

Figure 18.19. Distribution of Mesozoic sauropoda plotted on a mid-Cretaceous plate tectonic small Earth model. Sauropoda data are shown as red dots in relation to climate zones and distribution of ancient continental seas and modern oceans. Palaeontological data after PaleoBioDB, 2015.

The theropod dinosaurs are a diverse sub-group of the bipedal saurischian dinosaurs. They include the largest terrestrial carnivores ever to have existed on Earth, ranging in size from the crow-sized microraptor to the tyrannosaurus rex.

The distribution of theropod dinosaurs is shown on the late-Cretaceous plate tectonic small Earth model in Figure 18.20. In this figure, by the late-Cretaceous the theropods had become widely dispersed, populating all continents and extending from high northern to high southern latitudes. This distribution also included theropods in remote islands such as New Zealand and Madagascar. Theropods in these remote islands may represent a holdover of species from early beginnings during the late-Triassic times when New Zealand, for example, was connected to Central America and Madagascar was connected to India and South Africa.

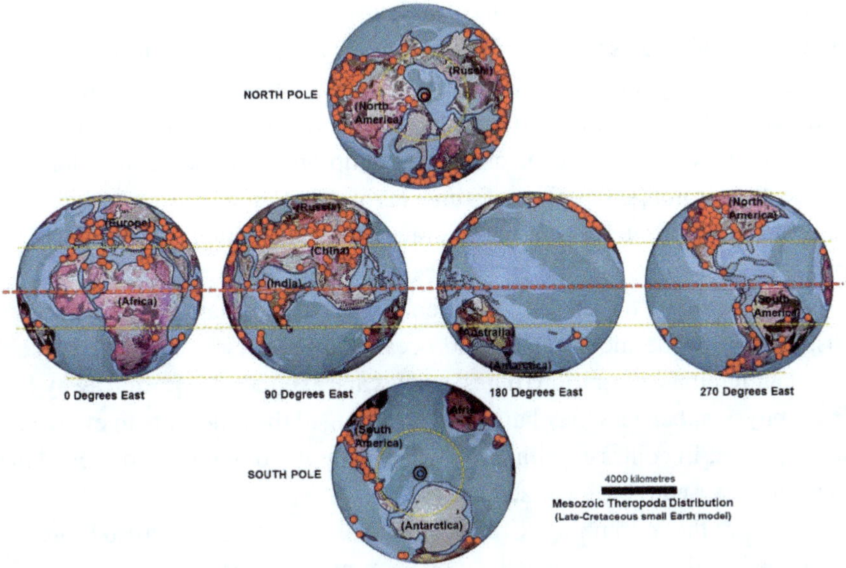

Figure 18.20. Distribution of Mesozoic theropoda plotted on a late-Cretaceous plate tectonic small Earth model. Theropoda data are shown as red dots in relation to climate zones and distribution of ancient continental seas and modern oceans. Palaeontological data after PaleoBioDB, 2015.

Theropods first appeared during the late-Triassic period, some 230 million years ago, and included the sole large terrestrial carnivores from the early-Jurassic until at least the close of the Cretaceous. During the Jurassic, birds evolved from small specialized theropods. Among the features linking theropod dinosaurs to birds are the three-toed foot, a wishbone, air-filled bones, brooding of the eggs, and often feathers.

Unlike the sauropod saurischians, all of the theropods were obligate bipeds. Their hind legs provided support and locomotion while the short forelimbs and mobile hands were adapted for grasping and tearing prey. Theropods were carnivorous, although they exhibited a wide range of diets, from insectivores to herbivores and carnivores.

The distribution and diversification of dinosaurs shown on plate tectonic small Earth models highlights the evolution of all Mesozoic species during breakup of the ancient Pangaean supercontinent to form the modern continents and opening to form the modern oceans.

In addition to the breakup of Pangaea, the dinosaurs were subjected to a number of multiple extinction events with catastrophic outcomes to most species. The post-Permian extinction resurgence of life-forms then enabled previously subdued species to rapidly diversify and populate the entire Earth.

On an increasing radius plate tectonic Earth, the distribution, migration, and eventual demise of the dinosaurs can be visualized in conjunction with this very involved period of crustal breakup and new ocean development. While the prolonged period of time involved during this crustal breakup allowed the dinosaur species to migrate to more equitable locations, it also presented new physical and environmental challenges.

The draining of ancient continental seas, for instance, influenced the distribution of the modern seas and oceans and adversely affected available dinosaur habitats, migration routes, and escape avenues. Migration away from these physical barriers may have then encouraged the dinosaurs to evolve into new species. In contrast, failing to evolve, migrate, or being land-locked may have also caused their localized demise.

During the remaining Cretaceous times, continental crustal breakup continued unabated as each of the modern continents became well established. By then, all of the intervening modern oceans, except for the Southern Ocean, had opened. Breaching of the North American and Australian land connection had already occurred, giving rise to merging of the North and South Pacific Oceans, which resulted in the end-Triassic extinction event.

During this end-Triassic extinction event, most of the non-dinosaurian archosaurs, as well as most of the terrestrial vertebrates, the ancestors of mammals, and most of the large amphibians became extinct. This extinction enabled the surviving dinosaurs to then migrate and extend their coverage well beyond their original ancestral provinces. Because this end-Triassic extinction event mainly affected coastal environments, those dinosaurs, and other species adapted to more inland habitats, may have been well suited to survival.

This unique provincial distribution of the earlier Permian ancestral reptiles and Mesozoic dinosaurs on an increasing radius plate tectonic Earth demonstrates consistent evolutionary links between Permian, Triassic, Jurassic, and Cretaceous species. Many of these links were then permanently disrupted during the Triassic and Cretaceous Periods as a result of ongoing crustal breakup and continental dispersal, draining of the ancient seas, variable sea-levels, and disruption to existing climate zones.

18.7. Cenozoic Life

The Cenozoic Era is known as the Age of Mammals. It was a time when the end-Cretaceous extinction of many groups had allowed mammals to greatly diversify. The Cenozoic is divided into three periods: The Paleogene, Neogene, and Quaternary; and seven epochs: the Paleocene, Eocene, Oligocene, Miocene, Pliocene, Pleistocene, and Holocene. The era commenced with the end-Cretaceous extinction event and the Holocene Epoch is current to the present-day.

Early in the Cenozoic, the Earth was dominated by relatively small fauna, including small mammals, birds, reptiles, and amphibians. It did not take long for mammals and birds to diversify in the absence of large carnivorous reptiles that had dominated during the Mesozoic.

Mammals came to occupy almost every available niche, both marine and terrestrial, and some were able to grow very large. The Cenozoic is also renowned for the dominance of savanna-type environments, co-dependent flowering plants and insects, and the birds. Grass played a very important role in this era, shaping the evolution of birds and mammals that fed on it.

Climate-wise, the Earth had begun a drying and cooling trend, culminating in the glaciations of the Pleistocene Epoch. The configuration of continents was also looking familiar by this time, and all continents were progressively moving into their current positions.

Mammals are a life-form group capable of maintaining their bodies at a metabolically favourable temperature. They are four-footed animals distinguished from reptiles and birds by the possession of hair, three middle ear bones, mammary glands, and a neocortex region of the brain. The mammalian brain regulates body temperature and the circulatory system, including a four-chambered heart.

The mammals include the largest animals existing today, as well as some of the most intelligent, such as elephants, primates, whales, dolphins, and primates. The basic body type is a four-legged land-borne animal, but other mammals were adapted for life at sea, in the air, in trees, and on two legs.

The early mammalian ancestors were sphenacodont pelycosaurs, a dinosaurian group that also included the dimetrodon. During the Permian, this group diverged from the dinosaur line that led to today's reptiles and birds. The first mammals then appeared during the early-Mesozoic era. The modern mammalian orders arose during the early- to mid-Cenozoic Era, after extinction of the non-avian dinosaurs during the end-Cretaceous extinction event.

The distribution of mammals is shown on the Eocene plate tectonic small Earth model in Figure 18.21.

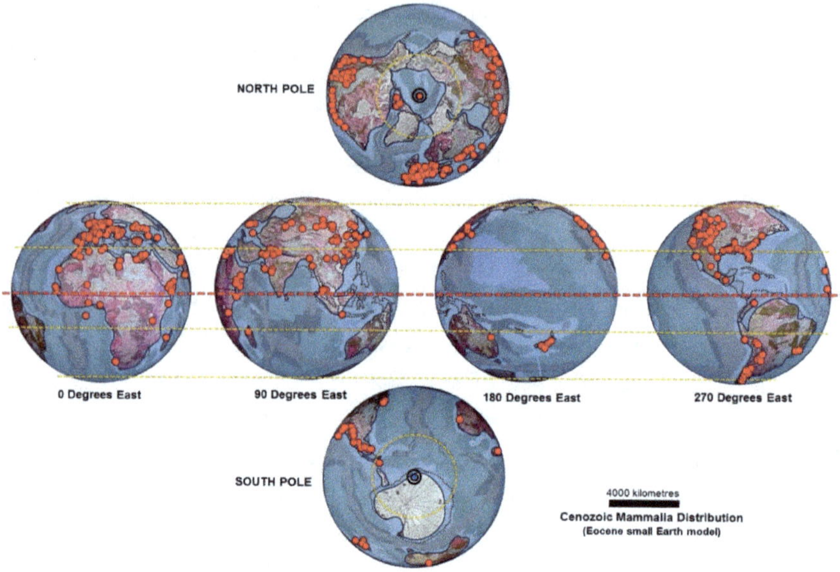

Figure 18.21. Distribution of Cenozoic mammalia plotted on an Eocene plate tectonic small Earth model. Mammalia data are shown as red dots in relation to climate zones and distribution of modern oceans. Palaeontological data after PaleoBioDB, 2015.

This figure shows a preferred distribution of mammals within equatorial and temperate regions, with minor occurrences within the north and south Polar Regions. This distribution may have been attributed primarily to their metabolic requirements and the presence of favourable vegetation.

On an increasing radius plate tectonic Earth, the diversity of modern mammalian life-forms is also attributable to the relative isolation of modern continents, as compared to earlier eras where land connections existed between most continents. Ancestral land connections favoured interactive migration between continents, while isolation of the continents favoured adaptive radiation and diversification of the mammalian species to the present-day.

An example of adaptive radiation in isolation is the marsupials, Figure 22.

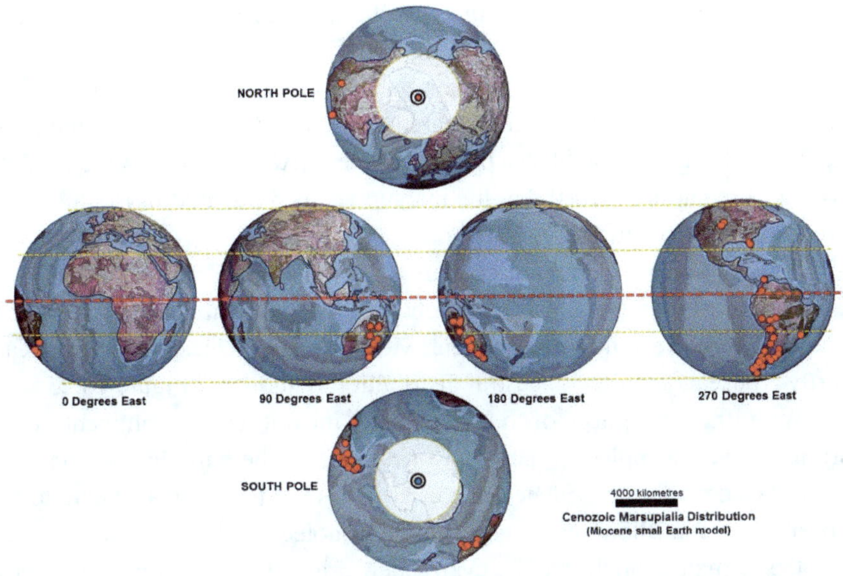

Figure 18.22. Distribution of Cenozoic marsupialia plotted on a Miocene plate tectonic small Earth model. Marsupialia data are shown as red dots in relation to climate zones and distribution of modern oceans, as well as North and South Polar ice caps, shaded white. Palaeontological data after PaleoBioDB, 2015.

Marsupials are an infraclass of mammals now living primarily in Australia and the Americas. A distinctive characteristic, common to most species, is that the young are carried in a pouch. The ancestors of marsupials are thought to have split from the mammals during the mid-Jurassic period, originating with the last common ancestor of extant metatherians.

The oldest metatherian fossils are found in present-day China. Metatherians are thought to have then spread eastward into modern North America, where the earliest true marsupials are found. In South America, opossums evolved and developed a strong presence, and the Paleogene also saw the evolution of shrew opossums alongside non-marsupial metatherian predators.

The marsupials are thought to have reached Australia via Antarctica at least 50 to 65 million years ago, suggesting a single dispersion event of just one species, most likely a relative of the opossums from South America.

On an increasing radius plate tectonic Earth, a common link between South America, Antarctica and Australia existed during the Cretaceous and Paleocene Epochs within the southern temperate region (Figure 18.22). At that time ancestors of the marsupials existed in southern South America and the first fossils in Australia existed during the Eocene.

During that same time, Australia separated from Antarctica and has since remained as an island continent. Antarctica also migrated into the South Polar Region and began to establish a permanent ice cap during the Oligocene. With no prior predators in Australia, the marsupials then radiated in isolation into the wide variety of life forms seen today.

Primates arose from mammalian ancestors that lived in the trees of tropical forests, and many primate characteristics are thought to represent adaptations to life in this environment. Most primate species remained at least partly arboreal, and their ancestral distribution is shown in Figure 18.23.

With the exception of humans, who inhabit every continent, most primates live in tropical or subtropical regions of the Americas, Africa and Asia. Based on fossil evidence, the earliest known true primates date to the Eocene—55.8 million years ago—although molecular clock studies suggest that the primate branch may be even older, originating near the Cretaceous-Paleocene boundary around 63 to 74 million years ago.

Primates are characterized by large brains relative to other mammals, as well as an increased reliance on stereoscopic vision, at the expense of smell, the dominant sensory system in most mammals. Three-colour vision developed in some primates and most also have opposable thumbs and prehensile tails. These primate features are more developed in monkeys and apes and noticeably less so in lorises and lemurs. Primates also have slower rates of development than other similar sized mammals and reach maturity later but have longer lifespans. Depending on the species, adults may live in solitude, in mated pairs, or in groups of up to hundreds of members.

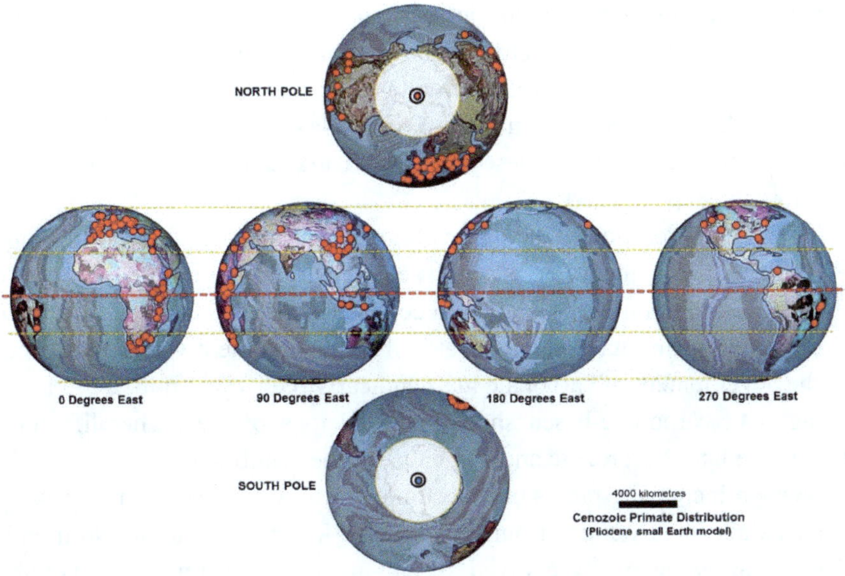

Figure 18.23. Distribution of Cenozoic primates plotted on a Pliocene plate tectonic small Earth model. Primate data are shown as red dots in relation to climate zones and distribution of modern oceans, as well as North and South Polar ice caps, shaded white. Palaeontological data after PaleoBioDB, 2015.

On an increasing radius plate tectonic Earth, ancestors to the primates existed in southern South America, North America, Europe, and China during the Paleocene and Eocene Epochs. From there, colonization rapidly extended into east and south Africa, Europe and China, with a decline in the Americas to the present-day. This distribution is consistent with land-based migration routes existing between each of these continents and excludes colonization of the island continents, such as Australia and New Zealand, prior to modern-day colonization.

18.8. Summary of Life on Earth

On increasing radius plate tectonic small Earth models, the timing and development of ancient continental seas and supercontinents, along with modern continents and oceans, is considered the prime cause of evolution of all life forms on Earth. The ancient continental seas, in particular, provided an ideal setting for the primitive Precambrian microbe's effectiveness as nurseries of evolution and to markedly drive evolutionary changes in all life forms.

Similarly, extinction of species is shown to be a by-product of sea-level changes, disruption to the distribution of established seas, and the inability of species to keep pace with these environmental changes. It is further considered that evolution of species during these times was subsequently driven by the need to survive amongst these ever-increasing changes to the Earth and environment.

With the evolution of oxygenic photosynthesis during the early- to mid-Proterozoic times, the stable platform sedimentary basins and shallow seas of the Precambrian times provided ideal conditions for chemical precipitation of iron, calcium carbonates, and silicates. This environment was also reflected in the development of carbonate mats and early reefs by primitive life forms, along with carbonate-based shells and skeletons of the higher life forms during the late-Proterozoic and beginning of the Cambrian.

On an increasing radius plate tectonic Earth, warm sea waters during the early-Palaeozoic extended from equatorial regions through to the North Polar Region, allowing newly evolved species to readily colonize and populate throughout much of the interconnected ancient Tethys, Iapetus, and Panthalassa seaways. The warm seas enabled stable carbonate shelf environments to develop adjacent to the ancient lands, along with deeper anoxic basins further away from the coastlines. The extensive algal mats and early reefal mounds present throughout much of the Precambrian eventually gave way to predation and bioturbation, and planktonic life forms were able to readily spread throughout the deeper seas.

Mid-Palaeozoic life forms on an increasing radius plate tectonic Earth continued to show that the network of relatively shallow continental seas provided an ideal environment for early reef development, and for marine creatures to continue to thrive and proliferate. Warm sea waters during the mid-Palaeozoic continued to extend from equatorial regions through to the North Polar Region, allowing evolved species to readily colonize and populate throughout much of the ancient seas.

Changing Earth surface curvature was, however, becoming increasingly prevalent, in particular later in the Palaeozoic. These changes fragmented and disrupted the distribution of ancient seas, including changes to surficial areas, sea levels, and warm water circulation patterns. All of these changes adversely affected species development, in particular during devastating mass extinction events. Changes to the environment also provided the impetus to drive colonization of the lands, sparking new adaptive terrestrial radiations never before seen.

During the late-Palaeozoic, increasing geosynclinal activity and orogenesis severely disrupted the distribution of continental seas, forming isolated, often relatively deep seaways. Warm sea waters during these times were mainly confined to the equatorial and northern hemisphere regions, with a well-developed South Polar icecap, and possible seasonal icing within the Northern Polar Region. Disruptions to the continental seas, along with elevated topography, further increased erosion, with extensive coal-bearing swamps confined to low-lying regions.

The late-Palaeozoic is well noted for the colonization and proliferation of plant and insect species, along with early vertebrates, such as the reptiles. This colonization was severely disrupted during the late-Permian with onset of breakup of Pangaea to form the modern continents and opening to form the modern oceans, coincident with the end-Permian extinction event. The post-extinction resurgence of life-forms then enabled previously subdued species to rapidly diversify and again repopulate the entire Earth.

While the prolonged period of time involved during this post-Permian crustal breakup allowed the dinosaur species to migrate to more equitable locations, it also presented new physical and environmental barriers. Draining of the ancient continental seas, for instance, influenced distribution of the modern seas and oceans and hence adversely affected available dinosaur habitats, migration routes, and escape avenues. Migration away from these physical barriers then encouraged the dinosaurs to evolve into new species. In contrast, failing to evolve, migrate, or being land-locked may have also caused their localized demise.

During the Mesozoic times, continental crustal breakup continued unabated as each of the modern continents became well established. By then all of the intervening modern oceans, except for the Southern Ocean, had opened. Breaching of the North American and Australian land connection had occurred during the Triassic Period, giving rise to merging of the north and South Pacific Oceans, which resulted in the end-Triassic extinction event. During this end-Triassic extinction event most of the non-dinosaurian archosaurs, as well as most of the terrestrial vertebrates, the ancestors of mammals, and most of the large amphibians became extinct.

The unique provincial distribution of the earlier Permian ancestral reptiles and Mesozoic dinosaurs on an increasing radius plate tectonic Earth demonstrates consistent evolutionary links between Permian, Triassic, Jurassic, and Cretaceous species. Many of these links were permanently disrupted during the Triassic and Cretaceous Periods as a result of crustal breakup and continental dispersal, draining of the ancient seas, variable sea-levels, and disruption to existing climate zones.

The Cenozoic Era saw the modern continents and oceans continue to migrate and open to their current configurations, and studies here show that this will continue unabated into the near future. The era was dominated by species development being increasingly isolated within island continents, with many species now endemic to specific continents or islands.

In recent times this isolation has, of course, been severely disrupted by the colonization activities of mankind, with all its attendant implications for renewed species diversification and secondary extinctions.

Chapter 19

A Paradigm Shift in Metallogeny

"*Major peaks in the abundance of metal deposits that formed, or were preserved in convergent-margin orogenic belts in the Late Archaean and metal deposits associated with either anorogenic magmatism or continental sedimentation in the Proterozoic are difficult to reconcile with the idea that Plate Tectonic processes have operated since the Late Archaean.*" Barley and Groves (1992).

Metallogeny is a branch of geology concerned with the study of the formation and distribution of mineral deposits. Metallic mineral deposits are defined as *naturally occurring accumulations or concentrations of metals or minerals of sufficient size and concentration that might, under favourable circumstances, have economic value.* The distribution of mineral deposits on the present-day Earth is determined by the geological processes that formed them. Mineral deposits are "*...commonly clustered into geological provinces, with some provinces being strongly endowed in particular mineral commodities*" (Jaireth and Huston, 2010).

The distribution of mineral deposits is also determined by the timing of ore forming processes. Ore deposits have formed at various periods over geological time from at least 3,500 million years ago to the present-day where mineral deposits are forming as part of natural processes both on land and on the seafloor.

Advances in dating geological processes have shown that many ore deposits formed over a relatively short period of time. Deposits have also been commonly formed through a combination of geological factors closely related in time, many of which have been repeated throughout the geological record, and most can be related to tectonic activity.

The distribution of metallic minerals on plate tectonic small Earth models is considered in the context of metallogenic epochs and metallogenic provinciality. A metallogenic epoch is "*...a regularly recurring sequence of events that gives rise to a cyclicity in mineral deposition during geological* *history*" (Smirnov, 1984). A metallogenic province is "*...a specific region possessing a notable concentration of a certain metal or metals*" (Evans, 1984) which are linked by common characteristic features.

Metallogenic provinces are often unified into global-scale tectonic belts, divided into sub-provinces based on common occurrences of individual kinds or groups of minerals, or subdivided into metallogenic regions indicated by the existence of structural or tectonic controls.

19.1. Metallogenic Epochs

Since the introduction and widespread application of radiometric age dating techniques carried out on rocks and minerals, it has been recognized that igneous and metamorphic mineral dates are periodic and roughly cyclic. The maxima indicated by these dates were considered to be closely related in time to periods of orogenic activity. The distribution of metal deposits and the nature and styles of mineralization in time and space suggests there has been an evolutionary trend in the concentration of metals, as well as a diversity in the form and structure of mineral occurrences.

Mineralization distribution patterns occurring over time were considered from a recycling perspective by Veizer *et al.* in 1989, suggesting that the observed patterns represent an evolutionary process of continental lithosphere, atmosphere, hydrosphere, and biospheric development. In 1981, and again in 1988, Meyer considered the distributions in relation to various groups of metal deposits. Gaál in 1992 and Vanecek in 1994 considered metallogeny in relation to plate tectonic cycles while in 1992 Barley and Groves attributed the distribution patterns to conventional plate tectonic cyclical aggregation and breakup of large supercontinents.

Meyer recognized three main transitional periods for the metallogenic distribution of major groups of metal deposit types, representing changes in the characteristics and frequency of metal occurrences with time. Characteristic peaks in the abundance of specific styles of metallic mineralization were also recognized, and Barley and Groves considered the uneven temporal distribution in time—to be related to three main factors, including the evolution of the hydrosphere and atmosphere, a decrease in global heat flow, and long-term tectonic trends.

This systematic temporal variation was considered by Barley and Groves in relation to a conventional plate tectonic model of cyclical aggregation and breakup of large continents throughout Earth history. In particular, metal deposits that formed in continental basins or were associated with anorogenic magmatism—magma that formed between orogenic events—were shown to be abundant in the mid-Proterozoic, corresponding to an inferred assemblage of a large proto-continent. Peaks in abundance of continental metal deposits were also shown to coincide with a late-Proterozoic Rodinian supercontinental assemblage and a near maximum extent of a Pangaean assemblage during the late-Palaeozoic.

Barley and Groves further considered that metal deposits formed or preserved in convergent-margin orogens were abundant in the late-Archaean, corresponding to periods of high global heat flow and rapid stabilization of continental crust. This abundancy was also considered to be occurring during the past 200 million years, corresponding to a present tectonic cycle, and similar mineralization styles were shown to be present in early-Proterozoic, late-Proterozoic, and Phanerozoic orogens.

The relationship between the abundance of specific metal deposit styles and long-term tectonic trends was less clear. Barley and Groves, and similarly Barnicoat *et al.* in 1991, considered that major peaks in the abundance of metal deposits that formed, or were preserved, in convergent-margin orogenic belts during the late-Archaean and metal deposits associated with either anorogenic magmatism or continental sedimentation in the Proterozoic are difficult to reconcile with the idea that conventional plate tectonic processes have operated since the late-Archaean.

Metallogenic epochs on an increasing radius plate tectonic Earth are considered in relation to the continental and seafloor crustal development history of the Earth. It was established in Chapter 7 that the primary mechanism for pre-Triassic crustal development on an increasing radius Earth was a process of crustal extension within an established global network of crustal weakness.

Similarly, orogenesis occurred as a result of intracratonic—sedimentary basins located in the middle of stable continental regions—motion during periods of crustal compression, and mountain building was related to crustal uplift along continental margins with an associated collapse of continental interiors during changing Earth surface curvature over time.

On an increasing radius plate tectonic Earth, these tectonic processes result in characteristic inter-related regional or tectonic settings, each reflecting a characteristic metallogeny which may crosscut present-day continental boundaries.

The correlation of Precambrian metallogenic epochs and tectonic development on an increasing radius plate tectonic Earth is categorized in the following tectonic settings (adapted from Barley and Groves, 1992; and Gaál, 1992):

1. Late-Archaean stabilization of continental crust (2.9 to 2.6 billion years ago) characterized by high global heat flow and magmatism.
2. Intracratonic rifting, extension and tensile fragmentation of the primitive Archaean crust (2.5 to 2.0 billion years ago) consisting of stages of early crustal fragmentation (2.5 to 2.3 billion years ago) with gold-conglomerate basins, banded iron formation basins, layered intrusions hosting platinum group elements, chromium, vanadium-titanium-iron and nickel-copper; early-Proterozoic greenstone belts (2.5 to 2.0 billion years ago) with vein gold, copper-gold sulphide deposits; and late crustal fracture (2.1 to 1.9 billion years ago) with layered intrusions hosting nickel-copper, platinum group elements and chromium.
3. Basement crustal extension and rifting (1.95 to 1.8 billion years ago) generating ophiolite associated copper-cobalt-zinc massive sulphide deposits (approximately 1.95 billion years ago), oceanic island arcs (1.93 to 1.85 billion years ago) with nickel-copper deposits and continental arcs (1.9 to 1.8 billion years ago) with zinc-copper-lead-silver-gold volcanogenic massive sulphide, porphyry copper-gold-molybdenum, banded iron formation and Kiruna-type iron deposits.
4. Cratonization, basement crustal extension and intracratonic rifting (1.8 to 1.6 billion years ago) characterized by rifted basins with sediment-hosted zinc-lead-copper and granitoids with tin, molybdenum, tungsten, copper, iron and rare Earth elements.
5. Formation of vast, elongate, slowly sinking basins (0.8 to 0.6 billion years ago) characterized by metalliferous carbonaceous argillites and marls, often associated with bedded phosphorites, copper-lead and zinc-lead base metals in a predominantly oxygenated hydrosphere.

On an increasing radius plate tectonic Earth, each of these tectonic settings are global scale and metallogenic epochs correlate globally. Crustal development during the Precambrian era had a common heritage of both crust and metals and crustal development was dominated by horizontal motion, including crustal extension and continental orogenesis accompanied by high heat flow.

Small Earth modelling suggests that the Archaean was an epoch where the crust had stabilized into discrete cratons which experienced limited internal deformation until the Proterozoic. Crustal extension during the Proterozoic was localized within a network of crustal weakness surrounding established cratons forming elongate platform sedimentary basins. These basins and zones of crustal weakness also hosted geosynclinal and orogenic activity.

Peaks in the abundance of specific styles of metallic mineralization on an increasing radius plate tectonic Earth are then seen as reflecting evolution of the continental crust, hydrosphere and atmosphere, the secular decrease in global heat flow and changing crustal conditions with time. In particular, metal deposits abundant in the early- to late-Proterozoic were formed during a period of increasing crustal extension within a network of platform sedimentary basins associated with continental magmatism and orogeny. By the late-Palaeozoic, peaks in continental metal deposits coincided with initiation of continental breakup, and during the Mesozoic and Cenozoic Eras with continental dispersal and orogenesis.

19.1.1. Archaean Metallogenesis

Laznicka in 1993 considered the early- and mid-Archaean to be an early stage of cratonic development where metal mineralization was absent or poorly developed. The amphibolite-facies greenstone rocks—metamorphosed volcanic rocks—hosted some banded siliceous iron deposits, minor stratabound massive copper sulphide deposits, layered and non-layered chromite, and gold, arsenopyrite, minor chalcopyrite and locally significant antimony and mercury mineralization. This metallic mineralization was considered to have resulted from mainly seafloor exhalations, hydrothermal convection—mineralization associated with high-temperature aqueous solutions, and metasomatism—the chemical alteration of rocks by hot fluids.

In contrast, the late-Archaean was considered by Laznicka to be a period where granite-greenstone and greenstone belts in particular achieved their peak development. Greenstone assemblages throughout the world are important host rocks for iron as banded iron formation; manganese as bedded ores; titanium, vanadium and chromium in differentiated mafic—a volcanic rock rich in magnesium and iron—and ultramafic intrusive rocks; nickel and cobalt associated with komatiites—volcanic rocks with a high to extremely high magnesium content; copper, zinc and lead as volcanogenic massive sulphide deposits; tin, mercury, uranium and gold in altered shear zones; silver associated with volcanogenic massive sulphides and to a lesser extent in shears and veins; as well as platinum group elements associated with nickel sulphides.

The proportion of ore deposits dependent on volcanic activity was considered by Meyer to be greatest in the Archaean than during later geological times. The only sedimentary ore types were the banded iron formation and volcanic-hosted massive sulphide deposits. The abundance of gold, volcanic-hosted massive sulphide, and komatiite-associated nickel sulphide mineralization in late-Archaean greenstone belts and sedimentary basins was suggested by Barley and Groves to have resulted from a rapid growth and stabilization of continental crust combined with high global heat flow.

It is significant to note, however, that modern geological evidence shows that a significant amount of rock material has been eroded from all of the ancient cratonic and orogenic regions exposed throughout the world today. This erosion has been estimated by others to vary between 10 to 40 kilometres vertical thickness which has, in turn, been further deposited in surrounding sedimentary basins.

It is considered that this extreme amount of erosion will significantly bias the interpretation of metallogenic settings during the Archaean and Proterozoic times, where most of the metallogenic settings may have since been eroded away.

On an increasing radius plate tectonic Earth, the Archaean comprised a complete assemblage of the most ancient cratonic and orogenic crusts preserved on Earth today. As such, many of these observed metallogenic settings were associated with zones of crustal fragmentation and extension localized within a global network of crustal weakness, surrounding stabilized cratons.

Small Earth models show that global-scale cratonization had occurred during a prolonged period of stability during the Archaean to mid-Proterozoic. Mineralization associated with this stable tectonic setting was then reflected in the predominance of ultramafic to felsic—rocks rich in silica, oxygen, aluminium, sodium, and potassium—magmatic associations, as well as an increase in accumulation of sediment and volcano-sediment associations during the late-Archaean.

19.1.2. Proterozoic Metallogenesis

The Archaean-Proterozoic boundary was considered by Meyer in 1981, as well as Lambert and Groves in 1981, to be a significant interface between two substantially different crustal-forming regimes. These differences were considered to reflect a primitive Archaean and a more evolved Proterozoic. In 1994, Vanecek described the late-Archaean to early-Proterozoic transition as an episodic event characterized by a change in the composition of granite rocks to potassium-rich granites displaying a negative europium-anomaly.

These granite rocks are characteristic of this era and belong to the tonalite-trondhjemite-granodiorite group of rocks. During that time, the ancient cratonic regions were also yielding eroded sediment and chemically precipitated material into the surrounding platform sedimentary basins.

Laznicka considered that the bulk of the Proterozoic greenstone belts and their depositional metallogeny had retained the same characteristics as the Archaean greenstones, with some signs of maturity such as higher lead contents in some volcanic massive sulphide deposits. Other volcano-sedimentary terranes departed substantially from the Archaean model, with a higher proportion of felsic volcanic rocks, metamorphosed carbonate sediments, iron formations, and granites. Laznicka attributed this to a significant involvement of continental crust as a source of some magma, to rifting regimes, to syn- and post-depositional regional metasomatism and to shallow water or subaerial deposition of sediments.

Meyer also considered that vast, elongate, slowly sinking basins were the dominant feature of the late-Proterozoic, with metalliferous carbonaceous sediment, often associated with bedded phosphorite, reaching a peak between 800 to 600 million years ago. Sedimentation in these continental basins and seas provided ideal environments for concentration of metals. The principle metals during that time were copper and lead, and zinc associated with lead, formed by local biogenic reduction of newly abundant sulphate supplied from a predominantly oxygenated hydrosphere.

The effect of abundant free oxygen on Earth's surface chemistry during the early-Proterozoic was considered by Meyer to be gradual but irreversible. This oxygen had major effects on chemical sedimentation, including the introduction of new types of sediment-hosted base metal sulphide ores and sediment affiliated uranium.

The Proterozoic Era on an increasing radius plate tectonic Earth was a period of steady to steadily accelerating increase in Earth radius and surface area resulting in increased crustal extension and sediment accumulation within a global network of platform sedimentary basins.

The Proterozoic small Earth modelling agrees with the observations of

Meyer where vast, elongate, slowly sinking basins were the dominant tectonic feature of the late-Proterozoic. Exposed Archaean cratonic regions existing at the time yielded eroded sediment to these platform sedimentary basins, and sedimentation in continental basins and seas provided ideal environments for concentration of base metals and chemically precipitated and anaerobic sediments.

19.1.3. Phanerozoic Metallogenesis

In most places of the world, there was no break between the Proterozoic and Phanerozoic rock associations. The late-Proterozoic was characterized by heat dissipation through rifts and mobile zones which evolved into Phanerozoic orogenic belts producing a variegated range of orogenic-related ore deposits. Meyers noted that most Phanerozoic mineral deposits lie within orogenic belts, and Mesozoic and Cenozoic deposits, in particular, were aligned with present plate boundaries, supporting the suggestion for an ascendancy of new mechanisms of crustal behaviour during the Palaeozoic.

Volcanogenic massive sulphide deposits were abundant during the Phanerozoic, locally containing lead as the principle metal and hosted by deformed volcanic rocks making up the cores of Phanerozoic mountain chains. Copper, silver and zinc deposits in black carbonaceous shales and lead in carbonate rocks were widespread during the Palaeozoic, becoming less abundant during later eras. Uranium formed in continental sandstones within oxidation-reduction weathering fronts. Copper-pyrite deposits and podiform chromite deposits in ophiolite assemblages first appeared in the Phanerozoic.

Large base and precious metal concentrations associated with various granites, including porphyry deposits containing copper, silver, molybdenum, tin, and tungsten were prominent during the Phanerozoic, reflecting a near-surface chemistry with abundant sulphate and moderate to strong hydrogen ion metasomatism.

On an increasing radius plate tectonic Earth, during the Precambrian and Palaeozoic all continental crusts were assembled together on a smaller radius Earth. Metal distributions continued to be primarily focused within a global network of crustal weakness and coincident sedimentary basins. Continental crustal extension and deposition of sediments within these zones of weakness culminated with the progressive breakup and dispersal of continents during the late-Permian Period.

Phanerozoic sedimentation then shifted from continental settings to marginal basins located around the perimeters of the newly formed continents. Similarly, long chains of volcanic island arcs located around the margins of the continents were associated with initiation of mid-ocean-rifting, as well as rift-bordered and back-arc basins and large shallow seas.

As noted by Meyers, most Phanerozoic mineral deposits lie within orogenic belts and Mesozoic and Cenozoic deposits in particular are aligned with present plate boundaries. On an increasing radius plate tectonic Earth, this alignment is due to breakup of the Pangaean supercontinental crusts and formation of the modern continents. This alignment supports the observations of Meyers for the ascendancy of new mechanisms of crustal behaviour.

19.2. Metallogenic Provincial Modelling

The global distribution of a selection of metals outlined in this section is based on the USGS Mineral Resource Data Set (MRDS) 2015, which lists the locations of metallic and non-metallic mineral resources throughout the world. Most commodities are listed as metal associations, e.g. gold-copper-silver-molybdenum, where, in this example, gold is the dominant commodity. It should be noted that, because the MRDS is a USGS data set, there is a strong bias in the presence and distribution of metals in the Americas, as compared with data from the rest of the world.

The MRDS data represents locational data and is not age dated. In order to model this data on small Earth models, the data are modelled only for the present-day, Permian, and Cambrian models. This modelling reflects metal distributions and associations persisting at the end of the Precambrian—shown on the Cambrian model, just prior to the end-Permian supercontinental breakup—shown on the Permian model, and where these same metals are located today—shown on the Pliocene to recent model.

Because there is no age dating, the plotted data represent either where the particular metals were located on each of the small Earth models, or where they resided within the crust or mantle prior to formation and preservation as mineral deposits. This plotting then reflects where the particular metal originated from, that is, the location of its original mantle or crustal source. Deposits within continental Greenland and Antarctica beneath present-day ice-sheets are unknown and are not shown.

A number of distinct metallogenic provinces are highlighted within the plotted data, whereby provinces from adjoining continents or tectonic regimes coincide, extending across present-day geographical boundaries. Many of these provinces are now separated by sedimentary basins or large oceans during post-Permian continental breakup and dispersal.

Examples of metallogenic provinces on plate tectonic small Earth models include:

1. The western North American Proterozoic gold (e.g. Homestake mine) and base metal (e.g. Sullivan mine) province located adjacent to the north Australian Proterozoic uranium-gold (e.g. Rum Jungle and Kakadu mines) and base metal (e.g. McArthur River and Mt Isa mines) provinces. Each of these metal deposits are sediment hosted and together form part of a distinct ancient sedimentary basin located between the South American Guyana Craton and Canadian Superior Provinces (Figures 19.1 and 19.2).

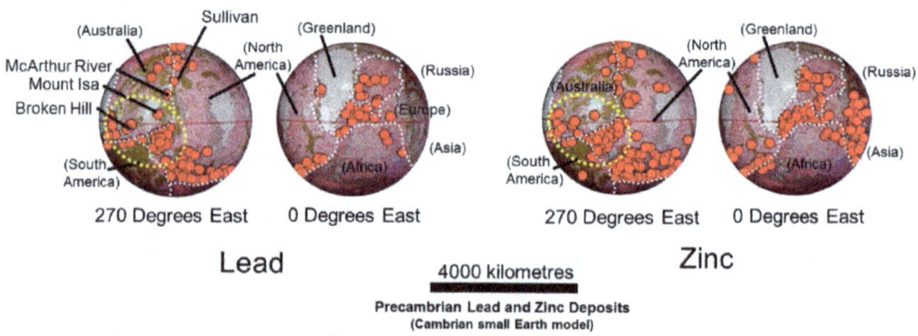

Figure 19.1. Distribution of lead and zinc plotted on the Cambrian plate tectonic small Earth model. Data are shown as red dots in relation to continental crustal assemblages highlighted as dashed white lines. Lead and zinc metallogenic province is highlighted by dashed yellow shape. Data after USGS Mineral Resource Data Set 2015.

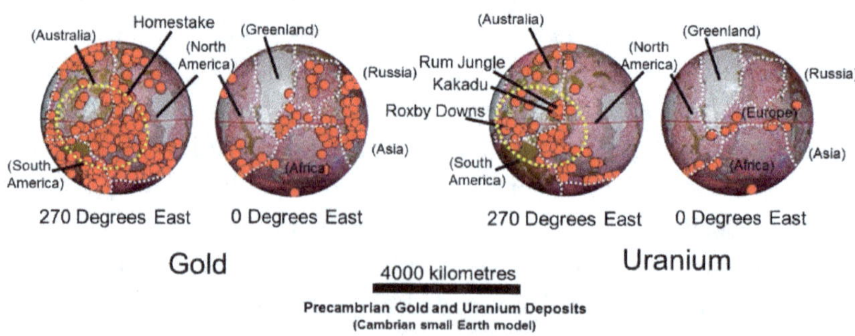

Figure 19.2. Distribution of gold and uranium plotted on the Cambrian plate tectonic small Earth model. Data are shown as red in relation to continental crustal assemblages highlighted as dashed white lines. The gold and uranium metallogenic province is highlighted by dashed yellow shape. Data after USGS Mineral Resource Data Set 2015.

2. The Proterozoic sedimentary banded iron formations of the Hamersley Basin, Western Australia, are located adjacent to Archaean to Proterozoic high-grade metamorphic and sedimentary iron deposits in northern China and south-eastern Russia (Figure 19.3).

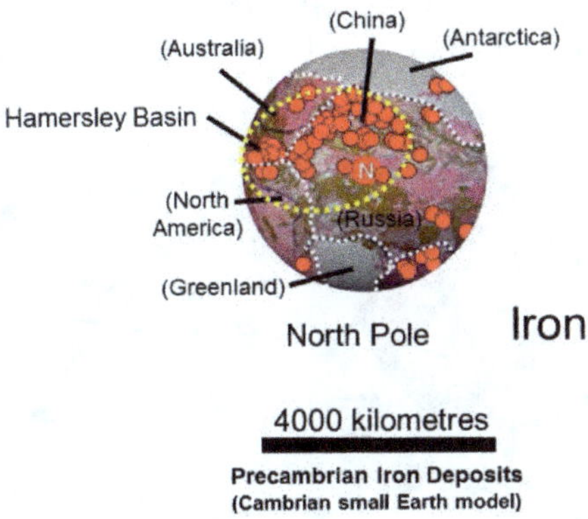

Figure 19.3. Distribution of iron plotted on the Cambrian plate tectonic small Earth model. Data are shown as red dots in relation to continental crustal assemblages highlighted as dashed white lines. Iron metallogenic province is highlighted by dashed yellow shape. Data after USGS Mineral Resource Data Set 2015.

3. Palaeozoic granite-associated tin deposits in southern Asia and China coincide with the assembled continental margins of East Antarctica and Australia and are located adjacent to Archaean tin-tantalum-tungsten pegmatite deposits in southwest Australia (Greenbushes mine) and the Pilbara Craton (Figure 19.4).

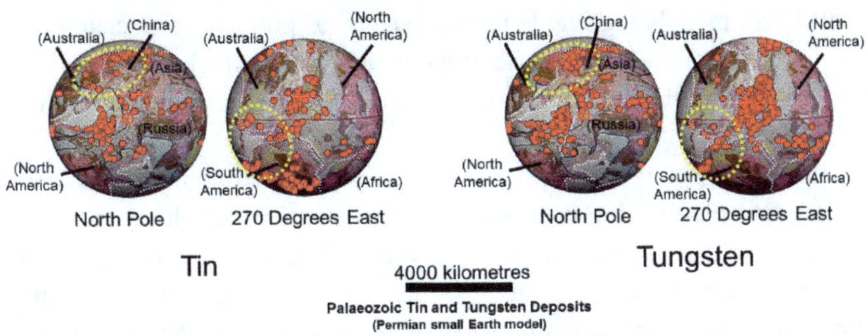

Figure 19.4. Distribution of tin and tungsten plotted on the Permian plate tectonic small Earth model. Data are shown as red dots in relation to continental crustal assemblages highlighted as dashed white lines. The tin and tungsten metallogenic province is highlighted by dashed yellow shape. Data after USGS Mineral Resource Data Set 2015.

4. Palaeozoic tin deposits in Tasmania and Eastern Australia are located adjacent to the Palaeozoic Bolivian tin deposits in South America (Figure 19.4).
5. Host rocks to nickel-cobalt-base metals in Queensland (Greenvale) and Central America (Honduras and Guatemala) coincide with ophiolite -hosted and lateritic deposits in Cuba and New Caledonia (Figure 19.5).

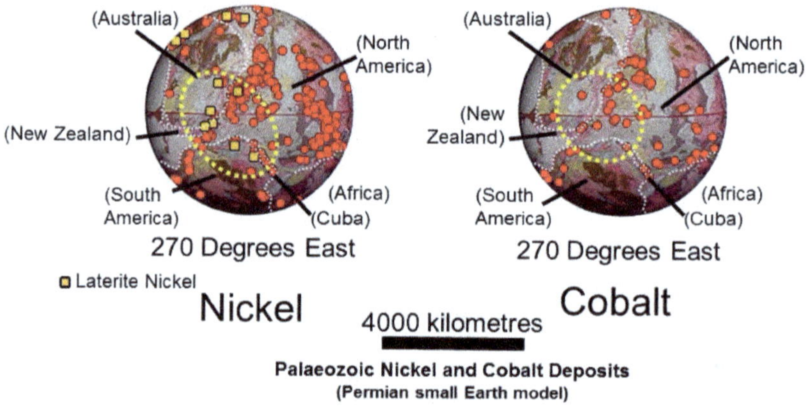

Figure 19.5. Distribution of nickel and cobalt plotted on the Permian plate tectonic small Earth model. Data are shown as red dots in relation to continental crustal assemblages highlighted as dashed white lines. Lateritic nickel deposits are shown as yellow squares. The nickel and cobalt metallogenic province is highlighted by dashed yellow shape. Data after USGS Mineral Resource Data Set 2015.

The Mesozoic and Cenozoic metallogenic distribution on plate tectonic small Earth models highlights the abundance of porphyry-associated metals concentrated within Phanerozoic orogenic belts, in particular the Cordilleran-Andean tectonic belt along the west coasts of the Americas, shaded yellow on Figure 19.6.

On an increasing radius plate tectonic Earth, these orogenic belts crosscut and displace pre-existing Palaeozoic and Precambrian metallogenic provinces and the mechanism for emplacing of metals is considered to be related to isostatic uplift of continental margins during change in Earth surface curvature. During breakup and dispersal of the continents a common mantle source for some metals is also evident, forming metallogenic provinces linking the original continents across each of the modern oceans.

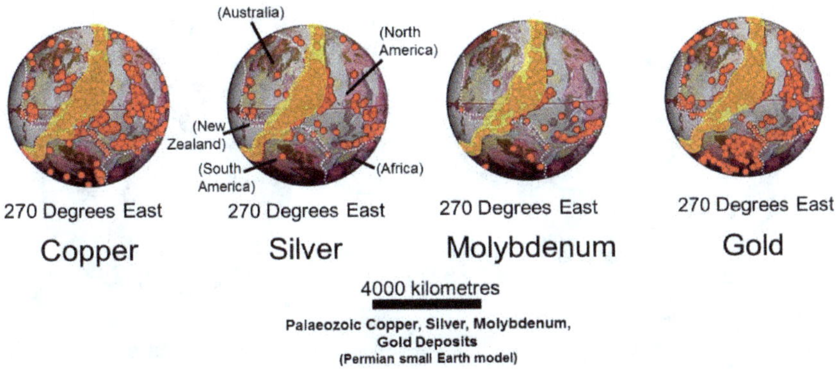

Figure 19.6. Distribution of orogenic copper, silver, molybdenum, and gold plotted on the Permian plate tectonic small Earth model. Data are shown as red dots in relation to continental crustal assemblages highlighted as dashed white lines. Cross-cutting orogenic plate boundary metal deposits located along the west coasts of the Americas are shaded yellow. Data after USGS Mineral Resource Data Set 2015.

The distribution of selected Mesozoic and Cenozoic orogenic-hosted metals is shown plotted on the Permian plate tectonic small Earth model (Figure 19.6). The Permian model highlights the distinct alignment of Mesozoic and Cenozoic orogenic metals with present plate boundaries prior to breakup of the continental crusts. The orogenic metal deposits shaded in yellow are shown to crosscut pre-existing Precambrian metallogenic provinces, prior to complete separation during opening of the Pacific Ocean (Figure 19.7).

Metals in Figure 19.6 include:

1. Porphyry-associated copper, molybdenum, silver, gold. Additional metals include tungsten and tin mineralization previously shown plotted in Figure 19.4, which are abundant within Phanerozoic orogenic belts.

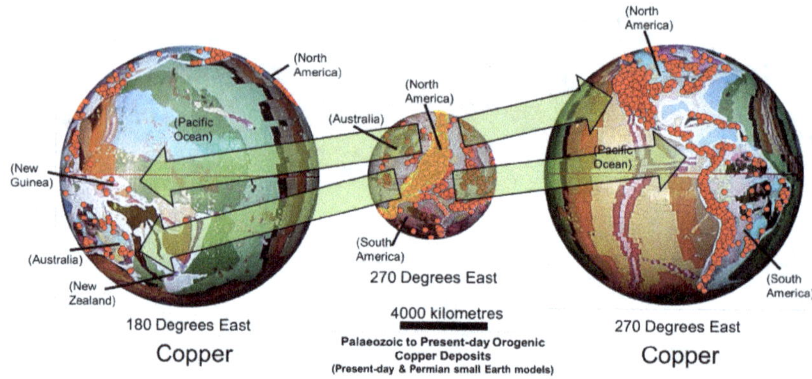

Figure 19.7. Distribution of orogenic copper in relation to opening of the Pacific Ocean plotted on the Permian and present-day plate tectonic small Earth models. Data are shown as red dots in relation to continental crustal assemblages highlighted as dashed white lines. Cross-cutting orogenic metal deposits located along the west coasts of the Americas and east coasts of Australia and New Guinea are shaded yellow prior to opening of the Pacific Ocean. Data after USGS Mineral Resource Data Set 2015.

The selection of metallogenic epochs and mineralization settings highlighted in this section are essentially identical to those identified within conventional studies. The difference being that, on an increasing radius plate tectonic Earth, prior to the early-Triassic Period, all continental crusts were assembled together on a smaller radius supercontinental Earth. The increasing radius assemblages then enable pre-Triassic metallogenic epochs and provinces from otherwise remote locations to be assembled together as unique, inter-related provinces on smaller radius Earth models.

The assemblage of continents and crustal elements on an increasing radius plate tectonic Earth then provides a means to investigate the spatial and temporal distribution of metals across adjoining continents and crustal regimes. Recognition and understanding of past metal distributions on the present Earth then potentially enable mineral search and genetic relationships to be extended beyond their known type localities.

Chapter 20

A Paradigm Shift in Hydrocarbons

More than 85 percent of the world's energy still comes from fossil fuels. Despite centuries of growing use, these fuels remain abundant. Powerful economic and political interests are organized around the fossil-energy system, as are complex social arrangements.

Fossil fuel is a general term for *buried combustible geologic deposits of organic materials, formed from decayed plants and animals that have been converted to crude oil, coal, natural gas, or heavy oils by exposure to heat and pressure in the earth's crust over hundreds of millions of years.* Fossil fuels include shale oil and shale gas, coal, petroleum, and natural gas and these are generally considered to have been formed by natural processes such as anaerobic decomposition of buried organisms.

Fossil fuels contain high percentages of carbon which range from volatile materials with low carbon to hydrogen ratios, such as methane and liquid petroleum, and non-volatile materials composed of almost pure carbon, such as anthracite coal. Petroleum and natural gas, in particular, are formed by the anaerobic decomposition of remains of organisms including phytoplankton and zooplankton that settled to the sea or lake floor in large quantities under anoxic conditions.

Over geological time, this organic matter was then buried under heavy layers of sediment. The resulting high levels of heat and pressure caused the organic matter to chemically alter, first into a waxy material known as kerogen which is found in oil shales, and then, with more heat, into liquid and gaseous hydrocarbons by a process known as catagenesis.

Studies of the organic content of sedimentary rocks were shown by North in 1990 to be at a maximum during the Cambrian-Ordovician, Carboniferous, Jurassic and Cenozoic times, and at a minimum in Silurian-Devonian and Permian-Triassic rocks containing concentrations of evaporite deposits. Hydrocarbon occurrences were considered to correlate with source sediment maxima, with Tiratsoo in 1984 showing that the worldwide percentage-by-weight distribution of crude oil and natural gas is greatest in Mesozoic reservoir rocks, followed by the Cenozoic and Palaeozoic.

A major Phanerozoic marine transgression—a geological event where sea levels rise relative to the land—was noted by Hallam et al. in 1983 and Exxon (Figure 20.1) to have occurred during the Cambrian. This occurred before marine life was sufficiently abundant or diversified to provide rich accumulations of organic matter.

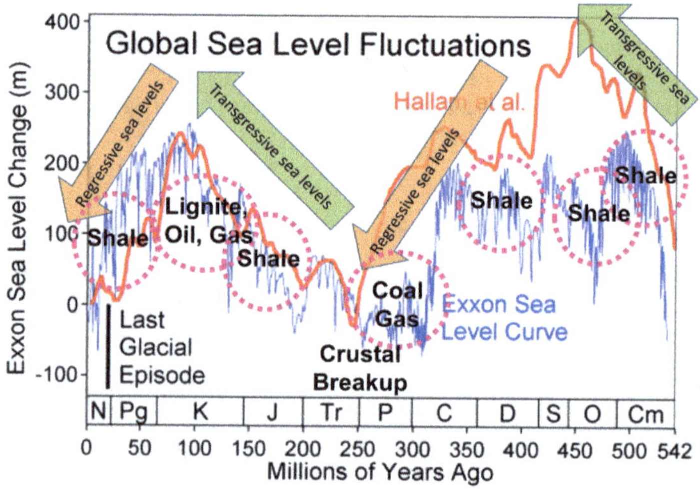

Figure 20.1. Major peaks in fossil fuel occurrences in relation to sea-level changes during the past 542 million years. Sea level curves after Hallam *et al.*, 1983, and Exxon.

North also described lesser transgressive events during the Ordovician, contributing effective source sediments in North America, and a late-Devonian transgression contributing significant source sediments in the Ural-Volga and western Canadian basins, plus lesser basins in eastern and southern North America and North Africa. This transgressive sea-level was followed by a fluctuating period of marine sea-level regression—a geological event where areas of submerged seafloor were exposed during lowering of sea levels—throughout the remaining Palaeozoic, culminating in breakup of Pangaea during the late-Permian.

An extensive marine transgressive period again occurred during the mid-Jurassic and continued with minor setbacks until near the end-Cretaceous. This transgression formed the principal source sediments for fossil fuels in the Middle East, western Siberia, North Sea and Central America and Peru. North also records smaller transgressions during the Oligocene and Miocene which formed source sediments in the Californian, Caucasian, Carpathian and Indonesian provinces.

20.1. Shale Oil-Gas

Oil and gas shales are fine-grained sedimentary rocks containing significant amounts of kerogen which belong to the group of sapropel fuels. Oil and gas shale formation takes place in a number of depositional settings and has considerable compositional variation. The shales can be classified by their composition—including carbonate minerals such as calcite or detrital minerals such as quartz and clays, or by their depositional environment—including large lakes, shallow marine, or lagoonal settings. Much of the organic matter in oil and gas shale is of algal origin but may also include remains of vascular land plants. Some shale deposits may also contain metals such as vanadium, zinc, copper, and uranium.

As a sapropel fuel, oil and gas shales differ from humus fuels in their lower content of organic matter. The organic matter in shales forms a complex macromolecular structure which is insoluble in common organic solvents. The organic portion of shale consists largely of a pre-bitumen to bituminous groundmass, made up of remains of algae, spores, pollen, plant cuticles and corky fragments of herbaceous and woody plants, and cellular debris from other lacustrine, marine, and land plants.

Terrestrial shales may contain resins, spores, waxy cuticles, and corky tissues of roots and stems of vascular terrestrial plants, while lacustrine shales include lipid-rich organic matter derived from algae. Similarly, marine shales are composed of marine algae, acritarchs, and marine dinoflagellates. Organic matter in oil and gas shales may also contain organic sulphur and a lower proportion of nitrogen.

Most oil and gas shale formation took place during the mid-Cambrian, early- and mid-Ordovician, late-Devonian, late-Jurassic and Paleogene periods (Figure 20.1). These shales were formed by the deposition of organic matter in a variety of depositional environments including freshwater to highly saline lakes, epicontinental marine basins and subtidal shelves and were often restricted to estuarine areas such as oxbow lakes, peat bogs, limnic and coastal swamps, and peatlands. Continuous burial and further heating and pressure often resulted in the formation of petroleum and natural gas from the shale source rock.

The global distributions of assessed and non-assessed oil and gas shales are plotted on the Permian plate tectonic small Earth model (Figure 20.2). Data were sourced from the 2013 U.S. EIA report "*Technically Recoverable Shale Oil and Shale Gas Resources: An Assessment of 137 Shale Formations in 41 Countries outside the United States*".

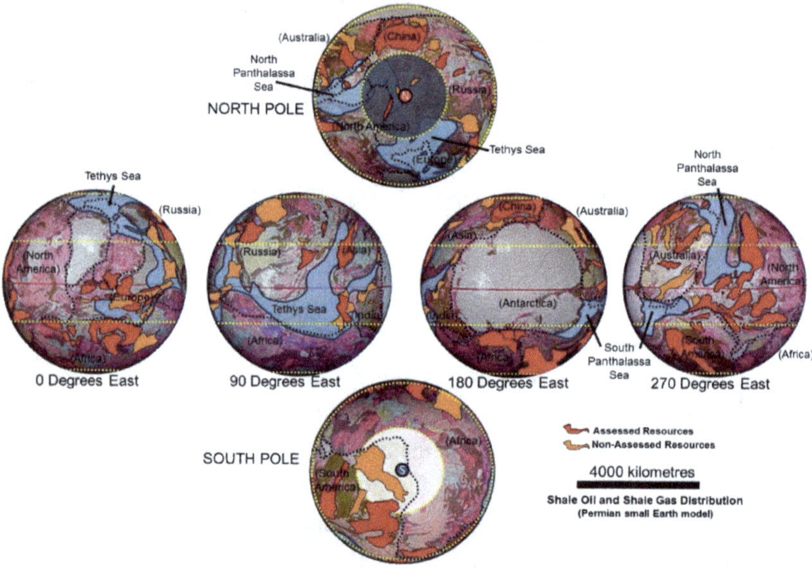

Figure 20.2. Permo-Carboniferous shale oil and shale gas distributions plotted on a Permian plate tectonic small Earth model in relation to climate zones, the late-Palaeozoic south polar icesheet, shaded white, and the distribution of ancient continental seas. Data after U.S. EIA report, 2013.

The distributions of oil and gas shales on the Permian plate tectonic small Earth model represent distributions covering the entire Palaeozoic Era. The plotted data shows a global distribution extending from high northern to high southern latitudes. While a preference for terrestrial locations is evident in this figure, the changing configuration of continental seas during the earlier Palaeozoic times suggests that the majority of resources in Figure 20.2 were shallow marine with lesser terrestrial environments.

20.2. Permo-Carboniferous Coal

In 1994, Calder and Gibling described an extensive late-Palaeozoic peat forming ecosystem flourishing across the southern, equatorial, and to a lesser degree, northern temperate lowlands of Europe and America. Preserved deposits of these tropical peat lands were said to record the effects of climate, tectonism, and global sea-level changes existing during these times, with many peats forming in foreland—basins that develop adjacent and parallel to a mountain belt—and intermontane basins—basins located between mountains or mountain systems—associated with equatorial orogenic belts or adjacent cratonic basins.

In 1994, Faure *et al.* noted that early-Permian to mid-Triassic coal deposits of southern Africa, Australia, Antarctica, South America and China were deposited predominantly in foreland basins. Similarly, Russian and Siberian coals occur in basins, flanked in the north, east and south by regions of mountain building.

The late-Palaeozoic coal age of Calder and Gibling followed the evolution of vascular plants during the late-Silurian. Repeated cyclotherms—alternating sequences of marine and non-marine sediments—of coal, sandstone and carbonate strata were inferred to have been deposited during periods of glacial-related changes to global sea level or associated shifts in climate. Evidence from stable isotope measurements suggested to Faure *et al.* that there was a decrease in the atmospheric $^{13}C/^{14}C$ ratio during the Permian-Triassic transition, which coincided with an abrupt world-wide coal hiatus.

The late-Palaeozoic was characterized by profound changes to life on land, including evolution of insects and colonization of broad lowland swamps by spore-bearing trees. Floral experimentation during the early-Carboniferous gave rise to the emergence of highly successful late-Palaeozoic flora of the northern coal swamps and adjacent habitats. The most important northern hemisphere coal swamp genera were the spore-bearing Lepidodendron and Sigillaria plants confined to swamps which contributed much of the log and foliage that was later compressed and buried to form coal.

Similarly, Glossopteris, a seed fern with a tree-like appearance, and other lesser plants played an important role in forming coal deposits within the southern supercontinent Gondwana. These plants became a dominant part of the southern flora throughout the rest of the Permian, though they dwindled to extinction by the end of the Triassic Period. The rapid appearance, expansion, and relatively quick extinction of this group, as well as the large number of species, made the fossil group very important for understanding paleobiogeography since the breakup of Pangaea.

The distribution of late-Palaeozoic to Triassic anthracite and bituminous coal is plotted on the Permian plate tectonic small Earth model in Figure 20.3, shown in relation to important coal swamp flora. The distribution of coal data is taken from the "*Major Coal Deposits of the World*" map (2010), and fossil data are from the PaleoBioDB, (2015).

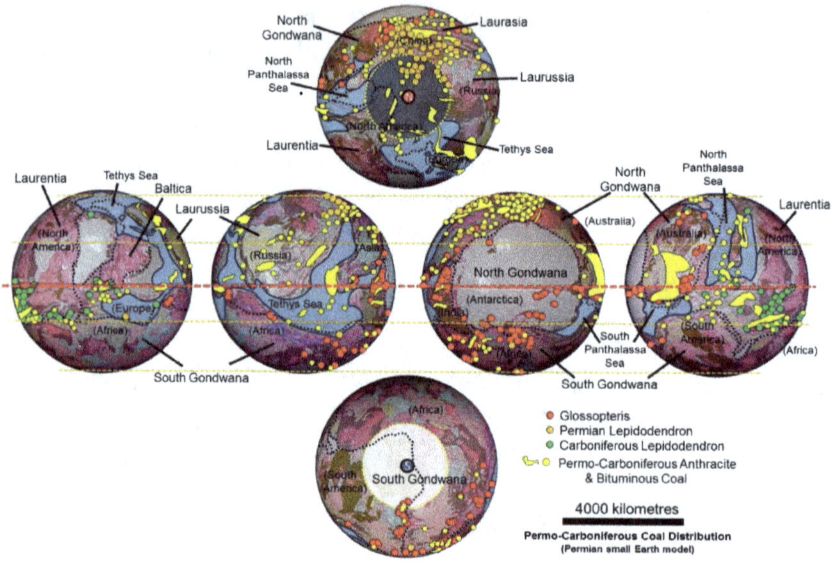

Figure 20.3. Permo-Carboniferous anthracite and bituminous coal distributions shown as yellow dots and shapes plotted on a Permian plate tectonic small Earth model. The distribution of late-Palaeozoic Glossopteris data is shown as red dots, Permian Lepidodendron data as orange dots, and Carboniferous Lepidodendron data as green dots in relation to climate zones, the late-Palaeozoic south polar icesheet, shaded white, and the distribution of ancient continental seas. Coal data after Major Coal Deposits of the World map, 2010. Plant data after PaleoBioDB, 2015.

The coals, highlighted in yellow in Figure 20.3, are predominantly located within the northern temperate zone, but also extend from high northern to mid-southern latitudes.

The distribution of Permian and Carboniferous lepidodendron, as well as glossopteris plant species in Figure 20.3 highlight the concentration of Permian lepidodendron within Laurasia—now China (orange dots), and Carboniferous lepidodendron within Laurentia—now North America (green dots), and a broad distribution of glossopteris species within both north and south Gondwana (red dots).

Very little integration of these species is shown in this figure, possibly representing the presence of physical or environmental barriers such as climate, ancient seas, or topography. In all cases the distributions of these plant species are now constrained to known coal deposits within the northern and southern hemispheres.

20.3. Cretaceous Lignite Coal

Unlike the Palaeozoic Era, Cretaceous lignite coals accumulated in either basins or passive continental margins with extensive deposits occurring in western North America and eastern Russia. Other areas include Europe, Japan, northern China, eastern Australia, New Zealand, northern Africa, Peru, Columbia and Argentina. The largest Cretaceous coal resources are located in foreland basins that stretched along the western margin of the Western Interior seaway of North America.

Cretaceous lignite coals were considered by McCabe and Parrish in 1992 to be distributed in a similar fashion to modern peats. These were further considered to form in mires—wet, soggy, muddy ground—within coastal regions, particularly near the equator where rainfall was presumably higher, or in high mid-latitudes where precipitation may have been relatively high and evaporation low.

Examination of the conventional plate tectonic global distribution of coals throughout the Phanerozoic indicated to McCabe and Parrish that there was a latitudinal shift in coal deposition, from principally equatorial in the Carboniferous, to high latitudes in Permian and later times. This shift was thought to have resulted from the onset of a Pangaean monsoonal climate, which disrupted climatic circulation in low- to mid-latitudes and led to drying of the continents within the equatorial regions.

Cretaceous lignite coal distribution is plotted on the late-Cretaceous plate tectonic small Earth model in Figure 20.4.

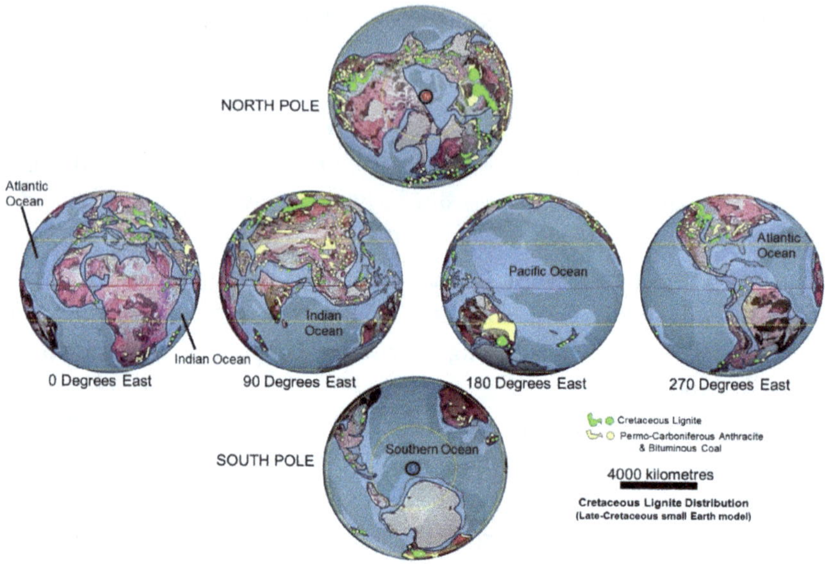

Figure 20.4. Cretaceous lignite coal distributions, highlighted in green, on a late-Cretaceous plate tectonic small Earth model. Lignite distributions are shown in conjunction with Permo-Carboniferous to Triassic coals, shown as pale yellow. Coal data after Major Coal Deposits of the World map, 2010.

Coal data was plotted after the "*Major Coal Deposits of the World*" map (2010). The distribution of lignite coal highlights the predominance of deposits located in the northern hemisphere, as noted by McCabe and Parrish, and these can be compared with the distribution of Permo-Carboniferous coals, shown as pale-yellow shapes in Figure 20.3.

On an increasing radius plate tectonic Earth, during the late-Cretaceous, continental breakup, dispersal, and opening of the modern oceans was well advanced and the spatial configuration of continents was similar to the present-day. During that time sea levels and transgression of continental seas was at its maximum, prior to regression of the seas during the Cenozoic to the present-day configuration and distribution of oceans.

On small Earth models, the latitudinal shift in Phanerozoic coal deposition, as noted by McCabe and Parrish, reflects the rapid opening of each of the modern oceans, along with a northward migration of continents during the Mesozoic and Cenozoic Eras. The predominance of coal deposits in the northern hemisphere is attributed to the greater extent of available landmasses influencing rainfall, as well as the extent of remnant continental basins existing during periods of high sea levels suitable for coal formation in these areas.

20.4. Petroleum and Natural Gas

Petroleum and natural gas deposits are generated by thermal processes from organic material entombed in sedimentary rocks. For petroleum and natural gas to be generated and preserved within a depositional environment there must be suitable organic-rich source rocks, such as oil and gas shale, and suitable reservoir rocks, such as porous sandstone.

In 1990, North considered the most favoured source sediments to be early basin deposits, forming at the peak of transgression of the seas following a trans-basinal unconformity—buried erosional surfaces that can represent a break in the geologic record of hundreds of millions of years or more. These source sediments were considered by North to be older than the main reservoir rocks and may be separated from the source rocks by entire rock formations devoid of recoverable hydrocarbons.

When viewed in context with global and transgressive-regressive sea-level changes (Figure 20.1), oil and gas development coincides with periods of peak sea-level transgression and maximum areas of continental seas and oceans. These conditions were considered by North to generate the most favoured early basin source sediments, forming at the peak of transgression following a trans-basinal unconformity. The Cretaceous, in particular, coincided with a period of post-late-Palaeozoic glacial melting, rapid opening of modern oceans, generally warm climatic conditions and rapid plant and animal diversification.

The widest distribution of natural gas not associated with petroleum occurs in Cretaceous molasse sediments—sandstone, shale, and conglomerate sediments deposited in front of rising mountain chains—with important deposits located in north-western Siberia and the western interior of North America. Natural gas also occurs during the Permian to Triassic periods, associated with the spread of Pennsylvanian coal deposits. The Cretaceous is included within the time of greatest development of oil-source sediments and the Permian to Triassic periods coincide with an oil-source minimum.

The distribution of petroleum and natural gas resources is shown on the Permian plate tectonic small Earth model in Figure 20.5. Cretaceous oil and gas resources are shown for context only. Petroleum and gas data were sourced from the "*World Oil and Gas Map*" (4th edition) produced by the Petroleum Economist (2013).

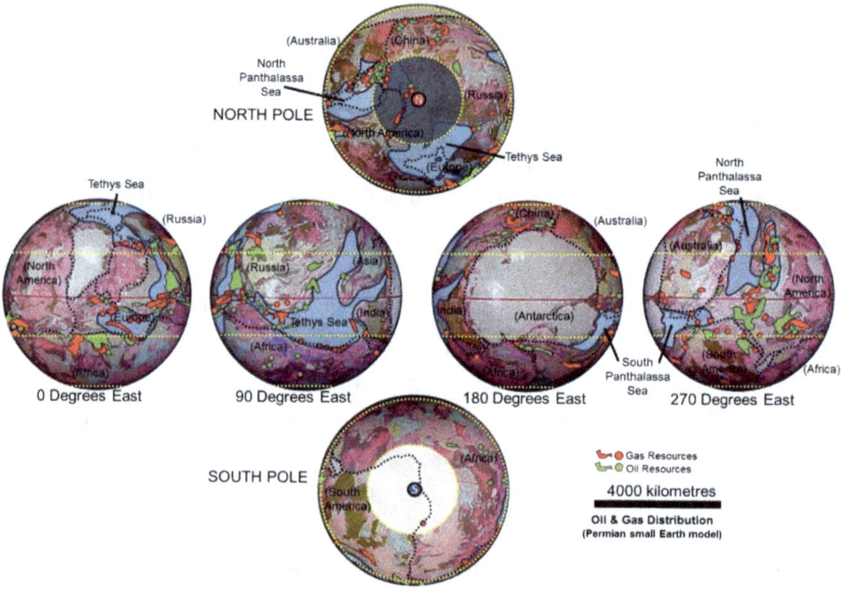

Figure 20.5. Oil (red shapes) and gas (green shapes) resource distributions shown on a Permian plate tectonic increasing radius small Earth model. Data sourced from the World Oil and Gas Map (4th edition) produced by the Petroleum Economist (2013).

Figure 20.5 confirms the distribution of both petroleum and natural gas coinciding with development of major Phanerozoic continental to marginal basin settings, with resources extending from high northern to high southern latitudes. A broad zonation of resources is evident in this figure, straddling the ancient equator and extending from low southern to mid-northern latitudes. This distribution highlights a northward shift in continents, and hence climatic zonation, over time.

20.5. Fossil Fuel Global Distributions

Plotting fossil fuel data distributions on plate tectonic small Earth models provides a unique opportunity to display the data holistically in relation to the distribution of ancient continental seas and supercontinental lands. The data for Phanerozoic shale oil and gas, coal, petroleum, and natural gas are summarized from previous Figures 20.2 to 20.5 and are layered onto the Permian plate tectonic small Earth model in Figure 20.6.

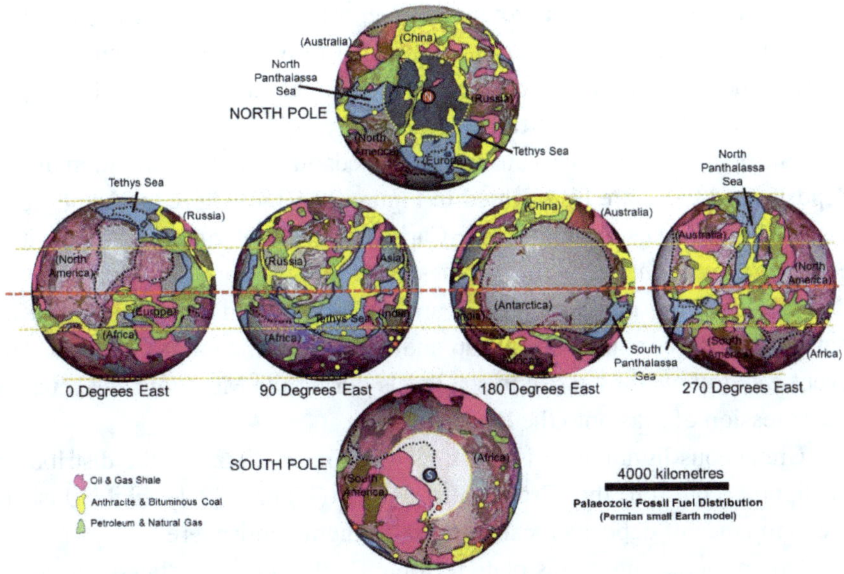

Figure 20.6. Compilation of oil and gas shale (magenta), coal (yellow), and petroleum and natural gas (green) distributions shown on the Permian plate tectonic increasing radius small Earth model. possible source there.

In Figure 20.6, oil and gas shale distributions are shown as pink shapes, anthracite and bituminous coal distributions as yellow shapes, and petroleum and natural gas distributions are shown as green shapes. Cretaceous lignite coal distributions are not shown. Petroleum and natural gas distributions include major occurrences from the Cretaceous Period and are shown in Figure 20.6 for context only.

The distribution of fossil fuels plotted on the Permian plate tectonic small Earth model coincides with the documented period of minimum sea levels occurring during the late-Palaeozoic (Figure 20.1).

On an increasing radius Earth, the Permian small Earth model represents a time of low sea levels prior to rupture of the Pangaean supercontinent to form the modern continents and opening of the modern oceans. This period was accompanied by regression and draining of continental seas from the supercontinental lands.

Most oil and gas shale deposits had already formed during the mid-Cambrian, early- and mid-Ordovician, and late-Devonian times. These deposits accumulated during a time of early colonization of the lands by primitive plants and were formed by the deposition of organic matter in a variety of depositional environments. Depositional environments included freshwater to highly saline lakes, continental to marine basins, and subtidal shelves and were often restricted to estuarine areas such as oxbow lakes, peat bogs, coastal swamps, and peatlands.

The declining sea levels during the Palaeozoic favoured formation of coal deposits and these are highlighted in Figure 20.6 by the predominant global distribution of coals during the late-Carboniferous to Permian coal era within equatorial to high northern latitudes. The Permo-Carboniferous coal era terminated during the late-Permian and Triassic Periods during onset of the Pangaean supercontinental breakup and opening of the modern oceans. This event was followed by a rise in sea levels during the Mesozoic and renewed transgression of seas onto the lands.

Cretaceous lignite coal is not shown in Figure 20.6, but the distribution of lignite coal on the Cretaceous model (Figure 20.4) highlights the predominance of deposits located in the northern hemisphere.

On an increasing radius plate tectonic Earth, during the late-Cretaceous, continental breakup, dispersal, and opening of the modern oceans was well established and the overall continental configuration was similar to the present-day. During that time, sea levels and transgression of continental seas was at its maximum, followed by a regression of the seas during the Cenozoic to the present-day configuration of oceans. Similarly, peak petroleum and natural gas development coincides with the Cretaceous period of sea-level transgression and maximum surficial areas of epi-continental seas.

The distribution of all fossil fuels on plate tectonic small Earth models highlights the global interrelationships of resources coinciding with Palaeozoic continental seas and low-lying terrestrial environments. The transition from deposition of oil and gas shale to coal to petroleum and natural gas is shown to be consistent with the various periods of maximum and minimum sea level changes occurring during periods of marine transgression and regression.

Part Three

Speculating on the Geological Rock Record

"*The history of science demonstrates…that the scientific truths of yesterday are often viewed as misconception and, conversely, that ideas rejected in the past may now be considered true*". Naomi Oreskes, 1999.

Chapter 21

A Continental Example

"*Evidence suggests that west Australia, and much of Australia, has been largely intact since about 2.5 Ga [2,500 million years ago] and that most subsequent orogenic activity has been ensialic.*" Etheridge et al. (1987).

In this chapter, the geological evolution of Australia is briefly summarized in order to demonstrate the application of increasing radius plate tectonic Earth models to the stratigraphic and tectonic history of a continent. The geological history and evolution of Australia is extensively covered in various technical and non-technical publications, including Blewett, 2012, and was briefly introduced in Chapter 9.

The following descriptions are summarized in part from Rutland *et al.* (1990) and Solomon and Groves (1994) and are supplemented by my own observations and work experiences throughout much of Australia. This example is not intended to be an in-depth history or study, as would be expected from a professional geologist, but only as an overview.

When describing the geological evolution of Australia, geographical orientations relative to the ancient equator will be given in lower case, for example, the long axis of Australia was orientated north-south throughout the Precambrian—as distinct from its current east-west orientation. In contrast, current geographical descriptors and names of continents relative to the present-day equator will be given in upper case, for example West Australia.

In 1990, Myers, in summarizing the Precambrian geological history of Australia, concluded that West Australia has had a long record of continental growth and fragmentation. Myers considered West Australia to have formed from a collage of small, accreted crustal fragments. Crustal generation was then seen as a consequence of the amalgamation of small "*rafts*" of continental crust and the destruction of intervening oceanic crust. In other words, as complex, random, amalgamation-dispersal-amalgamation plate tectonic cycles.

This interpretation of Myers contrasts strongly with the geological evidence, whereby field mapping and structural interpretation by others suggests that the older parts of both West Australia, and much of Central Australia, have been largely intact since at least 2,500 million years ago. Most subsequent tectonic activity has then been confined to basins located on the Australian continental crust, rather than within inferred intervening oceanic crust.

In contrast, on an increasing radius plate tectonic Earth, each of the modern continents have only existed in their current form since breakup of the ancient Pangaean supercontinental crust commenced some 250 million years ago. The older parts of the Australian continent had their beginnings as part of an ancient Precambrian supercontinental crustal assemblage (Figure 21.1).

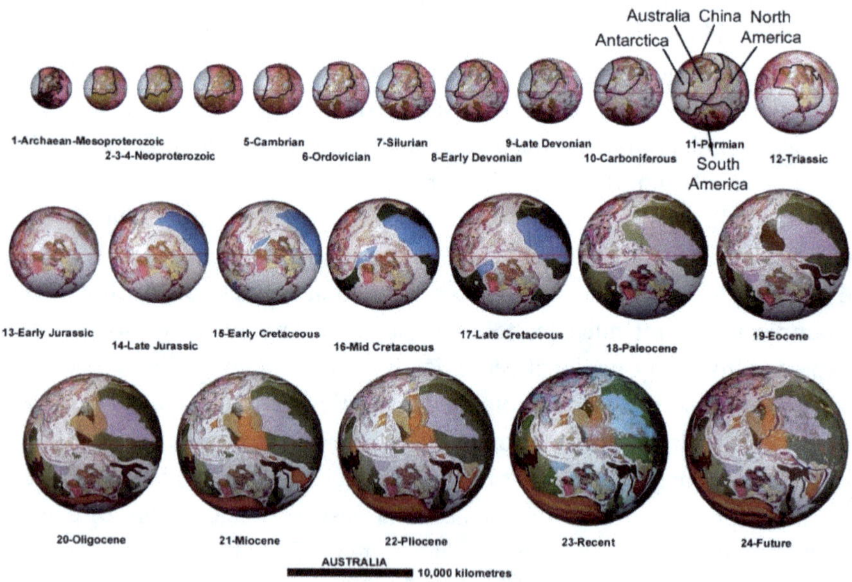

Figure 21.1 Continental development of Australia. The outline of the pre-Jurassic Australian continent is highlighted as a black line. Adjacent continents are named and highlighted on the Permian plate tectonic small Earth model only. The horizontal red line represents the location of the ancient equator. After Maxlow, 2001.

During these ancient times, the various Australian crusts abutted directly against ancient crusts from China to the north, Canada and North America to the east, South America to the south, and East Antarctica to the west—note, these orientations are relative to the ancient continental configuration. At that time, ancient Australia was located within mid- to high-northern latitudes, relative to the ancient equator, and the ancient equator passed through what is now Central and Northern Australia.

Once the ancient Pangaea supercontinent started to breakup and the modern Pacific Ocean commenced opening during the Triassic Period, the newly defined Australian continent then rotated counter-clockwise to its present east-west orientation and migrated south into its present location in the southern hemisphere.

21.1. Australian Crustal Development

Rutland *et al.* in 1990 subdivided Australia into three tectonic superprovinces, broadly based on the distribution of Archaean, late-Archaean to mid-Proterozoic, and late-Proterozoic and Palaeozoic crustal rocks (Figure 21.2).

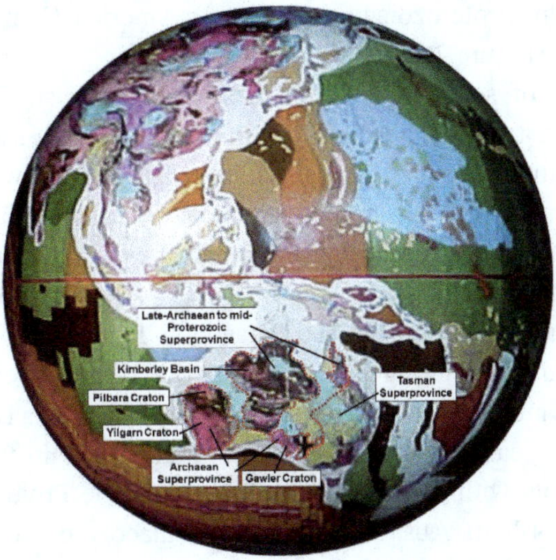

Figure 21.2. Australian Archaean, Proterozoic, and Phanerozoic Superprovinces named and highlighted by red dashed lines. Intermediate basins are Palaeozoic or younger in age.

The Australian Archaean Superprovince comprises an older Pilbara Craton—aged from 3,600 to 3,000 million years—and a relatively younger Yilgarn Craton—aged from 3,000 to 2,600 million years old.

These cratons consist mainly of granite, greenstone, sediment and iron formation rocks. The late-Archaean to mid-Proterozoic Superprovince includes provinces that are now widely separated, and fragments are presently located in West, North, Central and Southern Australia. These fragments consist mainly of sediments eroded from the older provinces, as well as interbedded volcanic rocks, and granite.

The late-Proterozoic and Palaeozoic Tasman Superprovince includes most of East Australia and extends north to include Mesozoic and Cenozoic provinces in New Guinea and south to include provinces in East Antarctica. The Tasman Superprovince is essentially sedimentary in origin and is made up of a number of distinct fold belts and orogenic zones.

On an increasing radius plate tectonic Earth, during pre-breakup supercontinental times, the Proterozoic sedimentary basins of Northern and Central Australia formed part of an extensive global network of sedimentary basins. These basins extended north into the Proterozoic basins of Alaska, Canada, Northern Russia, and Asia, and to the east and south these basins were linked to the Proterozoic basins of North America, Central America, and South America (Figure 21.3).

Deposition of sediments within these ancient Australian sedimentary basins was most active to the south—within what is now Eastern Australia—and this deposition extended into adjoining regions in New Zealand, South America, North America, and Antarctica. Breakup of this extensive sedimentary basin occurred during Permian times, during initial opening of the South Pacific Ocean.

Throughout the Proterozoic and Palaeozoic Eras, this global network of basins remained intact during an extended period of crustal extension, crustal rifting, crustal mobility and ongoing basin development. During this long period of geological activity, there was a relative shift in the deposition of eroded sediments within Australia, shifting from the Proterozoic basins of Central and Northern Australia, into new Palaeozoic sedimentary basins located in Eastern Australia.

Crustal rupture and development of the modern continents and surrounding oceans on an increasing radius plate tectonic Earth first commenced during the Permian Period. The first formed oceans were located adjacent to the Pilbara and Kimberley coastlines of northwest Australia, and along the south coast—now east coast of Australia. These two rupture sites represent the initiation and opening of the North and South Pacific Oceans respectively.

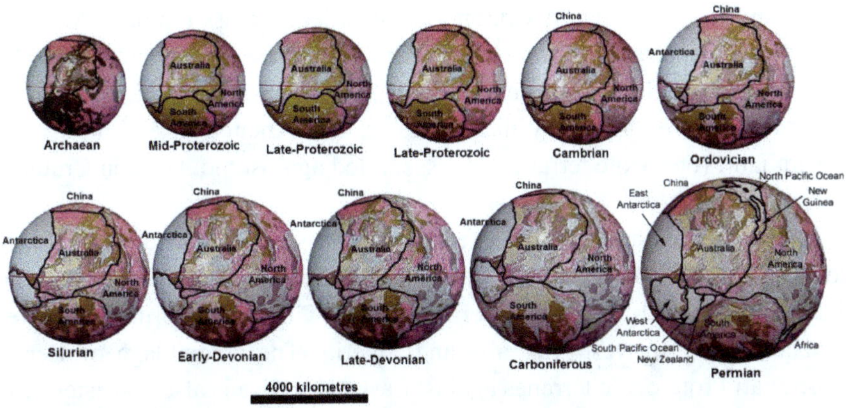

Figure 21.3. Detailed pre-breakup crustal history of Australia and adjoining continents on plate tectonic small Earth models. Red lines represent the ancient equator, black lines are continental outlines, pink and red colours represent distribution of Archaean and Proterozoic cratonic rocks, khaki colour represents distribution of Proterozoic sedimentary basins, brown colours represent Palaeozoic sedimentary basins, pale blue and yellow colours represent Mesozoic and Cenozoic sedimentary basins respectively. Maxlow, 2001.

The early North and South Pacific Oceans then rapidly opened as two separate oceans during the Permian and early-Triassic Periods. During the Triassic Period, Australia continued to migrate away, and finally separated, from North America, and the two oceans merged to form a single Pacific Ocean.

Similarly, opening of the Indian Ocean on an increasing radius plate tectonic Earth commenced during the Jurassic Period, separating Australia from China and Southeast Asia. Rifting between Australia and East Antarctica then commenced during the Paleocene Epoch, and all of these oceans have continued to open to the present-day.

21.1.1. Archaean Superprovinces

The assemblage of Australia on an Archaean to mid-Proterozoic increasing radius plate tectonic Earth is shown in Figure 21.3—top left small Earth models—in relation to ancient continental crusts from adjoining continents. This assemblage comprises the Yilgarn and Pilbara Cratons assembled against the Kimberley, North Australian and Central Australian Proterozoic terranes, and the Gawler Craton of South Australia—see Figure 21.2 for location of named cratons. These cratons, basins, and terranes are, in turn, assembled against the Canadian Slave, Churchill, and Superior Provinces.

Similarly, the Yilgarn Craton is assembled against remnant Archaean cratons and Proterozoic terranes from China, Southeast Asia, and Korea. The Yilgarn and Gawler Cratons and Central Australian Proterozoic terranes are in turn assembled against Wilkes Land in East Antarctica, and the Central Australian Proterozoic terranes are assembled against the Amazon Craton of South America.

This assemblage is consistent with the tectonic and stratigraphic correlations of Li *et al,* 1996, who suggested from palaeomagnetic evidence that early- to mid-Proterozoic terranes from Tibet and North China were attached to North-west Australia and Siberia. The assemblage of Central Australian Proterozoic terranes against East Antarctica is also consistent with Zhao *et al*, 1999, and consistent with a plate tectonic connection between west Canada, Australia and China during the Proterozoic Era.

Rocks older than 2,500 million years old are widely exposed in the Archaean superprovinces, and these rocks are also suspected to underlie terranes in each of the younger Proterozoic provinces. The granite and greenstone rocks from the Yilgarn and Pilbara Cratons are very similar, although the Pilbara Craton lacks a distinct directional grain, as seen in the Yilgarn Craton, and is older.

On an increasing radius plate tectonic Earth, this directional grain has a similar orientation to the directional grain shown in the adjoining Canadian Churchill and Superior Provinces. This grain implies that the two regions had a similar, and possibly related tectonic history, prior to ongoing basin extension during the Proterozoic Era.

The Pilbara Craton, and much of the Yilgarn Craton, consists of about 30 vertical kilometres of continental crust, and the lower part is made up of high-grade metamorphic rocks. Other Archaean terranes show similar crustal patterns, which suggests that Archaean high-grade metamorphic rocks are generally representative of the lower crust.

By the close of the Archaean Eon, the crust of the Pilbara and Yilgarn Cratons, and by inference the Gawler Craton from South Australia, underwent little change in volume. By the Early Proterozoic Era, basement extension was well established, forming the loci for ongoing extension, basin formation and orogenesis within each of the Northern and Central Australian Proterozoic terranes.

In 1989, Barley *et al.* suggested that deformation of the older greenstone belts of the Archaean Superprovince, at about 2,650 to 2,600 million years ago, reflected a westward propagation of stresses associated with compression of the greenstone belts in the southeast Yilgarn Craton. The younger greenstones of the Yilgarn Craton were then formed in linear belts related to this stress regime and included abundant felsic and ultramafic volcanic rocks.

On an increasing radius plate tectonic Earth, this compressional phase is a vector stress component related to oblique basement extension, and counterclockwise translation and rotational motion between Australia and North America. This crustal motion occurred between the Canadian provinces, the Pilbara Craton and Northern Australian terranes, relative to the Yilgarn Craton. This oblique continental basement extension and rotational stress regime is consistent with orogenic events affecting Proterozoic terranes throughout Northern Australia during the early-Proterozoic Era.

21.1.2. Archaean to Proterozoic Superprovince

The late-Archaean to mid-Proterozoic Australian Superprovince is described as a continental basin-forming episode, resulting from local extension of pre-existing Archaean continental crust. Rutland *et al.* in1990 noted that each of the provinces making up this superprovince show a uniformity in their older age patterns. There is also a consistency in orogenic and magmatic events occurring prior to about 1,800 million years ago. These periods of tectonic activity were considered to have been important in defining the present structure of the superprovince, which further controlled later tectonic events occurring during the late-Proterozoic and Palaeozoic Eras.

The Hamersley Province of the Pilbara Craton represents a preserved transition from the Archaean to early-Proterozoic Eras and displays abundant banded iron formation rocks dated at about 2,500 million years old. These rocks conformably overlie basalts and sandstones dated at 2,750 million years old. Early-Proterozoic rocks older than 2,000 million years ago are also recognized in the Pine Creek, Granites-Tanami and Gawler regions of Northern and Southern Australia. Elsewhere, Proterozoic rocks cover a large part of Central and West Australia, and they occur in a number of discrete provinces overlain in part by Phanerozoic sediments.

On an increasing radius plate tectonic Earth, this continental basin formation event and subsequent tectonism is consistent with oblique east-west—Precambrian orientation—crustal extension between the Australian Archaean Superprovinces and the Canadian Archaean Slave, Churchill and Superior Provinces. Translational motion and orogenesis accompanying this crustal extension also occurred during the early-Proterozoic Era, which was reactivated again during later orogenic events.

Much of the Pilbara and Yilgarn Cratons of Australia, and the Slave, Churchill and Superior Provinces of Canada, are here inferred to represent exposed land surfaces on an increasing radius Earth during the Proterozoic Era. These areas supplied sediment to each of the surrounding Proterozoic basins, which formed part of an extensive global network of continental seas.

This occurrence is supported by evidence from Proterozoic sedimentary rocks, as well as intrusive dolerite sills overlain by late-Proterozoic glacial rocks, in the Kimberley Basin of northern Australia. These sediments provide evidence for strong unidirectional longshore currents, which confirm that landmasses existed to the west and northeast of the present Kimberley region.

Similarly, throughout the Precambrian and Palaeozoic Eras the Central Australian basins on an increasing radius plate tectonic Earth were located in equatorial to low northern latitudes. These basins extended into mid- to high-northern latitudes in the West Australian regions. This location is supported by the occurrence of late-Proterozoic glacial marine deposits interbedded with warm-water carbonate sediments in the Central Australian basins. Palaeomagnetic evidence confirms deposition of these marine glacial deposits in equatorial latitudes and is also demonstrated by the presence of carbonate-rich sediments.

Each of the Proterozoic provinces of Australia are shown to have similar early basin forming histories, with very little evidence for tectonic activity between 2,500 and 2,100 million years ago. Most of these provinces underwent basin formation, initiated by crustal extension, between 2,100 and 1,900 million years ago. This basin formation event was then abruptly terminated by an orogenic event between 1,880 and 1,840 million years ago.

The small Earth models in Figure 21.3 suggests that these orogenic events in North and East Australian were aligned with the Grenville Orogenic event of Eastern North America. The Southern Australian Albany-Fraser and parts of the Central Australian Musgrave provinces are also considered by Solomon and Groves to form an extension of the Grenville fold belt, although Figure 21.3 suggests that these were separate parallel zones.

In Australia, this orogenic event was then followed by widespread intrusion and extrusion of felsic igneous rocks between 1,600 and 1,450 million years ago, and intrusion of diamond-bearing volcanic pipes in the Halls Creek fold belt in Northern Australia at around 1,200 million years ago.

21.1.3. Tasman Superprovince

The late-Proterozoic to Palaeozoic Tasman Superprovince involved a long history of orogenic fold belt activity along the Eastern margin of Australia. The Tasman Superprovince is made up of the Kanmantoo, Lachlan, Thomson, Hodgkinson-Broken River and New England fold belts (Figure 21.4). These fold belts form part of an orogenic belt extending south to the Borchgrevink Orogen in Antarctica, and north to Papua New Guinea.

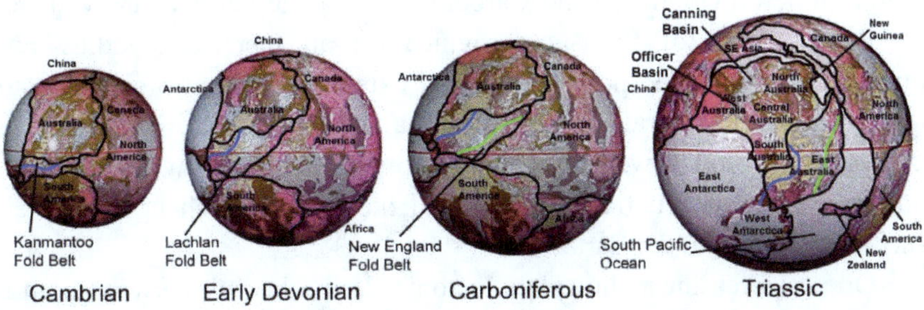

Figure 21.4. The late-Proterozoic and Palaeozoic Tasman Superprovince on an increasing radius plate tectonic Earth. Red, blue and green lines represent margins of the Kanmantoo, Lachlan and New England fold. Data after Walshe *et al.*, 1995.

Solomon and Groves described the Tasman Fold Belt system in Eastern Australia as an eastward growth of the Indo-Australian plate, which was considered to have been active from the early-Cambrian Period to the early-Mesozoic Era. During that time successive fold belts, island arcs or terranes were said to have been accreted onto the Australian craton from west to east, and also from the northeast.

It was suggested by Solomon and Groves that growth of the East Australian region was mostly by westward subduction and terrane accretion, preceded and followed by orogenic movements. Granite studies though provide evidence for the occurrence of Proterozoic basement rocks in the region, which suggests that orogeny could not be solely the consequence of lateral crustal accretion.

This conventional plate tectonic plate history for the Tasman Superprovince cannot be reconciled on an increasing radius plate tectonic Earth. On an increasing radius Earth, the Tasman Superprovince was located between the central Australian Proterozoic terranes, the Amazon Craton of South America and remnant terranes in Central America. The superprovince also extended east to include Palaeozoic basins in Southeast North America, and south to include basins in West Antarctica and South America.

In East Australia, the ages of these fold belts, as well as associated tectonic activity and granite intrusions throughout the Tasman Superprovince, become progressively younger when moving from northwest to southeast—relative to assemblage on the small Earth models. Crustal deformation in the Kanmantoo and related Adelaide fold belts, for instance, was late-Cambrian in age, and in the Lachlan Fold Belt the major deformation events were late-Ordovician, mid-Devonian and mid-Cretaceous, while in the New England Fold Belt deformation was mid-Carboniferous to late-Permian and younger.

Development of the superprovince on an increasing radius Earth represents an extended period of crustal extension and basin sedimentation located between Australia, North America, and South America. Orogenesis and eastward growth were then related to periodic pulses within the basin, during ongoing gravitational collapse of the basin and changes in overall surface curvature of the Earth.

On an increasing radius plate tectonic Earth, the link between both the Cordilleran and Andean orogenic belts of North and South America was maintained throughout the late-Proterozoic and Palaeozoic Eras, forming a common link between the Tasman Superprovince of Australia and the Americas. Rifting between Australia and the Americas during the Permian Period, and rapid opening of the Pacific Ocean during the Mesozoic Era, then terminated this established link.

During opening of the Pacific Ocean, the Andean and Cordilleran orogenic belts were fragmented. The resultant mountain belts then remained as part of South America and North America, with the New England fold belt remaining as a small remnant within East Australia, and smaller remnants remain in New Zealand.

Chapter 22

A Paradigm Shift in the Rock Record

"*...the past history of our globe must be explained by what can be seen to be happening now.*" Hutton, 1788.

The opportunity is now taken to speculate on the geological implications raised in this book, after considering global tectonic observational data modelled on an increasing radius plate tectonic Earth.

It has been shown that, by using modern geological mapping of the seafloor and continental crusts to constrain plate assemblages, each plate assembles together precisely, with one plate-fit option, on smaller radius Earth models. Similarly, the extensive range of plotted modern global data substantiates these small Earth models and shows that distribution of this data is better suited to an increasing radius plate tectonic Earth model. As Shields noted in 1997 "*Ultimately world reconstructions must be congruent not only with the data from geology and geophysics, but also with palaeobiogeography, palaeoclimatology, and palaeogeography.*"

22.1. Primitive Atmosphere and Hydrosphere

Fundamental to the concept of an increasing radius plate tectonic Earth is the premise that ocean waters and atmospheric gases have been accumulating throughout much of geological time, in sympathy with the formation of ancient supercontinental crusts and new seafloor volcanic crusts.

It was considered by Bailey and Stewart in 1983 that "*...for an Earth undergoing expansion with time, the bulk of the oceans would have to be outgassed since the Palaeozoic, requiring fundamental changes in atmosphere, climate, biology, sedimentology and volcanology.*" It was also considered by Carey in 1988 that, "*as the generation of the ocean floors depends fundamentally on the out gassing of juvenile water, it would therefore be expected that the volume of seawater* [and atmospheric gases] *and capacity* *of the ocean basins both increased, but not necessarily precisely in phase, in a related way.*"

The primitive atmosphere and hydrosphere—the combined mass of all water and atmospheric gases found on Earth—was considered by Lambert in 1982 to have been formed largely from elements and molecules degassed from the Earth's interior, and subsequently modified by physical, chemical, and biological processes. As early as 1975, Rubey proposed that degassing—the removal of dissolved gases from liquids [inclusive of molten magma]—has been a continuous or recurrent process, which is still occurring today. Rubey further suggested that "*the whole of the waters of the oceans have been exhaled from the interior of the Earth, not as a primordial process, but slowly, progressively and continuously throughout geological time.*"

This consideration contrasts strongly with what current mainstream science says about the origin of water on the earth where: "*Earth's water originally accumulated during planetary accretion* (Wood et al., 2010) *and/or soon afterward* (Albarede, 2009) *by delivery from asteroids* (Morbidelli et al., 2000) *and comets* (Hartogh et al., 2011). *Ever since, a changing portion of this water has been stored within our planet's surface oceans.*"

Studies of melted igneous rocks carried out since the 1970s and 1980s have shown that the solubility of water in melted rocks increases with increasing pressure and temperature, until a maximum value is reached in the mantle. Quoted examples range from 14 to 21 percent by weight of water dissolved in volcanic rocks at temperatures varying between 1,000 to 1,200 degrees Celsius, accompanied by high pressures. For silica-rich magmas, carbon dioxide was also shown to be readily dissolved, in particular under high pressures.

It was concluded from these studies that, if water and carbon dioxide were available, they would both be highly soluble in magmas normally generated in the upper mantle.

Eggler in 1987 further considered that the elements carbon, oxygen, hydrogen, and sulphur would also exist in the Earth's mantle, within volatile-bearing minerals—minerals capable of retaining these elements in their crystal lattices. It was also considered possible that solution of elements in crystalline minerals represents a significant repository for each of these elements within the Earth's mantle and crust. These volatile-bearing minerals occur in most volcanic and metamorphic rocks, as well as in carbonate rocks and sulphide ores. Similarly, each of the volatile minerals such as water, carbon dioxide, carbon monoxide, methane, hydrogen, sulphur dioxide, and hydrogen sulphide are soluble in silica-rich melted rocks, with water and methane being more soluble than the rest.

The main controls on the chemistry of the Archaean hydrosphere were considered by Lambert to be reactions with hot igneous rocks, seafloor weathering, and additions of new water degassed from the crust and mantle, with minor input from river systems. Oxygen isotope studies from Archaean sediments suggest that fluids in existence at that time had a similar composition to modern seawater, with prevailing seawater temperatures of up to 70 degrees Celsius.

One of the most important events occurring during the Proterozoic Era was the accumulation of oxygen in the Earth's atmosphere. Though oxygen may have been released by photosynthesis or chemical processes well back in Archaean times, it was unable to build to any significant degree until chemical sinks—the presence of unoxidized sulphur and iron—had been filled.

Until about 2,300 million years ago, it had been estimated by others that oxygen was probably only 1 percent to 2 percent of its current level today. Banded iron formation rocks, which now provide much of the world's iron ore, were also a prominent chemical sink where most iron accumulation ceased after 1,900 million years ago. Red beds—sediments coloured by the iron oxide mineral hematite—indicate an increase in atmospheric oxygen after 2,000 million years ago as they are not found in older rocks.

This oxygen build-up is considered in the literature to be due to two main factors: a filling of the chemical sinks, and an increase in carbon burial, which sequestered organic compounds that may have otherwise been oxidized by the atmosphere.

On an increasing radius plate tectonic Earth, it is suggested that degassing of the dissolved volatile elements and minerals from a primitive mantle and crust commenced during early-Archaean times. This degassing first commenced once the primitive crust had cooled sufficiently to start crystallizing and expelling excess volatile elements. It is envisaged that, once degassed, these elements would have been retained as water-rich gases in the atmosphere until such time as temperatures had cooled sufficiently to retain liquids on the surface of the Earth.

Evidence suggests that a reduced atmosphere was well established during Archaean times, comprising mainly water and carbon dioxide, with lesser carbon monoxide, nitrogen, sulphuric acid, sulphur dioxide, hydrochloric acid, and small amounts of nitric acid and carbonic acid. The absence of an oxygen source implied to Lambert that oxygen was not a stable component of the early atmosphere, but instead, was generated later by secondary processes, such as photosynthesis and microbial action.

22.2. Archaean Crust-Mantle

The Archaean extends in time from around 4,000 to 2,500 million years ago. On an increasing radius plate tectonic Earth, this represents a period of time where surface temperatures were becoming sufficiently low to allow crystallization and solidification of a thin crust over a mantle that still retained a relatively high temperature gradient.

As cooling progressed, the early crust thickened and extended in area as primitive volcanism became widespread. Volcanic activity covered the entire primordial Earth surface, and this activity was associated with a complex system of mantle plumes, where hot volcanic rock was rising through the Earth's mantle to form the Earth's crust.

During this process, hot gases were also degassed from the mantle to form a primitive atmosphere while later, once surface temperatures had cooled sufficiently, condensed fluids started to pool in low-lying depressions.

The primitive early-Archaean Earth had a radius of approximately 1,700 kilometres. It is suggested that, by that time differentiation—a chemical processes whereby molten magmas undergo bulk chemical change—of a molten pre-Archaean Earth silica-rich magma was sufficiently advanced to form a distinct core and mantle prior to cooling and crystallization to form a primitive surface crust.

The presently known Archaean geological history commenced around 4,000 million years ago, with the stabilization and preservation of a thin outer primitive crust prior to onset of increasing Earth radius and fragmentation of the crust during mid- to late-Archaean times (Figure 22.1).

The formation of localized stable patches of juvenile crusts, as suggested by Moorbath in 1982, is consistent with an early-Archaean increasing radius plate tectonic Earth model. During the early-Archaean, as the primitive silica-rich crust progressively cooled and crystallized, these patches of juvenile crusts eventually coalesced to enclose the entire primordial Earth with stabilized crust. Because of the small size of the primordial Earth, this event may have been globally synchronous, occurring at the same time all over the primordial Earth.

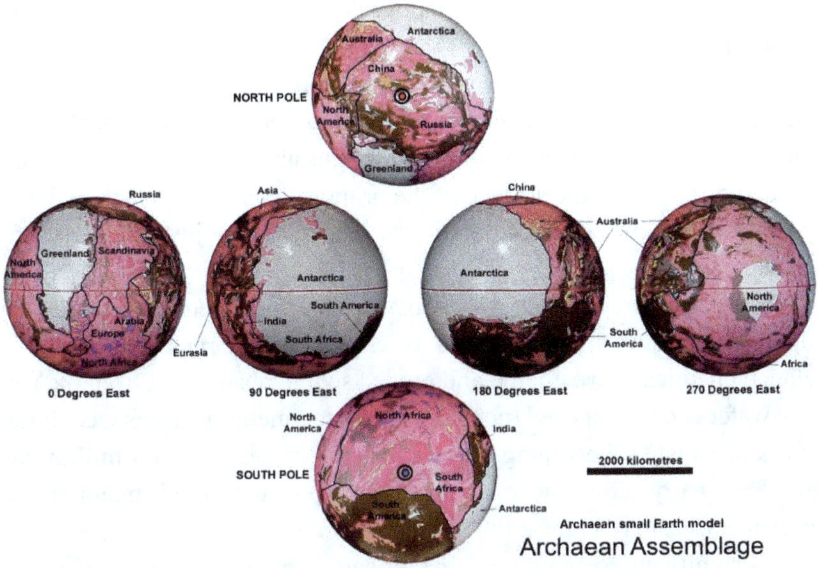

Figure 22.1. Primordial Archaean to Proterozoic increasing radius plate tectonic small Earth model. Cratons are shown as pink and red, Proterozoic basement rocks are shown as khaki. Outlines of the remnant present-day ancient crusts are shown as black lines. (Present-day Antarctic and Greenland icesheets are shown as pale blue areas). Maxlow,2001.

The following proposal for crust-mantle evolution and development of volcanic, granite, and metamorphic crusts during the Archaean is based on a similar, albeit conventional proposal put forward by Kröner in 1982.

After the Earth-Moon system had formed and prior to the early-Archaean cooling and crystallization phase, it is suggested that the pre-Archaean Earth was molten and volcanically turbulent. At that time, patches of primitive juvenile crusts may not have survived the extensive volcanic plume activity and may have been, in part, re-absorbed back into the mantle, further reducing mantle temperatures in the process.

By about 4,000 million years ago, mantle temperatures and heat flow had decreased sufficiently to maintain a granite and volcanic lava-dominated crust, allowing patches of juvenile volcanic rock to rapidly increase in height, area, and thickness. Growth of this crust may have also occurred through chemical change and crystallization of the underlying upper mantle, giving rise to further cooling, thickening, and stabilization of the overlying crusts.

In contrast, regions away from the main volcanic centres may have maintained a relatively thin upper crust, which may have ultimately formed the observed loci for on-going crustal weakness, crustal extension, and basin formation during latter eras

During early-Archaean times, this primitive crust was made of predominantly silica and alumina-rich rocks, comprising a mixture of intruded granites and an assortment of iron and magnesia-rich volcanic rocks. Once cooled and stabilized, the surface of the primordial Earth was then exposed to erosion by wind and waters, in particular erosion by hot juvenile waters from volcanic eruptions.

This erosion process initially deposited greywacke sediments in low lying sedimentary basins and comprised volcanic ash and rock debris and later included chemically weathered and eroded sediments derived from the action of the waters on the exposed rocks. The early sedimentary basins were initially small and isolated, becoming more extensive by about 3,800 million years ago. This early crust now forms the oldest preserved magmatic and sedimentary basin remains known on Earth today.

It is significant to again note that modern geological evidence shows that all Precambrian cratons throughout the world have been eroded by the order of 10 to 40 kilometres vertical thickness. This eroded material now forms the surrounding younger sedimentary basins and orogenic belts making up much of the continents on Earth today and may have been recycled many times throughout Earth history.

The implication from this evidence is that any discourse on Archaean crusts on present-day Earth only represents the deeper ancient crusts, and not necessarily the exposed ancient rocks, or topography, present on the original ancient Archaean land surface.

The formation of silica and alumina-rich granite and related volcanic rocks from a dense iron-magnesia-rich upper mantle source suggested to Lambert that the primordial molten crustal material may have initially formed as a surface layer of not more than two to three kilometres thick. Lambert considered that the high thermal gradient existing during pre- to early-Archaean times suggested that the crustal material would have continued to segregate to also form a lower region of iron-magnesia-rich rocks extending from depths of 30 to 100 kilometres thick.

The present-day survival of a 30- to 40-kilometre-thick Archaean crust further implied the presence of a thick stable sub-crustal zone, which may have ultimately extended still further to greater than 200 kilometres thick.

Kröner envisaged that by about 3,600 million years ago, continental crust had attained sufficient thickness and rigidity to support large sedimentary basins. On an increasing radius plate tectonic Earth, once cooled, this continental crust was subjected to the initial phases of crustal extension and fracturing during early increase in Earth radius. The magnitude of this increase in Earth radius, and crustal extension, during these times was of the order of microns per year and may have initially been represented by simple cooling cracks.

Over an extended period of time—hundreds of millions of years—this small crustal extension process ultimately gave rise to crustal fracture, fissuring, and rifting, generating large sedimentary basins or sharply bounded, fault-controlled landscapes. These low-lying basin areas collected water, shallow-water sediments, and extruded volcanic lava originating from the upper mantle.

A small proportion of this volcanic lava reached the surface as magnesia-rich komatiite lava, which is now known to be unique to the high mantle temperature conditions existing during Archaean times. A much larger volume of magma was also retained near the base of the crust to become the source for later volcanic eruptions and granite intrusives.

It is envisaged that on-going crustal extension of the stabilized Archaean crust during onset of an increase in Earth radius resulted in early crustal fracturing and initiation of a global network of crustal weakness. Faulting and crustal extension was focused within this network and was accompanied by deposition and accumulation of sequences of volcanic lava and eroded sedimentary rocks.

The Archaean is also renowned for the presence of high-grade metamorphic terranes, as preserved throughout the ancient crusts today. These terranes represent largely pre-volcanic silica and alumina-rich rocks, together with early sedimentary deposits and granites produced during early stages of crustal fissuring and basin formation.

Crustal extension and sub-crustal mantle movements during on-going increases in Earth radius and surface area then provides a mechanism for stretching and extension of these rocks within the Archaean crust. Crustal extension, accompanied by high mantle temperatures also provided the necessary heat-flow for recrystallization and metamorphism of these high-grade gneissic terranes.

The Archaean Eon extended over an extremely long period of time, and the increase in Earth radius during that time was less than the thickness of human hair per year. It is envisaged that, over the subsequent 1,500-million-year history, the crustal rocks were then fractured, intruded by volcanic lava, eroded during weathering, and sediments were deposited in low-lying basins, all of which were complexly contorted and folded as a result of the steadily changing surface curvature.

Because of the extremely long time involved, as well as the minute rate of increase per year, it is further envisaged that on-going erosion of the surface of the entire Earth ultimately maintained a very flat landscape throughout the latter part of this eon, where erosion kept pace with changing surface curvature. This landscape may have been barren, rocky, and inhospitable by present-day standards.

The latter phase of Archaean crustal development eventually gave rise to formation of very large, stable sedimentary platform basins and shallow seas with a low elevation contrast between the continents and seas.

Continental sedimentary input during that time was limited to mainly chemically precipitated sediments, such as calcium and silica-rich chert and banded iron formation rocks, with lesser fine-grained sediments. These rocks were common during mid- to late-Archaean times and formed laterally extensive deposits which are now known throughout most of the ancient crusts on the present-day Earth.

During the Archaean Eon, the ancient magnetic poles were located in what are now Mongolia-China and west Central Africa, and the ancient equator passed through North America, the Baltic region, Antarctica, and Australia. There was also a primitive climate zonation, which persisted for the entire duration of the eon. It is significant that primitive microbial life forms first evolved in the seas during this eon, possibly evolving from within carbon-rich sediments formed during the early anoxic atmosphere and seawater phase, prior to onset of an oxygen-rich atmosphere.

22.3. Proterozoic Earth

The Proterozoic Eon extended in time from around 2,500 to 541 million years ago. The eon was characterized by comparative crustal stability, dominated by stable exposed continental lands, extensive sedimentary platform basins and shallow seas during slowly increasing changes in surface curvature.

The change in Earth radius throughout the Archaean to mid-Proterozoic times amounted to approximately 60 kilometres increase in radius spread over a 3,000-million-year history, and around 400 kilometres increase in Earth circumference. The increase in circumference during that time was reflected by the steadily increasing surface areas of the sedimentary basins surrounding each of the ancient cratons.

On a Proterozoic increasing radius plate tectonic Earth, there was not a lot of difference between late-Archaean and Proterozoic times. The location of the poles and equator were essentially the same, and the distribution of exposed ancient supercontinental lands was also much the same. The distinguishing feature of Proterozoic history was that increasing changes in surface curvature were slowly starting to take effect by increasingly elevating the exposed lands.

Like the Archaean, the rate of change in surface curvature during the Proterozoic was still small enough for erosion to continually maintain a relatively flat landscape. In time, this increasing elevation contrast began to slowly increase the rate of erosion of the exposed lands, which also influenced changes to the distribution of seas and corresponding coastal outlines of the established supercontinental lands. These changes to the coastal outlines then formalized the existence and configuration of both the ancient seas and the ancient Rodinia supercontinent (Figure 22.2).

After the mid-Proterozoic—about 1,700 million years ago—the steadily increasing Earth radius and crustal conditions were reflected in the gradual onset of increasing surface curvature and increasing elevation contrast between the lands and seas. This increasing elevation contrast gave rise to changes to, and distribution, of sediments deposited in the established network of shallow seas during these times, changing from predominantly chemically precipitated sediments to predominantly course grained sand and silt. The course grain-size of these sediments, in turn, reflected the increased erosional energy available to both erode and move these sediments. This energy, in turn, is confirmation of the increased elevation contrast of the land surface during these times.

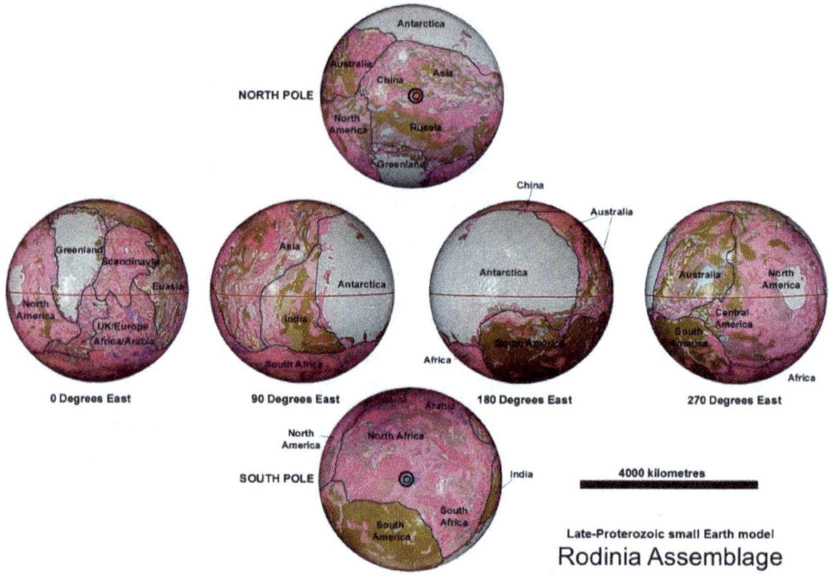

Figure 22.2. Late-Proterozoic Rodinia plate tectonic small Earth supercontinent. The model shows the distribution of Precambrian cratons (pink and red) inferred to represent exposed land, and a network of basins (khaki and brown) inferred to represent continental seas. The black lines represent outlines of remnant present-day continents. Maxlow,2001.

Another distinguishing feature of Proterozoic times was the changing atmospheric conditions. These conditions gradually changed from chemically reduced atmospheric conditions throughout much of the earlier Archaean, through to the accumulation of atmospheric oxygen during the Proterozoic. This transition was marked in the rock-record by the increasing accumulation of iron oxide-rich banded iron formation, calcium and magnesium carbonate rocks, and hematite-rich red beds during this time.

These rock types are now found within most of the ancient terranes located on the present-day Earth and they signify globally synchronous conditions existing throughout Archaean and Proterozoic times. This transition also coincided with development and preservation of the earliest life forms, soon to explode in diversity during the following Phanerozoic Eon.

Throughout the Proterozoic Eon, the ancient North and South Poles remained within what is now northern China-Mongolia and West Africa respectively. The ancient equator continued to pass through what is now North America, East Antarctica, Australia, Greenland, and Scandinavia, as well as

Europe, Asia, and India, essentially the same as during Archaean times. This continuity of magnetic pole locations, as well as continuity of the equator and associated climate zonation is testament to the extremely long period of crustal stability existing during the Rodinia supercontinental configuration.

On subsequent, younger small Earth models, this early Rodinia supercontinental assemblage represents a precursor to the better known Gondwana supercontinent, and ultimately to the Pangaea supercontinental configuration. This transition from one supercontinent to the next was intimately related to the variation in coastal outlines during changes in sea levels, and the passive nature of this transition is unique to an increasing radius Earth. On an increasing radius plate tectonic Earth, it was only Pangaea that eventually broke up and dispersed to form the modern continents and modern oceans at around 250 million years ago.

Tectonism during the Proterozoic was essentially confined to an established global network of crustal weakness, which was coincident with the network of sedimentary basins. This tectonism was intimately related to increases in surface curvature and was marked by increased jostling of the ancient cratons. Tectonism was also accompanied by on-going crustal extension within the sedimentary basins.

This tectonism generated long, narrow, elongated geosynclinal zones and may have also been accompanied by renewed volcanic activity and granite intrusion. Each of these zones coincided with the established network of sedimentary basins and in turn continued to represent a global network of crustal weakness.

Proterozoic crustal extension was confined to this network of sedimentary basins and was also accompanied by high heat flow from the mantle. Multiple phases of crustal extension-mobility-extension are a prime feature of increasing radius plate tectonic Earth models, continuing through to Palaeozoic times, and prior to continental crustal rupture during the early-Permian and continental breakup and dispersal during the Mesozoic and Cenozoic to the present-day.

22.4. Palaeozoic Earth

The Palaeozoic Era represents the early part of the Phanerozoic Eon and extends in time from about 541 to 252.6 million years ago. The boundary between the Proterozoic Eon and the Palaeozoic Era is located at the base of the Cambrian Period, coinciding with a time when fossils of hard body marine animal species became noticeably abundant. Compared to the preceding Precambrian times, the Palaeozoic was a time of increasingly dramatic geological, climatic, and species evolutionary change.

The Cambrian Period, in particular, witnessed the most rapid and widespread diversification of life in Earth's history in which relatives of most modern life forms, including fish, arthropods, amphibians, and reptiles first appeared and evolved. Life during that time began in the seas but eventually extended onto the land, and, by the late-Palaeozoic, great forests of primitive plants covered the continents, many of which formed the coal beds of Europe, Russia, Australia, and North America. Towards the end of this era, large reptiles became dominant and modern plants also appeared.

On an increasing radius plate tectonic Earth, the transition from late-Proterozoic to the Palaeozoic represents an accelerating phase of increasing Earth radius, along with accelerating increases in surface area and surface curvature. During the Palaeozoic, distinct elevated ancient land surfaces were becoming prevalent, an erosional cycle was prevalent, and rivers were actively carving valleys and modifying landscapes.

These land surfaces were, in turn, surrounded by a network of relatively shallow continental seas, which, in turn, defined the outline of the ancient Palaeozoic supercontinents punctuated by major extinction events during changes to both sea levels and coastal outlines.

At that time, Gondwana (Figure 22.3) was subdivided into a North and South Gondwana, separated in part by an early Panthalassa Sea—the precursor to the modern Pacific Ocean.

North Gondwana was made up of Australia, East and West Antarctica, and India, and also possibly included Tibet and Afghanistan. South Gondwana, joined in Madagascar, comprised Africa, Arabia, and South America. The ancient Laurentia and Baltica sub-continents were made up of North America, Greenland, and Scandinavia, as well as smaller islands making up England and Ireland. Similarly, the ancient Laurussia sub-continent was centred on the Precambrian regions of Mongolia and northern Russia.

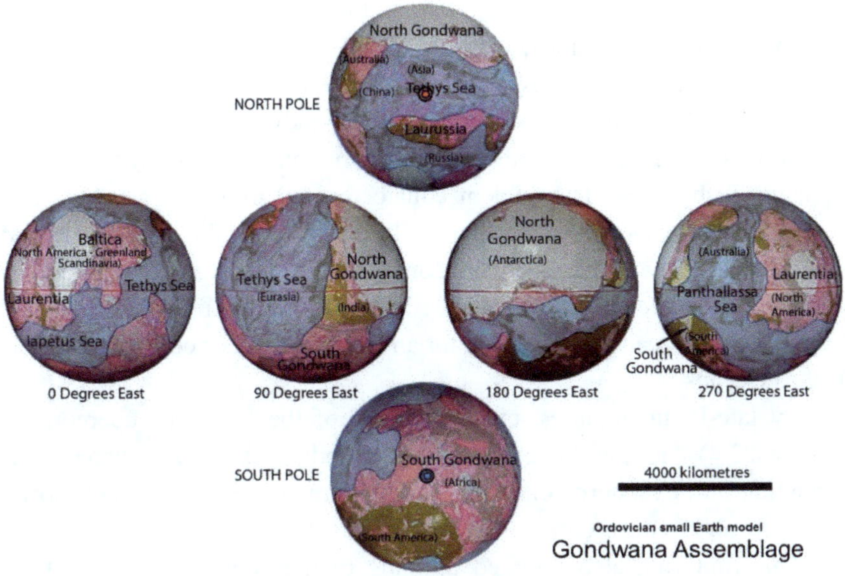

Figure 22.3. The Ordovician Gondwana plate tectonic small Earth assemblage. This model shows the ancient coastline distribution (blue lines) defining North and South Gondwana in relation to Laurentia, Baltica, and Laurussia. The ancient Tethys, Iapetus, and Panthalassa Seas form part of a global network of continental seas (shaded areas) surrounding each of the exposed supercontinents. Maxlow,2001.

These exposed Gondwanan land surfaces were surrounded by the ancient continental Tethys, Iapetus, and Panthalassa Seas. Remnants of these seas are now preserved and represented by many of the ancient sedimentary basins that are located in Eastern Australia, North and South America, Europe, Asia, and Africa. These continental seas continued to maintain a global network and their outline and distribution were dictated by the changing coastal outlines and sea-levels.

At that time, accelerating changes to surface curvature gave rise to a marked increase in erosion of the lands and deposition of new sediments within the network of sedimentary basins. The increasing changes in surface curvature, and super-elevation of the lands, also initiated localized compression of the geosyncline-related sedimentary basins to form long linear fold mountain belts, further disrupting established seaways.

The presence of mountain glaciers and ice sheets during a number of Palaeozoic glacial events is testament to the marked increase in land surface elevation contrast. The presence of marine glacial debris deposited within equatorial regions is also testament to the presence of continental seas extending into polar-regions.

During Palaeozoic times, the South Pole continued to be located within central West Africa in what was then South Gondwana. Similarly, the North Pole was located within Northern China-Mongolia in what was part of the northern Tethys Sea. The ancient equator passed through East Antarctica, central Australia, North America, central Eurasia, and India through what was then North Gondwana. This crustal configuration approximates conventional plate tectonic reconstructions in part but differs substantially in the South Pacific region because of the need for an expanse of inferred Panthalassa and Tethys Oceans.

By late-Permian times, crustal rupture of the Pangaea supercontinent (Figure 22.4) had commenced, which resulted in breakup of Pangaea and formation of the modern continents and modern oceans during the following Mesozoic and Cenozoic Eras.

This rupturing also initiated draining of the continental seas into the newly emerging marine sedimentary basins—the forerunners of the modern oceans—and was accompanied by expulsion of new waters from the newly established mid-ocean-rift zones. Disruptions to the continental seas, initiated exposure of the former sea-floors to the ravages of erosion, which resulted in an unprecedented accumulation of sediment and organic matter in the form of coal beds in newly emerging coastal environments.

These Palaeozoic times coincided with the rapid development and evolution of all modern life forms on Earth. On an increasing radius plate tectonic Earth, this evolution of life forms was driven by the need to keep pace with the rapidly changing environmental conditions and, similarly, extinction is seen as a by-product of rapidly changing sea levels, disruptions to the continental seas, and associated environmental changes. The degree of crustal and environmental changes during this interval of time is considered to have markedly influenced evolutionary change in all life forms.

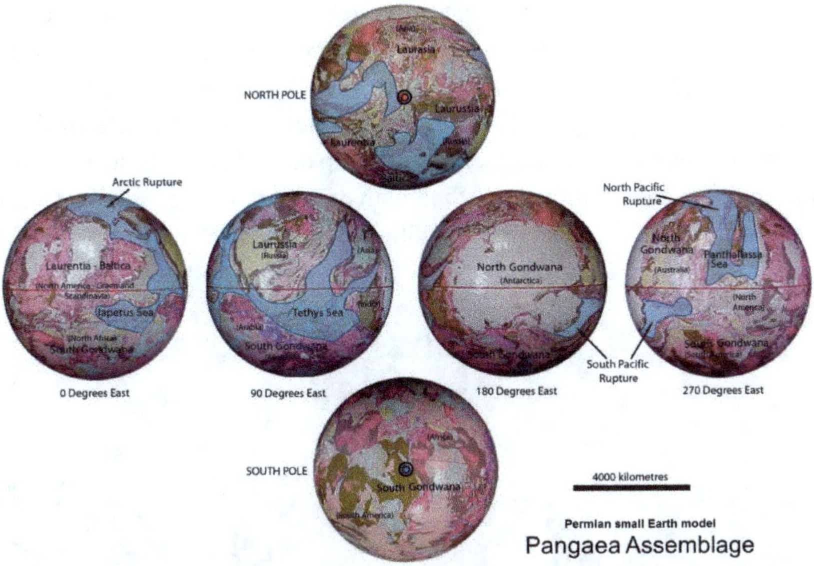

Figure 22.4. Permian Pangaea plate tectonic small Earth crustal assemblages. The model shows the ancient coastline distribution (blue lines) as well as the ancient Tethys, Iapetus, and Panthalassa Seas (blue shaded areas) forming part of a global network of continental seas. The figure also shows the locations of continental rupture commencing in the north and south Pacific and Arctic Ocean regions to form the modern oceans. Maxlow,2001.

The Palaeozoic Era then ended with rupture and breakup of the Pangaea supercontinental crust, and draining of the continental seas into the newly opening modern oceans—the end-Permian extinction event.

22.5. Mesozoic Earth

The Mesozoic Era was an unprecedented interval of geological time, extending from 252.6 million years ago to about 66 million years ago. The era began in the wake of the end-Permian extinction event, the largest mass extinction in Earth's history, and ended with the end-Cretaceous extinction event, which is best known for the demise of the dinosaurs along with many plant and animal species.

The Mesozoic is also recognized as a time of significant tectonic, climate, and evolutionary change. The climate of the Mesozoic was varied, alternating between warming and cooling periods. Dinosaurs appeared in the mid-Triassic and became the dominant terrestrial vertebrates early in the Jurassic, occupying this position for about 135 million years until their demise at the end of the Cretaceous. Birds first appeared in the Jurassic, having evolved from a branch of theropod dinosaurs. The first mammals also appeared during the Mesozoic but would remain relatively small in size until the Cenozoic Era.

On an increasing radius plate tectonic Earth, the Mesozoic Era commenced as a result of rupture and breakup of the Pangaean supercontinent to form the modern continents, along with draining of the continental Tethys, Panthalassa, and Iapetus Seas to initiate formation of the modern oceans (Figure 22.5).

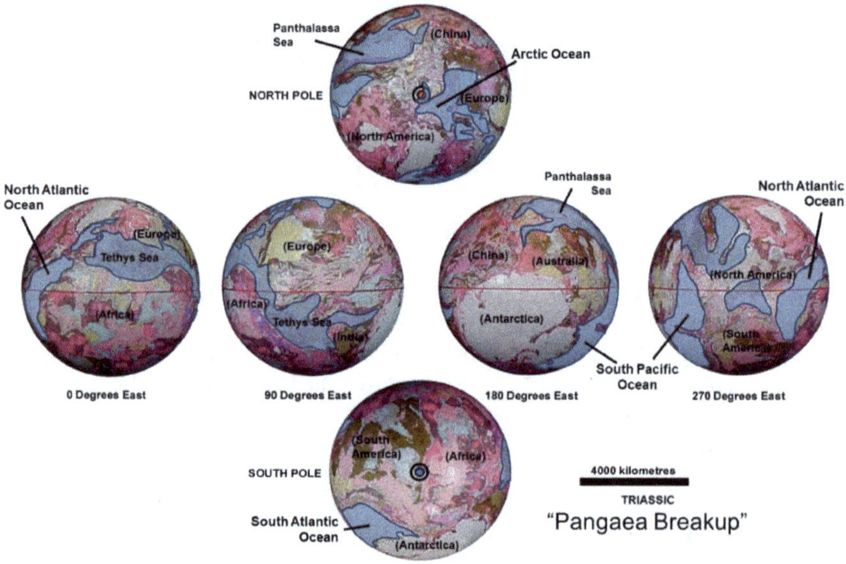

Figure 22.5. Triassic post-Pangaea breakup plate tectonic small Earth crustal assemblage. The model shows the ancient coastline distribution (blue lines) as well as remnants of the ancient Tethys, Iapetus, and Panthalassa Seas (blue shaded areas) forming part of a global network of separate continental seas. The figures also show the locations of the modern north and south Pacific, Atlantic, and Arctic Ocean regions. Maxlow,2001.

The subsequent migration history of the modern continents and seafloor crustal history is now preserved within the intruded volcanic seafloor lavas existing throughout all of the modern oceans. This post-Pangaea interval of time also saw large apparent shifts in the location of the North and South Poles and equator, occurring as a direct result of opening of the modern oceans and an apparent shift in the location of each of the modern continents.

Draining of the continental seas during breakup was accompanied by a shift in sedimentary deposition, changing from continental to shallow marine and marginal continental basin settings. Draining of the continental seas into the newly opening marine basins also gave rise to a number of separate and discrete seas and embryo oceans, each with their own separate sea-levels, salinities, ocean currents, temperatures, and marine species.

The location of the newly opening marine basins coincided with the precursors to the modern oceans and mid-ocean-rift zones. The influx of sediments within these basins, along with intruded magma and volcanism, also initiated formation of island-arcs, which were first initiated and located between the emerging modern continents.

The Mesozoic Era is also noted for its progressive breaching and merging of previously separate seas and oceans, resulting in a number of sea-level related extinction events. Breaching and merging of the North Pacific and South Pacific Oceans gave rise to the end-Triassic extinction event, and similarly breaching between Australia and Antarctica during the Paleocene Epoch initiated opening of the South Pacific Ocean and gave rise to the end-Cretaceous extinction event.

Breakup of the supercontinental crusts during the Mesozoic initiated apparent migration of the continents and opening of new oceans. Migration of the continents, in turn, affected the geographical location of the magnetic poles and severely disrupted established climate zones and influenced global climates. Breakup and disruption, in turn, affected the existing habitats and migration routes of both marine and terrestrial life forms.

This disruption forced species to imperceptibly migrate over time to keep pace with the migrating continents and climate zones, to evolve or better adapt to changing climate, or to simply perish if unable to keep pace with environmental change.

Continental breakup and opening of the modern oceans were accompanied by seafloor spreading along well-defined mid-ocean-rift zones, along with intrusion of basaltic seafloor lava and expulsion of new water and atmospheric gases. During this early continental breakup phase the seafloor spreading process was initially masked by input of sediments into the small marine basins as well as around the newly formed continental shelves.

As the opening of the modern oceans progressed further, beyond the reach of sedimentary deposition, the intruded seafloor volcanic lava was eventually exposed on the seafloor, as is now shown preserved in the Geological Map of the World.

22.6. Cenozoic Earth

The Cenozoic Era began in the wake of the end-Cretaceous extinction event some 66 million years ago and continues to the present-day. On an increasing radius plate tectonic Earth, the Cenozoic was marked by the establishment of symmetric seafloor spreading in all of the modern oceans. This seafloor spreading represents a preservation of the seafloor spreading and growth history of the Earth and is testament to opening and enlarging of all of the modern oceans.

The Cenozoic was also a time when all of the oceans had merged into a single global ocean, thus removing the threat of further sea-level-related mass extinction events and allowing for relative stability of marine and terrestrial life forms.

Crustal extension, as a result of increase in Earth radius, is now predominantly confined to the mid-ocean-rift zones, with minor basin extension and rifting within the continents during adjustment for change in surface curvature. Crustal breakup of both seafloor and continental crust continues to be closely related to lengthening and propagation of the mid-ocean-ridges, and includes focused seismic activity, heat-flow, and magmatism along well-defined zones of crustal weakness.

Lengthening of the East Pacific mid-ocean-ridge spreading axis is currently occurring as a northward extension of the spreading ridge passing through the Gulf of California. In the near future, this gulf will eventually rift and separate the Californian Peninsula from North America to form an island. A northward extension of the Red Sea Rift zone through the Gulf of Aqaba and Dead Sea region into Turkey will also result in rifting and separation of the Sinai Peninsula from Arabia. A northern extension of the Marianas spreading ridge will continue towards Japan and a southern extension of the Tongan spreading ridge will continue through New Zealand.

The Cenozoic Era has seen the modern continents and oceans continue to migrate and open to their current configurations. Studies here show that this will continue unabated into the near future.

The era was dominated by species development being increasingly isolated within island continents, with many species now endemic to specific continents or islands. Climatic changes occurring during this time also coincided with the Antarctician continent migrating from an equatorial position into the South Polar Region. A permanent continental icesheet was established in Antarctica about 33 million years ago and this migration may have subsequently had a direct influence on the long-term effects on climate.

Chapter 23

A Paradigm Shift in Cosmological Implications

"*A realistic attempt to reconstruct the early history of the solar system must necessarily be of a different character. It is essential to choose a procedure which reduces speculation as much as possible and connects the evolutionary models as closely as possible to experiment and observation*" Alfvén and Arrhenius, 2005.

The plate tectonic small Earth models presented in this book extend from the early-Archaean, approximately 4,000 million years ago, through to the present-day, with one model extended to 5 million years into the future. Age dating and petrological studies suggest that during early-Archaean times, around 4,000 million years ago, the ancient crusts were just starting to stabilize, implying that the first minerals on Earth were cooling and crystallizing to form recognizable rocks.

Similarly, age-dating from Moon rocks has returned ages extending to approximately 4,540 million years old, which potentially extends knowledge about the Earth-Moon system back a further 500 million years. Because of the age constraints of rocks on Earth, prior to 4,000 million years ago, there is no physical evidence in the Earth's rock-record to say what the Earth was like during these pre-Archaean times. There is enough subtle evidence preserved in the early Earth and Moon rock-record though to make some speculation.

Conventional cosmological studies consider that formation of the Earth was by accretion of interstellar debris during pre-Archaean times. Accretion was considered to have been completed by the early-Archaean and once stabilized, the size of the ancient Earth was assumed to be the same, or very similar to the size of the present-day Earth.

In order to account for the presence of continental crusts on a constant radius Earth, the ancient crusts were further considered to have either fully covered the Earth by about 4,000 million years ago, and have since been recycled through the mantle throughout geological time with no net change in crustal mass or, that the earliest continental crust nucleated as small localized patches and this juvenile continental crust has not been recycled back into the mantle.

Likewise, the prevailing hypothesis for formation of the Moon, although by no means definitive by empirical evidence, is that the Earth–Moon system formed as a result of a giant impact, whereby a Mars-sized body hit the newly formed proto-Earth and blasted material into orbit around it. This ejected material subsequently accreted to form the Moon.

Simulations of this event by others are said to show that most of the Moon material may have come from the impactor, not from ancient Earth. However, more recent simulations suggest that more of the Moon may have come from the Earth rather than the impactor. Both of these conventional Earth and Moon formation considerations are highly speculative and, as such, are open to further speculation and interpretation.

During 1982, Shuldiner presented evidence from ancient high-grade metamorphic rocks to suggest that Earth surface temperatures during the pre- to early-Archaean were 300 ± 100 degrees Celsius hotter than what it is now. This evidence implies that there was no liquid water, no oceans, and no rain on the ancient Earth, but probably some semblance of a primitive atmosphere. This evidence is also conceivable, because the bulk of the early-Archaean rocks surviving through to the present-day are granite and volcanic rocks, along with high-grade metamorphic rocks, which all require high temperatures to form.

Also in 1982, Kröner estimated that the upper mantle temperatures on Earth during the late-Archaean to early-Proterozoic times were up to 150 degrees Celsius hotter than the present. High temperatures during early-Archaean to Proterozoic times suggested to Kröner that these temperatures were attributed to a higher mantle geothermal gradient still prevailing during these ancient times.

Evidence from glacial studies suggested to Eyles and Young in 1994 that, by late-Archaean to early-Proterozoic times, the Earth surface temperatures had fallen significantly to maybe 30 degrees less than now. This surface temperature was inferred to have enabled atmospheric gases to condense to form surface water and ice. These low Archaean temperatures were further considered to be the result of an insulating effect, caused by the development and stabilization of a primitive crust in conjunction with a declining mantle geothermal gradient.

In contrast, increasing radius plate tectonic small Earth modelling studies presented here have shown that the size of the early-Archaean Earth was around 27 percent of the present Earth radius—a similar size to the present Moon. At that time, the entire primordial Earth crust comprised granite and volcanic rocks, with very little sedimentary rocks, and no large oceans.

It is speculated from the crustal and mantle temperature studies of others that the pre-Archaean Earth—times older than 4,000 million years ago—may have been incandescent, that is, hot enough to remain molten without crystallization to form a stable crust or dateable minerals and rocks.

In order for the primordial Earth, and by inference the Moon, to be incandescent during pre-Archaean times, it is speculated that the Earth-Moon system was located much closer to the primitive young Sun than it is now. The very much reduced Earth-Moon size and mass existing at that time would insist that centrifugal forces and angular momentum would have been vastly different to what they are now, and hence this suggestion may be plausible.

By Archaean to early-Proterozoic times, the Earth-Moon system may have then moved sufficiently far away from the influence of the Sun's surface temperature and gravity to cool and form a stable primitive crust. Sufficiently far to also lower Earth-Moon surface temperatures enough to retain liquid water on the primitive Earth surface, while still retaining a high residual geothermal gradient in the mantle.

During the distant pre-Archaean times—an indeterminate time span of maybe billions of years—it is further speculated that the Earth-Moon system may have also been originally combined as a single molten planet. By combining the volume of the present Moon and the volume of the early-Archaean Earth, the pre-Archaean Earth-Moon system is readily calculated to have been approximately 2,100 kilometres radius prior to separation. At, or sometime prior to 4,540 million years ago, the Earth and Moon may have then separated.

It is envisaged that this separation occurred as a result of gravitational instability of the more basaltic molten surface layer of the primitive incandescent Earth-Moon, possibly as a result of a high rotational velocity and centrifugal forces, forming a double planet. This mechanism is described in conventional literature as *fission of the Moon from the Earth's primitive basaltic crust through centrifugal forces occurring during a period of high angular momentum of the Earth–Moon system*. A high angular momentum and rotational velocity may again add support to the suggestion that the Earth-Moon system was originally located much closer to the primitive young Sun.

It is further speculated that, once separated, this double planet scenario would have originally comprised two bodies, the Earth and the Moon, of approximately equal size—around 1,700 kilometres radius—and in close proximity, co-rotating around each other as they do today. Currently, the Earth and Moon are continuing to move further apart and are now separating at a measured rate of 38 millimetres per year.

It is also feasible to consider that the Earth-Moon system is continuing to move further away from the Sun. This, more passive separation process, in contrast to a speculative impact event, may then go a long way to explain why the Moon is currently in synchronous, albeit retrograde rotation around the Earth, always showing the same face to the Earth.

This speculation is further supported by research in 2001, where a team at the Carnegie Institute of Washington reported precise measurement of isotopic signatures—variants of a particular chemical element—of lunar rocks. The team found that lunar rocks gathered during the Apollo program carried an isotopic signature that was identical with rocks from Earth and were similarly different from other known bodies in the Solar System.

In 2012, further analysis of titanium isotopes from surface lunar samples, and similarly in 2016 for oxygen isotopes, also indicated that the Moon has the same composition as the Earth, and the two bodies are indistinguishable. This research conflicts strongly with the current speculation that the precursor to the Moon originated far from Earth's orbit and was formed as a result of a giant impact between Earth and a Mars-sized body.

Not only is this isotopic signature data saying that the Moon did not originate from debris left over after, or fusion of, a giant impact between Earth and a Mars-sized body, it is also indirectly saying that the Moon and Earth may not have been involved in heavy bombardment from asteroids, or comets, colliding with the Earth and lunar surface. Debris from foreign asteroids or comets would have been readily detected in the research carried out by the Carnegie Institute by showing a distinctly separate bipolar isotopic signature, which it clearly doesn't.

In contrast, it is envisaged that the original molten basaltic material that may have separated from the combined Earth-Moon body to form the primitive Moon, was somewhat cooler than the remnant core-mantle material left over to form the primitive Earth.

The Moon is now known to have a distinct crust, mantle, and core, and rock samples collected from the surface of the Moon confirm that the composition is iron rich, similar to the seafloor crust and ancient volcanic rocks on Earth. The inner core of the present Moon is known to be small, about 20 percent the size of the Moon, in contrast to about 50 percent estimated for most other terrestrial bodies. This small core and mantle would also offer an explanation as to why the Moon has a limited surface gravity and magnetic field and why there is no atmosphere or hydrosphere.

During this speculated Earth-Moon separation process, it is envisaged that the hotter primitive core-mantle, as well as remnants of the surface layer of the former Earth-Moon body, remained intact, forming the primordial Earth (Figure 23.1).

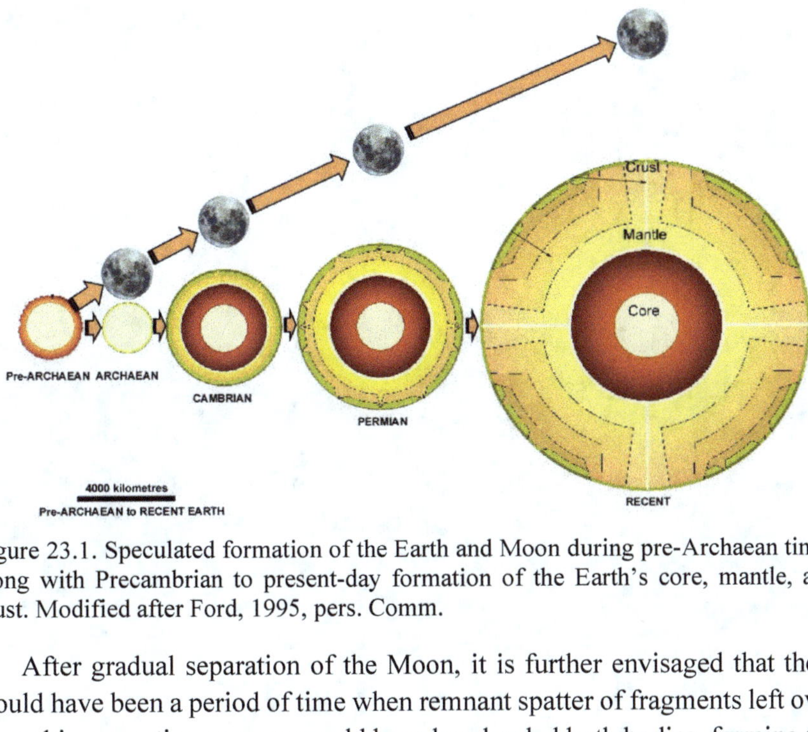

Figure 23.1. Speculated formation of the Earth and Moon during pre-Archaean times along with Precambrian to present-day formation of the Earth's core, mantle, and crust. Modified after Ford, 1995, pers. Comm.

After gradual separation of the Moon, it is further envisaged that there would have been a period of time when remnant spatter of fragments left over from this separation process would have bombarded both bodies, forming the now familiar cratering on the near side of the Moon's surface.

Much cooler temperature conditions on the Moon may also explain why the primitive Moon cooled and crystallized 500 million years earlier than the primitive Earth, thus preserving this period of surface bombardment and cratering.

Once the Earth-Moon body separated, it is further speculated that the remaining molten core-mantle material forming the primitive Earth began to differentiate, whereby the remaining molten elements and minerals separated into a distinct core and mantle. Over time, surface temperatures then gradually cooled sufficiently to form a new primitive silica-rich outer crust.

On Earth, during this differentiation process, the expulsion of fluids and gases from the mantle similarly began to accumulate to form shallow seas, and a primitive atmosphere, with an accompanying further reduction in mantle temperatures over time. Since then, without a strong magnetic core, the size of the Moon has remained relatively static while the Earth, with its strong magnetic core-mantle, has increased in size exponentially to the present-day.

Chapter 24

A Paradigm Shift in Thinking

The biggest problem facing geoscience today is accepting that Plate Tectonics may well be better suited to an increasing radius Earth scenario. Maxlow, 2018.

One of the prime aims of this book has been to encourage all scientists to broaden their research outlooks beyond conventional continental drift-based plate tectonic thinking in order to at least look at and consider what else modern global observational data has to offer.

Since the mid-1960s this modern data has traditionally been constrained to a constant Earth radius continental drift-based theory, yet nobody has bothered to test this data to see if the mechanism for plate tectonics is indeed better suited to an increasing radius Earth scenario.

It has been shown throughout this book that the extensive geological evidence and empirical data modeling studies more than adequately suggest that the decision to use continental drift as the basis of plate tectonic theory during the 1960s was poorly informed.

While an increasing radius Earth concept has historically been rejected, this rejection was based primarily on inconclusive palaeomagnetic studies made prior to 1976. This rejection was then followed by further emotive rejection of all proffered mechanisms and evidence subsequently proposed in support of an increasing mass and radius Earth over time.

It was emphasized throughout this book that, even if a mechanism for an increasing Earth mass and radius were previously not fully known, or understood, the extensive analysis presented here is based on readily available modern global observational data. The outcomes of this analysis are further based on reproducible empirical scientific observation, and it is again emphasized that a lack of a fully comprehended causal mechanism did not invalidate the need to at least test the concept of an increasing radius plate tectonic Earth model.

It was then proposed that, if a physically viable causal mechanism for an increasing Earth mass and radius were given, do we seriously consider this mechanism, test the new proposal, accept the empirical geological evidence, and revise current plate tectonic theory, or do we continue to reject the physical evidence and causal mechanism and instead remain supportive of a theory based on a presumed constant Earth radius premise?

The extensive empirical research presented in this book has been based on two important contributions to modern scientific understanding of the Earth, which have only been available since 1990 and 2000 respectively, including:

- Completion of bedrock geological mapping and age dating of all continental and seafloor crusts, and
- Recognition of the significance of solar wind related particles entering the Earth.

None of this data, or evidence, was available when the continental drift, plate tectonic, expanding Earth, or any of the alternative tectonic theories were first proposed during the mid- to late-twentieth century. In addition, none of this evidence, or data, has since been researched or modelled from anything other than a static radius Earth perspective.

By shifting our research perspective to an increasing radius Earth model, the successful integration of this data and evidence into plate tectonics must then constitute a paradigm shift in thinking, and the potential benefits to modern science are considered immeasurable.

The essence of all chapters presented in this book has been to utilize modern global observational data in order to test and quantify what this data is really telling us about crustal assemblages and subsequent tectonism on the Earth. Heavy reliance was initially made on using the published Geological Map of the World map to constrain assemblage of both the oceanic and continental plates back in time.

In order to achieve this aim, all preconceptions about Earth radius were simply ignored, in order to both measure the ancient radii of the Earth and to establish a formula to determine an ancient Earth radius at any moment in time.

This bedrock mapping and measured ancient radius data were then used to construct spherical plate tectonic small Earth models, extending from the early-Archaean Era, some 4,000 million years ago, to the present-day plus one model extended to 5 million years into the future—an unprecedented outcome.

The seafloor mapping shown on the Geological Map of the World was initially used to accurately constrain the location and assemblage of all seafloor crustal plates on smaller radius Earth models. Assemblage of these seafloor plates consistently showed that each plate assembles with a single unique fit, with all plates assembling with a high degree of precision along each of the mid-ocean-ridge spreading zones. This assemblage contrasts strongly with the numerous, relatively unconstrained, supercontinental assemblages proposed by conventional continental drift-based plate tectonic theory.

It was considered that, if these small Earth crustal plate assemblages were mere coincidence, it would be expected that the seafloor mapping, as well as geological and geographical evidence from adjoining continents, would not match across plate boundaries on any of the small Earth models constructed.

The evidence instead shows that seafloor bedrock mapping does indeed match across the plate boundaries, all continental sedimentary basins do merge to form a global network of ancient continental seas, orogenic and fold mountain belts coincide, and ancient crustal regions assemble together precisely.

This assemblage of seafloor crustal plates was shown to extend 200 million years back in time to the early-Triassic Period and demonstrated the viability and uniqueness of this post-Triassic small Earth modelling process. This assemblage contrasts strongly with conventional, continental drift-based plate tectonic reconstructions, for the same time interval, where assemblage of crustal plates is primarily based on palaeomagnetic apparent-polar-wander evidence, with poorly constrained multiple plate-fit options to contend with.

The uniqueness of small Earth plate tectonic assemblages also contrasts strongly with the conventional plate tectonic requirement to arbitrarily fragment continents in order to comply with the seafloor mapping data. It also contrasts with the requirement to dispose of large areas of inferred pre-existing crust beneath subduction zones in order to maintain a constant surface area premise.

Instead, what has been seen from the seafloor plate tectonic small Earth modelling studies is that all remaining continental crusts unite precisely to form a single complete global Pangaean supercontinental crust during the late-Permian Period, at around 50 percent of the present Earth radius. At that time the bulk of the seafloor volcanic crust, along with much of the atmosphere and hydrosphere, were returned to the mantle—and thence back to the Sun—from where they initially came from.

From this Pangaean supercontinental assemblage, continental sedimentary basins were shown to merge to form a global network coinciding with relatively shallow continental seas and the ancient supercontinents and seas were, in turn, defined by the variation in coastal outlines and sea-levels.

Quantification of an increasing radius plate tectonic Earth process back to the early-Archaean required an extension of the fundamental cumulative seafloor crustal premise to include continental crusts. Continental crust was reconstructed on pre-Triassic small Earth models by considering the primary crustal elements cratons, orogens, and basins. In order to construct these small Earth models, further consideration was given to an increase in Earth surface area occurring as a result of crustal stretching and extension within an established network of continental sedimentary basins.

Moving back in time, this crustal extension was progressively restored to a pre-extension, pre-stretching, or pre-rift configuration by simply removing young sedimentary and intruded magmatic rocks and reducing the surface areas of each of the sedimentary basins in turn, consistent with the empirical data shown on the Geological Map of the World.

During this process, the spacial integrity of all existing ancient cratons and orogens was retained until restoration to a pre-orogenic configuration was required. By removing all basin sediments and magmatic rocks, as well as progressively reducing the surface areas of the sedimentary basins in turn, an ancient primordial small Earth with a radius of approximately 27 percent of the present Earth radius was achieved during the early-Archaean—much the same size as the present-day Moon. This primordial Earth comprised an assemblage of the most ancient Archaean cratons and Proterozoic basement rocks: all other rocks, minerals, and elements were simply returned to their places of origin.

It is significant to reiterate that, on plate tectonic small Earth models, each of the established Precambrian and Palaeozoic sedimentary basin settings merge together to form a global network on a common pan-global continental crust. It was shown that the sedimentary basins from each of the continents also coincide with a global network of crustal weakness, operating throughout Precambrian and Palaeozoic times.

It was within this network of global crustal weakness that crustal extension, generated during on-going increase in Earth surface area, was focused, as well as on-going crustal mobility, mantle-derived heat flow, magmatic and metallogenic activity, followed by crustal rupture, continental breakup, and eventual opening of each of the modern oceans. The merging of each of these crustal settings back in time shows that all global crustal-related processes correlate precisely with the overall development of the continental crusts.

From the outcomes of this empirical plate tectonic small Earth modelling exercise, it was concluded that crustal modelling studies more than adequately quantify the validity of an increasing Earth radius plate tectonic process. The unique assemblage of all continental and seafloor crustal fragments on small Earth models demonstrates that an increasing radius Earth, extending back 4,000 million years to the beginning of Earth's geological past, is indeed both unique and viable.

What the full range of Archaean to present-day plate tectonic small Earth models also demonstrate is that, rather than being a random, arbitrary, amalgamation-dispersal-amalgamation cyclical crustal forming process, as we are currently led to believe, crustal development on an increasing radius plate tectonic Earth model is instead shown to be a simple, evolving, and predictable crustal process.

Applying this small Earth modelling evidence to palaeomagnetics showed that, when the imposed constant Earth surface area and constant Earth radius premises were removed from palaeomagnetic observations, these same observations demonstrate that the data is consistent with an increasing radius Earth. The application of palaeomagnetics to small Earth models showed that all ancient magnetic poles cluster as unique diametrically opposed north and south poles—as they should—and similarly, plotted palaeolatitude measurements coincide with and quantify predicted climate zones on each small Earth model constructed.

Similarly, additional geographical and biogeographical information aptly quantify the location of these ancient magnetic poles, equators, and climate zones as determined from unconstrained palaeomagnetic pole and latitudinal data.

When published coastal geography was plotted on each of the small Earth models, it was shown that large, conventional, Panthalassa, Iapetus, and Tethys Oceans were not required on a smaller radius Earth. Instead, this same coastal geography defined the presence of more restricted continental

Panthalassa, Iapetus, and Tethys Seas, which represent precursors to the modern Pacific and Atlantic Oceans and emergent Eurasian continents respectively.

From this coastal geography, the coastal outlines and emergent land surfaces on each small Earth model was then shown to define the ancient Rodinia, Gondwana, and Pangaea supercontinents and smaller sub-continents. This coastal geography demonstrated an evolutionary progression and development of each of the ancient seas and supercontinents throughout Earth history which was shown to be intimately related to changes in sea-level, changes to the outlines of continental sedimentary basins, changes incurred during crustal mobility, and changes to sea-levels once the modern oceans opened to the present-day.

The timing and development of these ancient continental seas and supercontinents, along with formation of the modern continents and oceans, was then shown to be the prime cause for evolution of all life forms on Earth. The ancient continental seas, in particular, provided an ideal setting for the primitive Precambrian microbe's effectiveness as nurseries of evolution and to markedly drive subsequent evolutionary changes in all life forms.

On each of the small Earth plate tectonic models, warm sea waters during much of the Palaeozoic extended from equatorial regions through to the North Polar Region allowing newly evolved species to readily colonize and populate throughout each of the interconnected ancient Tethys, Iapetus, and Panthalassa seaways. This distribution of warm seas also limited the presence of a polar ice cap in the North Polar Region and instead restricted the presence of ice to the exposed Gondwanan South Polar Region throughout much of that time.

The distribution of warm seas enabled stable carbonate shelf environments to develop adjacent to the ancient lands, along with deeper anoxic basins further away from the coastlines. The extensive algal mats and early reefal mounds present throughout much of the Precambrian eventually gave way to predation and bioturbation, and planktonic life forms were able to readily spread throughout the deeper seas.

Similarly, extinction of species was shown to be a by-product of sea-level change, disruption to the distribution of established seas, and the inability of species to keep pace with these environmental changes. It was further considered that evolution of species during these ancient times was subsequently driven by the need to survive amongst these ever-increasing changes to the Earth and environment.

During the end-Permian continental breakup and opening of the modern oceans, the existing intercontinental migration and dispersal routes of the various terrestrial and marine species were shown to be disrupted, enabling species endemic to the various regions to interact and extend their boundaries with time.

From the distribution of climate-dependent rocks and animal species it was also shown that these distributions coincide precisely with climatic zones anticipated on an increasing radius Earth. It was similarly shown that these climatic indicators display a distinct latitudinal zonation paralleling the ancient equator, suggesting that an inclined Earth rotational axis inclined to the pole of the ecliptic may have been well established during at least the Palaeozoic Era and has persisted to the present-day.

On an increasing radius plate tectonic Earth, the small Earth modelling studies show that, during early-Palaeozoic to present-day times there have been a number of drastic and prolonged changes to sea-levels which coincide with known extinction events. On these models, major changes in sea-levels were shown to occur as a result of separation or merging of previous ancient continental seas, as well as onset of geosynclinal activity and orogenesis, breakup of the ancient supercontinents, opening of the modern oceans, and draining of the ancient continental seas.

Depending on the severity of these events, it was considered that sea-level changes may have also adversely affected regional to global-scale climate, as well as ocean-water circulation patterns, species habitats, and the type and location of sedimentary deposition.

Modelling metallogenic data and mineralization settings on small Earth models showed that the data and mineral settings where essentially the same as those identified within conventional studies. The difference being that, on an increasing radius plate tectonic Earth, prior to the early-Triassic Period, all continental crusts were assembled together on a smaller radius supercontinental Earth. The small Earth assemblages then enable pre-Triassic metallogenic provinces from otherwise remote locations to be assembled together as unique, inter-related provinces on smaller radius Earth models.

The assemblage of continents and crustal elements on small Earth models then provide a means to investigate the spatial and temporal distribution of metals across adjoining continents and crustal regimes. It was considered that recognition and understanding of past metal distributions on the present-day

Earth would then potentially enable mineral search and generic relationships to be extended beyond their known type localities.

Modelling the distribution of all fossil fuels on small Earth models highlighted the global interrelationships of resources coinciding with the distribution of Palaeozoic continental seas and low-lying terrestrial environments. The transition from deposition of oil and gas shale to coal to petroleum and natural gas was found to be consistent with the various periods of maximum to minimum sea level changes occurring during periods of marine transgression and regression, in particular after regression of the continental seas during the Palaeozoic time periods leading to crustal breakup and opening of the modern oceans.

After consideration of the global observational data modelling studies from each of the Earth sciences, the opportunity was then taken to speculate on the formation and early history of the Earth-Moon system, as well as the origin and formation of the hydrosphere and atmosphere. This opportunity was followed by a comprehensive summary of the geological history of the Earth and a summary of the geological history of a selected continent.

While speculative, these summaries were given in order to provide an alternative solution to avenues of research, as well as to stimulate further research into plate tectonics on an increasing radius Earth model.

With regard to a causal mechanism, irrespective of the infancy of space technology, and research into the input of solar wind related charged electrons and protons entering the Earth, it was considered here that we must at least ask the question, and seriously take note of, what is happening to these particles—the building blocks of all matter on Earth—once they enter the Earth. The causal mechanism for an increase in Earth mass and radius over time presented in this book was then based on, but not necessarily constrained to, this input of charged solar wind related electron and protons originating from the Sun.

It was suggested that these solar particles enter the Earth at the poles and recombine as new matter most likely within the 200 to 300 kilometres thick D" region, located at the base of the mantle directly above the core-mantle boundary. This matter generation process then represents a basis for formation of all new and existing elements and mineral species present on Earth.

It was further envisaged that formation of new matter within the Earth must then give rise to an increase in mass and volume of the mantle over time. Increase in mantle volume is then transferred to the outer crust as continental crustal extension, which is currently seen and preserved as extension along all mid-ocean-rift zones. Surface crustal extension within the mid-ocean-rift zones is further accompanied by intrusion of new basaltic lava and expulsion of new sea water and atmospheric gases.

In summary, it is concluded that the extensive global observational evidence and empirical modelling studies presented in this book more than adequately demonstrate that an increasing radius plate tectonic Earth is indeed a viable and demonstrable tectonic process. From a geological perspective, at no point have any fundamental physical law been violated. The commonly held presumption that Earth radius has remained constant throughout time was simply removed and instead, the Earth was allowed to tell its own story.

In order to create plate tectonic small Earth models, we simply removed from the Earth what was not previously there—intruded seafloor volcanic lava, intruded magma, and eroded sediments—to end up with a Pangaean Earth comprising an assemblage of continental crustal components, and ultimately to a primitive Archaean Earth comprising an assemblage of equally primitive cratonic and orogenic crustal components.

At all times the reader has been encouraged to keep an open mind, and to formulate their own opinions and conclusions based on the data modelling itself, rather than basing them on any preconceived conventional wisdom that may be present. By ignoring the empirical evidence presented here, and by continuing to accept any shortfalls conventional insistence imposes on the global observational data, it is considered that, this ignorance will continue to slow geologic progress by maintaining narrow, rigid viewpoints. This ignorance will continue to perpetuate scientific conundrum by discouraging alternative research and convincing students and scientists alike that the main plate tectonic problems have all been resolved.

Instead, the empirical evidence presented in this book suggests that plausible alternative models and mechanisms based on the extensive modern global observational data and evidence now available must be actively encouraged, not discouraged. This encouragement is desperately needed in order to fuel the traditional scientific method of multiple working hypotheses, so as to promote increased objectivity in the interpretation of all physical observations.

In this context, we should then at least consider that modern global plate tectonic observational data may well be better suited to an increasing radius

Earth scenario, and hence constitutes a paradigm shift in plate tectonic thinking, before continuing to unscientifically reject this proposal out of hand.

Bibliography

ADAMS N. See website at: http://www.xearththeory.com/

ALBAREDE F. 2009. Volatile accretion history of the terrestrial planets and dynamic implications: Nature, v. 461, no. 7268, p. 1227–1233, doi:10.1038/nature08477.

ALFÉN H. & ARRHENIUS G. 2005. Evolution of the Solar System. Scientific and Technical Information Office. NATIONAL AERONAUTICS AND SPACE ADMINISTRATION, Washington, D.C.

ANHAEUSSER C. R. 1981. The relationships of mineral deposits to early crustal evolution. Economic Geology, 75th Anniversary Volume, 42-62.

BAILEY D. K. & STEWART A. D. 1983. Problems of Ocean Water Accumulation on a Rapidly Expanding Earth. 67-69 in Carey S. W. (Ed): Expanding Earth Symposium, Sydney 1981, University of Tasmania.

BARLEY M. E., EISENLOHR B., GROVES D. I., PERRING C. S. & VEARCOMBE J. R. 1989. Late Archaean convergent margin tectonics and gold mineralisation: a new look at the Norseman-Wiluna Belt, Western Australia. Geology 17, 826-829.

BARLEY M. E. & GROVES D. I. 1992. Supercontinent cycles and the distribution of metal deposits through time. Geology 20, 291-294.

BARNETT C. H. 1962. A suggested reconstruction of the land masses of the Earth as a complete crust. Nature 195, 447-448.

BARNICOAT A. C., FARE R. J., GROVES D. J. & McNAUGHTON N. J., 1991. Synmetamorphic Code-gold deposits in high-grade Archaean settings. Geology 19, 921-924.

BARRON E. J. 1983. A warm, equable Cretaceous: the nature of the problem. Earth Science Reviews 19, 305-338.

BLATT H. & JONES R. L. 1975. Proportions of exposed igneous, metamorphic and sedimentary rocks. Geological Society of America Bulletin 86, 1085-1088.

BLEWETT R. S. 2012. Shaping of a Nation: A Geology of Australia. Geoscience Australia and ANU E Press, Canberra.

BLINOV V. F. 1983. Spreading rate and rate of expansion of the Earth. In: Carey S.W. ed. Expanding Earth Symposium, Sydney, 1981. University of Tasmania, 297-304

BOSS A. P. & SACKS I. S. 1985. Formation and Growth of Deep Mantle Plumes. Geophysical Journal of the Royal Astronomical Society 80, 241-255.

BRŐSSKE L. 1962. Wäschst die Erde mit Katastrophen? Düsseldorf.

BRUNNSCHWEILER R. O. 1983. Evolution of Geotectonic Concepts in the Past Century. 9-15, in Carey S. W. (Ed): Expanding Earth Symposium, Sydney 1981, University of Tasmania.

BURG J-P. & FORD M. 1997. Orogeny through time: an overview. In: Burg J-P. & Ford M. eds. Orogeny through time. Geological Society Special Publication 121, 1-17.

BUTLER R. F. 1992. Paleomagnetism: Magnetic Domains to Geologic Terranes. Blackwell Scientific Publications, USA.

CALDER J. H. & GIBLING M. R. 1994. The Euramerican coal province: controls on Late Paleozoic peat accumulation. Palaeogeography, Palaeoclimatology, Palaeoecology 106, 1-21.

CAREY S. W. 1958. The tectonic approach to continental drift. In: Continental Drift, a Symposium. University of Tasmania, Hobart, 177-355.

CAREY S. W. 1963. The Asymmetry of the Earth. Australian Journal of Science, 25, 369-383 & 479-488.

CAREY S. W. 1975. The expanding Earth - an essay review. Earth Science Reviews 11, 105-143.

CAREY S. W. 1976. The Expanding Earth. Elsevier, Amsterdam.

CAREY S. W. 1983a. Earth expansion and the null Universe. In: Carey S.W. ed. Expanding Earth Symposium, Sydney, 1981. University of Tasmania, 365-372.

CAREY S. W. 1983b. The necessity for Earth expansion. In: Carey S.W. ed. Expanding Earth Symposium, Sydney, 1981. University of Tasmania, 375-393.

CAREY S. W. 1988. Theories of the Earth and Universe: A History of Dogma in the Earth Sciences. Stanford University press, Stanford, California.

CAREY S. W. 1996. Earth, Universe, Cosmos. University of Tasmania, Hobart.

CARPENTER D. L. 1963. Whistler evidence of a 'knee' in the magnetospheric ionization density profile, J. Geophys. Res., 68, 1675–1682.

CEBULL S. E. & SHURBET D. H. 1992. Conventional plate tectonics and orogenic models. In: Chatterjee S. & Hotton N (III) Eds. New Concepts in global tectonic s. Texas Technical Press, USA, 111-117.

CHATTERJEE S. 1997. The Rise of Birds. Baltimore: Johns Hopkins University Press.

CGMW & UNESCO 1990. Geological Map of the World. Commission for the Geological Map of the World, Paris.

CHUMAKOV N. M. & ELSTON D. P. 1989. The paradox of late-Proterozoic glaciations at low latitudes. Episodes 12, 115-1120.

COX A. & DOELL R. R. 1961. Palaeomagnetic evidence relevant to a change in the Earth's radius. Nature 189, 45-47.

COX C. B. 1990. New geological theories and old biogeographical problems. Journal of Biogeography 17, 117-130.

CREER K. M 1965. An Expanding Earth. Nature 205, 539-544.

CREER K. M., IRVING E. & RUNCORN S. K. 1954. The direction of the geomagnetic field in remote epochs in Great Britain. Journal of Geomagnetics and Geoelectricity 6, 163-168.

CROWLEY T. J. 1994. Pangean climates. In: Klein G. D. ed. Pangea: paleoclimate, tectonics, and sedimentation during accretion, zenith, and breakup of a supercontinent. Geological Society of America Special Paper 288, 25-39.

CWOJDZINSKI S. 2003. The Tectonic Structure of the Continental Lithosphere Considered in the Light of the Expanding Earth Theory – A Proposal of a New Interpretation of Deep Seismic Data. Ed: Majewska A. Polish Geological Institute, Special Papers.

DEARNLEY R. 1965. Orogenic fold-belts, convection and expansion of the Earth. Nature 206, 1284-1290.

DENNIS J. G. 1962. Fitting the Continents. Nature 196, 364.

DODSON P. & TATARINOV L. P. 1990. Dinosaur extinction. In: Weishampel D. B., Dodson P. & Osmólska H. eds. 1990. The Dinosauria. University of California Press, Los Angeles, 55-62.

EGGLER D. H. 1987. Solubility of major and trace elements in mantle metasomatic fluids: experimental constraints. In: Menzies M. A. & Hawkesworth C. J. eds, Mantle Metasomatism. Academic Press, London. 21-41.

EGYED L. 1960. Some remarks on continental drift. Geofisica Pura Applica 45, 115-116.

EGYED L. 1963. The expanding Earth? Nature 197, 1059-1060.

EICHLER J. B. 2011. A New Mechanism for Matter Increase within the Earth. Ettore Majorana Foundation and Centre for Scientific Culture, 37th Interdisciplinary Workshop of the International School of Geophysics, Erice, Sicily, 4-9 October 2011, 159-162.

ERWIN D. H. 1993. The great dying: life and death in the Permian. Columbia University Press, New York.

ETHERIDGE M. A., RUTLAND R. W. R. and WYBORN L. A. I. 1987. Orogenesis and tectonic processes in the Early to Middle Proterozoic of Northern Australia. In: Kröner A. ed. Proterozoic lithospheric evolution, Geodynamics Series 17, Geological Society of America, 131-147.

EVANS A. M. 1984. An introduction to ore geology. Blackwell Scientific Publications, Oxford.

EYLES N. & YOUNG G. M. 1994. Geodynamic controls on glaciation in Earth history. In: Deynoux M., Miller J. M. G., Domack E. W., Eyles N., Fairchild I. J. & Young G. M. eds. International Geological Correlation Program Project 260: Earth's Glacial Record. Cambridge University Press, 1-28.

FAURE K., DeWIT M. J. & WILLIS J. P. 1994. Late Permian global coal discontinuity and Permian-Triassic boundary "events". Ninth International Gondwana Symposium. Hyderabad, India, 1994, 1075-1089.

FLÜGEL E. 1994. Pangean shelf carbonates: Controls and paleoclimatic significance of Permian and Triassic reefs. In: Klein, G. D. ed. Pangea: paleoclimate, tectonics, and sedimentation during accretion, zenith, and breakup of a supercontinent. Geological Society of America Special Paper 288, 247-266.

FRANCIS J. E. & FRAKES L. A. 1993. Cretaceous climates. Sedimentology Review 1993. 17-28.

GAÁL G. 1992. Global Proterozoic tectonic styles and Early Proterozoic metallogeny. South African Journal of Geology 95, 80-87.

GARFUNKEL Z. 1975. Growth, shrinking, and long-term evolution of plates and their implications for the flow pattern in the mantle. Journal of Geophysical Research 80, 4425-4432.

HAMBRY M. J. & HARLAND W. B. eds. 1981. Earth's Pre-Pleistocene glacial record. Cambridge: Cambridge University Press, London.

HARTOGH P., LIS D.C., BOCKELÉE-MORVAN D., de Val-BORRO M., BIVER N., KÜPPERS M., EMPRECHTINGER M., BERGIN E.A., CROVISIER J., RENGEL M., MORENO R., SZUTOWICZ S., and BLAKE G.A., 2011, Ocean-like water in the Jupiter-family comet 103P/Hartley 2: Nature, v. 478, no. 7368, p. 218–220, doi:10.1038/nature10519.

HEEZEN B.C. 1960. The rift in the ocean floor. Scientific America, 203, 98-110.

HILGENBERG O. C. 1933. Vom wachsenden Erdball. Selbstverlag, Berlin.

HILGENBERG O. C. 1962. Paläopollagen der Erde. Neues Jahrb. Geol. Und Paläontol, Abhandl 116, Struttgart.

HOFFMAN P. F., KAUFFMAN A. J., HALVERSON G. P. & SCHRAG D. P. 1998. A Neoproterozoic snowball earth. Science 281, 1342-1346.

HOTTON N. (III) 1992. Global distribution of terrestrial and aquatic tetrapods, and its relevance to the position of continental masses. In: Chatterjee S. & Hotton N. (III) Eds. New Concepts in global tectonic s. Texas Technical Press, USA, 267-285.

HURRELL S. W. 1994. Dinosaurs and the Expanding Earth. OneoffPublishing.com.

ITRF, 2008. International Terrestrial Reference Frame. Website: http://itrf.ensg.ign.fr/

JAIRETH S. & HUSTON D. 2010. Metal endowment of cratons, terranes and districts: Insights from a quantitative analysis of regions with giant and super-giant deposits. Ore Geology Reviews, 38, 288-303.

JEFFREYS H. 1962. A Suggested Reconstruction of the Land Masses of the Earth as a Complete Crust: Comment. Nature 195, 148.

KLEIN G. D. 1994. Pangea: paleoclimate, tectonics, and sedimentation during accretion, zenith, and breakup of a supercontinent. Geological Society of America Special Paper 288.

KOZIAR J. 1980. Ekspansja den oceanicznych I jej zwiazek z hipotaza ekspansji Ziemi. Sprawozdania Wroclawskiego Towarzystwa Naukowego, 35, 13-19.

KOZIAR J. 2002. Space geodesy and expanding Earth. 2002. (Unpublished).

KREMP G. O. W. 1992. Earth Expansion Theory versus Statical Earth Assumption. In: Chatterjee S. & Hotton N (III) Eds. New Concepts in global tectonic s. Texas Technical Press, USA, 297-307.

KRÖNER A. 1982. Archaean to early-Proterozoic tectonics and crustal evolution: A review. Revista Brasileira de Geociências 12, 15-31.

KUHN T. S. 1970. The Structure of Scientific Revolutions. University of Chicago Press.

LAMBERT I. B. 1982. Early geobiochemical evolution of the Earth. Revista Brasileira de Geociências 12, 32-38.

LAMBERT I. B. & GROVES D. I. 1981. Early Earth evolution and metallogeny. In: K. H. Wolf ed. Handbook of stratabound and stratiform ore deposits 8, Elsevier Co, Amsterdam, 339-447.

LARSON K. M., FREYMUELLER J. T. & PHILIPSEN S. 1997. Global plate velocities from the Global Positioning System. Journal of Geophysical Research 102, 9961-9981.

LARSON R. L., PITMAN W. C. (III), GOLOVCHENKO X., CANDE S. C., DEWEY J. F., HAXBY W. F. & LaBRECQUE (mapcompilers) 1985. The Bedrock Geology of the World. Freeman & Co., New York.

LAZNICKA P. 1993. Precambrian empirical metallogeny: precambrian lithologic associations and metallic ores. Developments in economic geology, 29. Elsevier. Amsterdam.

Le PICHON X. & HEIRTZLER J. R. 1968. Magnetic anomalies in the Indian Ocean and sea floor spreading. Journal of Geophysical Research, 73:2, 119-2,136.

LINDERMAN B. 1927. Kettengebirge, Kontinentale Zerspaltung und Erdexpansion. Gustav Fischer Publishers, Jena. 186p.

LI Z. X., ZHANG L. & POWELL C. McA. 1996. Positions of the East Asian cratons in the Neoproterozoic supercontinent Rodinia. Australian Journal of Earth Sciences 43, 593-604.

LOWMAN P.D. Jr. 1992. Plate tectonics and continental drift in geologic education. In: Chatterjee S. & Hotton N. (III) Eds. New Concepts in global tectonics. Texas Technical Press, USA, 3-9.

LUCKERT K. W. See website at: http://www.triplehood.com/

MAXLOW J. 1995. Global Expansion Tectonics: The geological implications of an expanding Earth. Unpublished Master of Science thesis, Curtin University of Technology, Perth, Western Australia.

MAXLOW J. 2001. Quantification of an Archaean to recent Earth Expansion Process using Global Geological and Geophysical Data Sets. Unpublished PhD thesis, Curtin University of Technology, Perth, Western Australia. http://adt.curtin.edu.au/theses/available/adt-WCU20020117.145715

MAXLOW J. 2005. Terra non-Firma Earth: Plate Tectonics is a Myth. Terrella Press. TerrellaPress@bigpond.com.au

MAXLOW J. 2014. On the Origins of Continents and Oceans: A Paradigm Shift in Understanding. Terrella Press. TerrellaPress@bigpond.com.au

McCABE P. J. & PARRISH J. T. 1992. Tectonic and climatic controls on the distribution and quality of Cretaceous coals. In: McCabe P. J. & Parrish J. T. eds. Controls on the distribution and quality of Cretaceous coals. GeologicalSociety of America Special Paper 267, 1- 15.

McCARTHY D. 2003. Biogeography and Scientific Revolutions. The Systematist 2005 No. 25, 3-12.

McELHINNY M. W. & BROCK A. 1975. A new paleomagnetic result from East Africa and estimates of the Mesozoic paleoradius. Earth and Planetary Science Letters 27, 321-328.

McELHINNY M. W. & LOCK J. 1996. IAGA paleomagnetic databases with Access. Surveys in Geophysics, 17, 575-591.

MEYER C. 1981. Ore-forming processes in geologic history. Economic Geology, 75th Anniversary Volume, 6-41.

MEYER C. 1988. Ore deposits as guides to geologic history of the Earth. Annual Reviews in Earth and Planetary Sciences 16, 147-171.

MEYERHOFF A. A., BOUCOT A. J., MEYERHOFF-HULL D. & DICKINS J. M. 1996. Phanerozoic faunal and floral realms of the Earth: The intercalary relations of the Milvinokaffric and Gondwana faunal realms with the Tethyan faunal realm. Geological Society of America, Memoir 189.

MORBIDELLI A., CHAMBERS J., LUNINE J.I., PETIT J.M., ROBERT F., VALSECCHI G.B., & CYR, K.E., 2000, Source regions and time scales for the delivery of water to the Earth: Meteoritics & Planetary Science, v. 35, no. 6, p. 1309–1320, doi:10.1111/j.1945-5100.2000.tb01518.x.

MYERS J. S. 1990a. Precambrian. In: Geology and mineral resources of Western Australia. Western Australia Geological Survey, Memoir 3, 737-749.

MYERS J. S. 1990b. Precambrian tectonic evolution of part of Gondwana, southwestern Australia. Geology 18, 537-540.

MILANOVSKY E. E. 1980. Problems of the tectonic development of the Earth in the light of concept on its pulsations and expansion. Revue de Geologique et de Geographie Physique 22, 15-27.

MOORBATH S. 1982. Crustal evolution in the Early Precambrian. Revista Brasileira de Geociências 12, 39-44.

MRDS, 2015. USGS Mineral Resource Data Set. Website: http://mrdata.usgs.gov/mrds/

NOËL D. 1989. NUTEERIAT: Nut Trees, the Expanding Earth, Rottnest Island and All That. Published for the Planetary Development Group TREE CROPS CENTRE. Cornucopia Press, Australia.

NORTH F. K. 1990. Petroleum geology. Unwin Hyman Inc., Boston.

OLLIER C. D. 1985. Morphotectonics of passive continental margins: Introduction. Z. Geomorphology Suppl. Bd. N. F. 54, 1-9.

OLLIER C. D. & PAIN C. F. 2000. The origin of mountains. Routledge, London.

OWEN H. G. 1976. Continental displacement and expansion of the Earth during the Mesozoic and Cenozoic. Philosophical Transactions of the Royal Society, London 281. 223-291.

ORESKES N. 1999. The Rejection of Continental Drift: Theory and Method in American Earth Science, Oxford University Press.

OWEN H. G. 1983. Atlas of Continental Displacement, 200 Million Years to the Present. Cambridge Earth Series, Cambridge University Press, Cambridge.

PARSONS B. 1982. Causes and consequences of the relation between area and age of the ocean floor. Journal of Geophysical Research 87, 289-302.

PBDB, 2015. Paleobiology Database, 2015. Website: https://paleobiodb.org/navigator/

PISAREVSKY S. 2005. Global Paleomagnetic Database (GPMDB V 4.6). Tectonics Special Research Centre of the University of Western Australia Web site (http://www.tsrc.uwa.edu.au/).

RICKARD M. J. 1969. Relief of curvature on expansion - a possible mechanism of geosynclinal formation and orogenesis. Tectonophysics 8, 129-144.

ROBAUDO S. & HARRISON C. G. A. 1993. Plate tectonics from SLR and VLBI global data. In: Smith D. E., & Turcotte D. L. eds. Contributions of Space Geodesy to Geodynamics: Crustal Dynamics. Geodynamics Series, Volume 23. American Geophysical Union.

RUBEY W. W. 1975. Geologic history of sea water; an attempt to state the problem. In: Kitano Y. ed. Geochemistry of Water. Dowden, Hutchinson & Ross Inc., Stroudsburg, Pennsylvania.

RUTLAND R. W. R., ETHERIDGE M. A. & SOLOMON M. 1990. The stratigraphic and tectonic setting of the ore deposits of Australia. In: Hughes F. E. ed. Geology of the mineral deposits of Australia and Papua New Guinea, 15-26.

SAULL V. A. 1986. Wanted: Alternatives to plate tectonics. Opinion. Geology 14, 1986

SCALERA G. 1988. Nonconventional Pangea reconstructions: new evidence for an expanding Earth. Tectonophysics 146, 365-383.

SCALERA G. 1990. General Clues Favouring Expanding Earth Theory. In: Critical Aspects of the Plate Tectonics Theory: Volume II, Alternative Theories. Theophrastus Publishers, Athens, Greece, 65-93.

SCHMIDT P. W., & EMBLETON B. J. J. 1981. A Geotectonic paradox: has the Earth expanded? Journal of Geophysics 49, 20-25.

SCOTESE C. R., GAHAGAN L. M. & LARSON R. L. 1988. Plate tectonic reconstructions of the Cretaceous and Cenozoic ocean basins. Tectonophysics 155, 27-48.

SCOTESE C. R. 1994. Paleogeographic maps. In: Klein, G. D. ed. Pangea: paleoclimate, tectonics, and sedimentation during accretion, zenith, and breakup of a supercontinent. Geological Society of America Special Paper, 288.

SCOTESE C. R. 2001. PALEOMAP Project http://www.scotese.com/

SHEN W., SUN R., CHEN W., ZHANG Z., LI J., HAN J., & DING H. 2011. Evidences of Earth Expansion from Space-Geodetic and Gravimetric Observations. Ettore Majorana Foundation and Centre for Scientific Culture, 37th Interdisciplinary Workshop of the International School of Geophysics, Erice, Sicily, 4-9 October 2011, 131-134.

SHIELDS O. 1979. Evidence for initial opening of the Pacific Ocean in the Jurassic. Palaeontology, Palaeoclimatology, Palaeoecology 26, 181-220.

SHIELDS O. 1996. Geologic significance of land organisms that crossed over the Eastern Tethys "Barrier" during the Permo-Triassic. Palaeobotanist 43, 85-95.

SHIELDS O. 1997. Is plate tectonics withstanding the test of time? Annali di Geofisica, Vol XL, 1-8.

SHULDINER V. I. 1982. The oldest high grade terranes: Possible relicts of primeval Earth crust. Revista Brasileira de Geociências 12, 45-52.

SMILEY C. J. 1992. Paleofloras, faunas, and continental drift: some problem areas. In: Chatterjee S. & Hotton N. (III) Eds. New Concepts in global tectonic s. Texas Technical Press, USA, 241-257.

SMIRNOV V. I. 1984. Periodicity of ore formation in geological history. Proceedings of the 27th International Geological Congress, Volume 12, Metallogenesis and mineral ore deposits, 1-15.

SMITH A. G., SMITH D. G. & FUNNELL B. M. 1994. Atlas of Mesozoic and Cenozoic coastlines. Cambridge University Press.

SOLOMON M. & GROVES D. I. 1994. The geology and origins of Australia's mineral deposits. Oxford University press, Melbourne.

SOUDARIN L., CRÉTAUX J-F. & CAZENAVE A. 1999. Vertical crustal motions from the DORIS space-geodesy system. Geophysical Research Letters 26, 1207-1210.

STEINER J. 1977. An expanding Earth on the basis of sea-floor spreading and subduction rates. Geology 5, 313-318.

STEVENS G. R. 1990. The influences of paleogeography, tectonism and eustacy on faunal development in the Jurassic of New Zealand. In: Pallini G., Cecca F., Cresta S. & Santantonio M. eds. Atti del secondo convegno internazionale. Fossili, evoluzione, ambiente. Pergola, 1987, 441-457.

STORETVEDT K. M. 1992. Rotating Plates: New Concepts in global tectonic s. In: Chatterjee S. & Hotton N. (III) (eds), New Concepts in global tectonic s. Texas Technical Press, USA, 203-220.

STRUTINSKI C. 2015. Discussion on the Cause of Earth Growth and its Consequences: A qualitative approach. Saarbrücken (In press).

TIRATSOO E. N. 1984. Oilfields of the World. Scientific Press Ltd, England.

TRÜMPY R. 2001. Why Plate Tectonics was not invented in the Alps. International Journal of Earth Sciences. (Geol Rundsch) 90:477-483.

U.S. EIA 2013. Technically Recoverable Shale Oil and Shale Gas Resources: An Assessment of 137 Shale Formations in 41 Countries outside the United States.

VAN ANDEL S. I. & HOSPERS J. 1968. Palaeomagnetic calculations of the radius of the ancient Earth by means of the palaeomeridian method. Physics of the Earth and Planetary Interiors 1, 155-163.

VANECEK M. 1994. Mineral deposits of the world. Ores, industrial minerals and rocks. Developments in economic geology, 28. Elsevier, Amsterdam.

VAN HILTEN D. 1963. Palaeomagnetic indications of an increase in the Earth's radius. Nature 200, 1277-1279.

VEIZER J., LAZNICKA P. & JANSEN S. L. 1989. Mineralization through geologic time: recycling perspective. American Journal of Science 289, 484-524.

VINE F. J. & MATHEWS D. H. 1963. Magnetic anomalies over oceanic ridges. Nature 199, 947-949.

VOGEL K. 1983. Global models and Earth expansion. In: Carey S.W. ed. Expanding Earth Symposium, Sydney, 1981. University of Tasmania, 17-27.

VOGEL K. 1984. Beitrage zur frage der expansion der Erde auf der grundlage von globenmodellen. Z. geol. Wiss. Berlin 12. 563-573.

VOGEL K. 1990. The expansion of the Earth - an alternative model to the plate tectonics theory. In: Critical Aspects of the Plate Tectonics Theory;

Volume II, Alternative Theories. Theophrastus Publishers, Athens, Greece, 14-34.

WARD M. A. 1960. On detecting changes in the Earth's radius. Geophysical Journal 8, 217-225.

WEGENER A. 1915. Die entstehung der kontinente und ozeane [The Origin of Continents and Oceans.]

WEIJERMARS R. 1986. Slow but not fast Global Expansion may explain the Surface Dichotomy of the Earth. Physics of the Earth and Planetary Interiors, 43. 67-89.

WEIJERMARS R. 1989. Global tectonics since the breakup of Pangea 180 million years ago: evolution maps and lithospheric budget. Earth-Science Reviews 26, 113-162.

WEISHAMPEL D. B. 1990. Dinosaurian distribution. In: Weishampel D. B., Dodson P. & Osmólska H. eds. 1990. The Dinosauria. University of California Press, Los Angeles.

WESSON P. S. 1973. The implications for geophysics of modern cosmologies in which G is variable. Quarterly Journal of the Royal Astrological Society 14, 9-64.

WOOD B.J., HALLIDAY A.N., and REHKAMPER M., 2010, Volatile accretion history of the Earth: Nature, v. 467, no. 7319, p. E6–E7, doi:10.1038/nature09484.

YUEN D. A. & PELTIER W. R. 1980. Mantle Plumes and the Thermal Stability of the "D" Layer. Geophysical Research Letters, 7, 625-628.

ZAREBSKI T. 2009. Toulmin's model of argument and the "logic" of scientific discovery. Studies in logic, Grammar and Rhetoric 16(29) 2009.

ZHAO G., CAWOOD P. A. & WILDE S. A. 1999. Reconstructions of global 2.1-1.8 Ga collisional orogens and associated cratons: implications for two pre Rodinia supercontinents? In: Watts G. R. & Evans D. A. D. eds. Two billion years of tectonics and mineralisation. Geological Society of Australia Abstracts 56, 60-64.

Appendix A

Kinematics

A mathematical equation for an exponential increase in Earth radius extending from the Archaean to present-day is derived by considering the mathematical equation for linear regression:

$$y = Ax + B \qquad A1$$

where y = the Y axis variable, x = the X axis variable, A = the gradient of a line, and B = the y-intercept of the line.

For a linear increase in ancient radius this equation can be written as:

$$R = At + B \qquad A2$$

where R = Earth radius, t = time before the present (negative), B = the y-intercept of the line.

To determine an exponential increase in ancient radius, equation A2 can be written as:

$$\ln R_a/R_0 = At + B \qquad A3$$

where ln = natural logarithm, R_0 = present Earth radius at time t_0, R_a = ancient Earth radius at time t_a.

Equation A3 simply expresses the exponential curve as a straight line suitable for analysis using the linear regression equation A2. Rearranging equation A3 for ancient Earth radius:

$$R_a = R_0^{e(At+B)} \qquad A4$$

where e = base of natural logarithm.

Mathematical modelling of the post-Triassic increase in Earth radius in Maxlow (1995) demonstrated that the y-intercept B is negligible and can be disregarded. The gradient of line A, representing the ancient Earth radius, is a constant k. equation A4 can then be written as:

$$R_a = R_0^{e(kt)} \qquad A5$$

The Archaean-Mesoproterozoic primordial Earth radius p, determined from empirical modelling, was found to be approximately 1,700 kilometres. This represents the radius of an Archaean to early-Proterozoic proto-Earth at formation or crustal stabilization. It is inferred that Earth radius then increased exponentially from a primordial proto-Earth to the present radius. An equation for exponential increase in Earth radius from the Archaean to present-day is expressed as:

$$R_a = (R_0 - R_p)e^{kt} + 1700 \quad \text{A6}$$

where $k = 4{\cdot}5366\text{x}10^{-9}$/year and the constant k is determined from equation A3 by using post-Triassic Earth radii derived from empirical seafloor data and 1,700 kilometres as the limiting primordial proto-Earth radius to solve for gradient A.

Equation A6 represents a mathematical expression for exponential change in the ancient radius of the Earth from the Archaean to the present-day and assumes that oceanic and continental crust is fully fixed in the rock record with little or no requirement for removal of crusts by subduction. This equation therefore represents the fundamental equation for determining ancient Earth radius at any time back or forward in time from the present-day.

Radius, Circumference, Surface Area, Volume

An investigation into the kinematics of an Earth increasing in radius over time uses equation A6 for ancient radius. Known parameters of the present-day Earth at time t_0 are:

Radius	$R_0 = 6.3708 \text{ x } 10^6$ m
Surface Area	$S_0 = 5.1000 \text{ x } 10^{14}$ m^2
Volume	$V_0 = 1.0830 \text{ x } 10^{21}$ m^3
Mass	$M_0 = 5.9730 \text{ x } 10^{24}$ kg
Mean Density	$D_0 = 5.5150 \text{ x } 10^3$ kg/m^3
Universal Gravity	$G_0 = 6.6732 \text{ x} 10^{-11}$ m^3 kg^{-1} sec^{-2}
Surface Gravity	$g_0 = 9.780317$ m sec^{-2}

Derived quantities for the variables radius, circumference, surface area and volume can then be calculated for the time interval t_0 (present-day) to t_{4000} (early-Archaean) using the following equations:

Ancient Radius	$R_a = (R_0-R_p)e^{kt}+R_p$
Ancient Circumference	$C_a = 2\pi R_a = 2\pi((R_0-R_p)e^{kt}+R_p)$

Ancient Surface Area $S_a = 4\pi R_a^2 = 4\pi((R_0-R_p)e^{kt}+R_p)^2$

Ancient Volume $V_a = (4/3)\pi(R_a)^3$

$= (4/3)\pi((R_0-R_p)e^{kt}+R_p)^3$

where R_a = past Earth radius, R_0 = present mean Earth radius = 6370.8 km, R_p = primordial Earth radius = 1,700 km, e = an exponent, k = 4.5366x10^{-9}/year, t = time before present (negative).

Derived quantities for radius, circumference, surface area and volume for the various time intervals used for each of the small Earth models are listed in Table A1.

Mass Density and Surface Gravity

The kinematics of mass, density and surface gravity, assuming a simplistic spherical Earth model with a homogeneous mass and density distribution is governed by the standard equations:

Density D = Mass /Volume

Surface gravity $g = GM/R^2$

Table A1. Derived quantities for the variable's radius, circumference, surface area and volume, calculated using radius equation A6. Maxlow, 2001.

Geologic Time Scale		Kinematics			
Event	Beginning Age	Radius	Circumference	Area	Volume
	(m.y.)	(km)	(kmx10^4)	(km^{2}x10^5)	(km^{3}x10^8)
Future	300	19916.4	12513.81	49845.91	330916.26
Future	200	13272.8	8339.54	22137.79	97943.42
Future	100	9052.2	5687.64	10297.07	31070.24
Future	5	6478	4070.22	5273.35	11386.84
Present-day	0	6370.8	4002.89	5100.32	10831.05
Quaternary	-1.9	6330.7	3977.7	5036.34	10627.87
Pliocene	-5.9	6247.4	3925.38	4904.72	10213.97
Miocene	-23	5908	3712.11	4386.23	8637.94
Oligocene	-37.7	5636.5	3541.54	3992.39	7501.08
Eocene	-59.2	5270.7	3311.68	3490.97	6133.29
Paleocene	-66.2	5159.1	3241.55	3344.69	5751.85
Late-Cretaceous	-84	4890.7	3072.94	3005.79	4900.19
Mid-Cretaceous	-118.7	4426	2780.93	2461.68	3631.79
Early-Cretaceous	-143.8	4132.6	2596.59	2146.14	2956.38
Late-Jurassic	-160	3960.2	2488.29	1970.84	2601.67
Early-Jurassic	-205	3542.9	2226.05	1577.32	1862.74
Triassic	-245	3237	2033.89	1316.75	1420.79
Permian	-286	2976.2	1869.98	1113.07	1104.22
Carboniferous	-360	2612.2	1641.32	857.5	746.66
Devonian	-408	2433.7	1529.16	744.31	603.82
Silurian	-438	2340.4	1470.5	688.3	536.96
Ordovician	-505	2172.5	1365.04	593.11	429.52
Cambrian	-570	2051.9	1289.22	529.06	361.85
Late-Proterozoic	-900	1778.7	1117.61	397.59	235.74
Mid-Proterozoic	-1,600.00	1703.3	1070.21	364.57	206.99
Early-Proterozoic	-2,500.00	1700.1	1068.18	363.19	205.82
Archaean	-4,000.00	1700	1068.14	363.17	205.8

In the density equation, volume is calculated from equation A6, and the variables mass M and density D are unknown. In the surface gravity equation, universal gravity G is assumed to be constant. Three possible scenarios exist for an expression representing the kinematics of mass, density and surface gravity with time:

1. Mass remains constant requiring density to decrease exponentially as volume increases throughout geological time.
2. Density remains constant requiring mass to increase exponentially as volume increases throughout geological time.
3. Both mass and density are variable as volume increases throughout geological time.

Derived quantities for the variables mass, density and surface gravity for conditions of a) constant mass and b) constant density are calculated by incorporating equation A6 for the time interval t_0 to t_{4000} using the following equations:

1. Constant Mass:

$$M = M_0$$

$$D = M/V = M_0/(4/3)\pi((R_0-R_p)e^{kt}+R_p)^3$$

$$g = GM/R^2$$

$$= GM_0/((R_0-R_p)e^{kt}+R_p)^2$$

2. Constant Density:

$$D = D_0$$

$$M = DV = D_0(4/3)\pi((R_0-R_p)e^{kt}+R_p)^3$$

$$g = GM/R^2$$

$$= GD_0(4/3)\pi((R_0-R_p)e^{kt}+R_p)^3/((R_0-R_p)e^{kt}+R_p)^2$$

$$= GD_0(4/3)\pi((R_0-R_p)e^{kt}+R_p)$$

Each of the constant density variables is shown graphically in figure A1 and is discussed further in Chapter 6.

Secular Rate of Change

The secular rate of change in Earth parameters is defined as an incremental increase in the physical dimensions of the Earth throughout geological time and may be quantified by considering the variables radius, circumference, surface area and volume. Mass, density and surface gravity are considered speculative, and mass is assumed to increase with time (Chapter

8). The application of equation A6 enables the secular rate of change to be determined for the following variables between time t_1 and t_2:

$$dR/dt = (R_1-R_2)/(t_1-t_2)$$
$$= (((R_0-R_p)e^{kt1}+R)-((R_0-Rp)e^{kt2}+R))/(t_1-t_2)$$

$$dC/dt = (C_1-C_2)/(t_1-t_2)$$
$$=((2\pi((R_0-R_p)e^{kt1}+R_p)-(2\pi((R_0- R_p)e^{kt2}+ R_p))/(t_1-t_2)$$

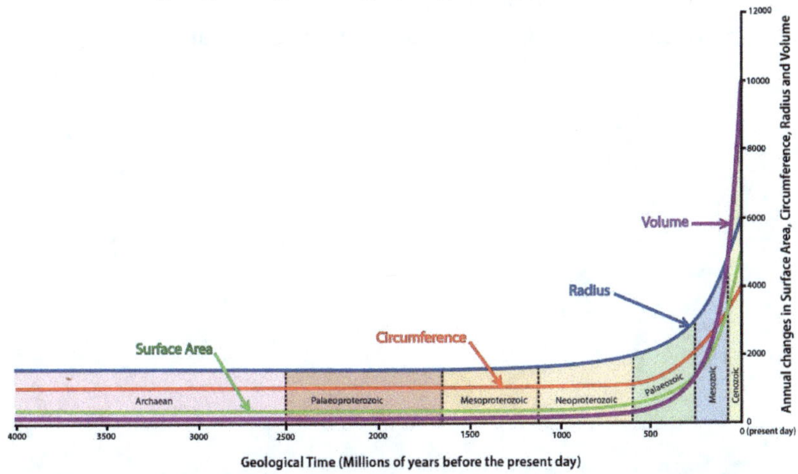

Figure A1. The annual changes in Earth radius, circumference, surface area and volume over time. The graphs range from the Archaean to the present-day and are derived using the established Earth radius formula. The annual change in Radius is in kilometres, Circumference is times 10 kilometres, Surface Area is times 100 million square kilometres, and Volume is times 100 million cubic kilometres. Maxlow, 2005.

$$dS/dt = (S_1-S_2)/(t_1-t_2)$$
$$= ((4\pi((R_0- R_p)e^{kt1}+ R_p)^2-(4\pi((R_0- R_p)e^{kt2}+ R_p)^2)/(t_1-t_2)$$

$$dV/dt = (V_1-V_2)/(t_1-t_2)$$
$$= ((4\pi/3)((R_0-R_p)e^{kt1}+ R_p)^3-(4/3)\pi((R_0- R_p)e^{kt2}+ R_p)^3/(t_1-t_2)$$

Assuming a constant Earth density with variable mass and surface gravity:

$$dD/dt = 0$$

$$dM/dt = (M_1-M_2)/(t_1-t_2)$$
$$= (D_0(4/3)\pi((R_0-R_p)e^{kt1}+R_p)^3-D_0(4/3)\pi((R_0-R_p)e^{kt2}+R_p)^3)/(t_1-t_2)$$

$$dg/dt = (g_1-g_2)/(t_1-t_2)$$
$$= (GD_0(4/3)\pi((R_0- R_p)e^{kt1}+ R_p)-GD_0(4/3)\pi((R_0-R_p)e^{kt2}+R_p))/(t_1-t_2)$$

The present secular rates of Earth expansion derived from the above formulae at time t0 are calculated to be:

1. Radius $dR/dt_0 = 22$ mm/year
2. Circumference $dC/dt_0 = 140$ mm/year
3. Surface Area $dS/dt_0 = 3.50$ km^2/year
4. Volume $dV/dt_0 = 11{,}000$ km^3/year
5. Mass $dM/dt_0 = 60 \times 10^{12}$tonnes/year
6. Density $dD/dt_0 = 0$
7. Surface Gravity $dg/dt_0 = 3.4 \times 10^{-8}$ $msec^{-2}$/year

Appendix B

Palaeomagnetic Formulae

The mathematical equation for an exponential increase in Earth radius extending from the Archaean to present-day (equation A6) is used to modify conventional palaeomagnetic formulae.

Magnetic Dipole Equations

The geocentric axial dipole model is central to the principles of palaeomagnetism (figure B1). In this conventional Earth model, a magnetic field is produced by a single magnetic dipole M, located at the centre of the Earth and aligned with the rotation axis.

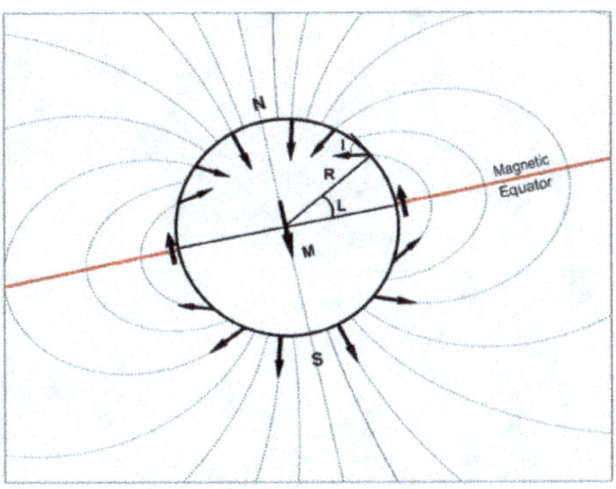

Figure B1. Geocentric axial dipole model. A magnetic dipole M is located at the centre of the Earth and aligned with the rotation axis. The geographic latitude is L, the mean Earth radius is R, the magnetic field directions at the Earth's surface, produced by the geocentric axial dipole, are schematically shown, inclination I is shown for one location and N is the North geographic pole. Maxlow, 2001.

Derivation of the conventional geomagnetic dipole equation is given in Butler (1992) and is defined as:

$$\tan I = 2 \tan L = 2 \cot P \qquad B1$$

Where: I is the mean inclination of the magnetic field at a particular site location, increasing from -90° at the geographic South Pole to +90° at the geographic North Pole, L is the geographic latitude determined from I, and P is the geographic colatitude determined from I.

Rearranging the dipole equation gives colatitude P as:

$$P = \cot^{-1} (\tan I/2) = \tan^{-1} (2/\tan I) \quad B2$$

Equation B2 represents a means of calculating a great-circle angular distance from the sample site to the magnetic pole and, like latitude, is independent of any Earth radius or time constraints imposed by the sample. This dipole equation is a general equation applicable to any sized magnetic sphere that obeys the geocentric axial dipole model (figure B1). On the sphere in figure B1 the magnetic lines of force behave in exactly the same manner, irrespective of the scale or size of the object, and, for a given site inclination, the colatitude (and latitude) calculated using the dipole equation will always remain the same.

Figure B2 demonstrates this very important characteristic of the dipole equation on a magnetized spherical object of different sizes.

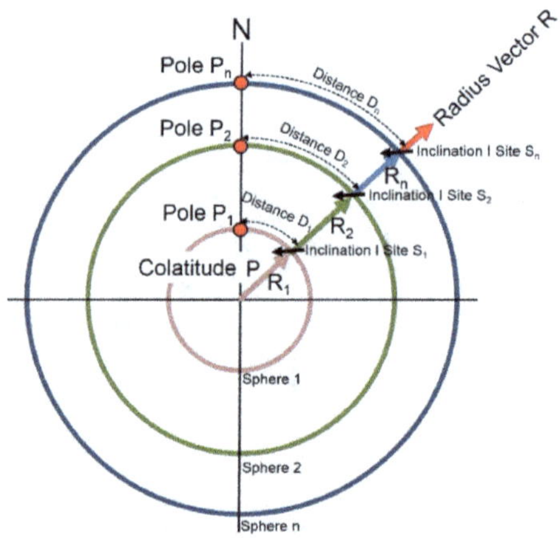

Figure B2. Cross sections of the Earth showing a number of geocentric axial dipolar spheres. The ancient magnetic pole locations determined using the conventional dipole equation are shown at Poles P_1 to P_n.

For a given site inclination value, I, the dipole equation remains true for an infinite number of sample sites along a radius vector R, passing from the centre of the Earth through the site location and beyond. Colatitude (equation B2) calculated from the inclination I at sites S_1 to S_n, located along the radius vector R, is, by definition, equal to a constant angular measurement P. As in figure B2, at site S_1 the colatitude P is also defined by a surface distance labelled Distance D_1 extending from the sample site to the ancient magnetic Pole P_1. This Distance D_1 is not equal to distances D_2 or D_n for sites S_2 and S_n using the identical values of inclination I and colatitude P.

To simulate a simplistic radial expansion of the Earth from Radius R_1 to R_2 to R_n the ancient magnetic pole positions Pole P_1, Pole P_2 and Pole P_n, calculated using the conventional dipole equation B2 (Figure B2). Because the conventional dipole equation uses angular measurements and has no provision for either a radial or time component to compensate for the shift in actual ancient pole position with increase in radius, the calculated pole positions at any moment in time will always coincide with the same geomagnetic pole—shown as red dots.

Determining the ancient pole position Pole Pa on the present-day Earth where radius varies exponentially with time, is shown schematically in figure B3. At Site Sn located on the present-day Earth of Radius R_n, using the conventional colatitude equation (equation B2) an inclination I gives a colatitude P which equates to the ancient colatitude locked in the rock record from its original location at site S_a at radius R_a.

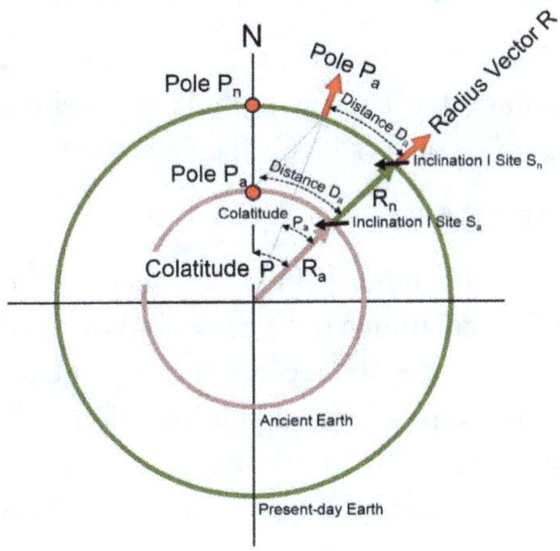

Figure B3. Determining the actual ancient pole position on the present-day Earth. The cross section shows an ancient Earth at radius Ra and a present Earth at present radius Rn.

At site S_a the surface distance D_a from the sample site to the ancient pole is equal to:

$$D_a = R_a P$$

where P is in radians. Rearranging:

$$P = (D_a/R_a)$$

To determine the ancient pole position Pa located on the present radius Earth, which equates to the ancient palaeopole position P_a determined from site Sa, colatitude:

$$P = (D_a/R_0) \times 180/\pi = (R_aP_a \times (\pi/180)/R_0) \times 180/\pi = R_aP_a /R_0$$

where P is in degrees.

The mathematical relationship for an exponential increase in the Earth's radius, derived from empirical measurements of seafloor and continental surface area data was previously found to be (equation A6):

$$R_a = (R_0 - R_p)e^{kt} + R_p \qquad \text{A6}$$

Incorporating equation A6 for R_a gives:

$$P = ((R_0 - R_p)e^{kt} + R_p)P_a/R_0$$

Incorporating equation B2 for P_a gives:

$$P = ((R_0 - R_p)e^{kt} + R_p))\ (\tan_{-1}\ (2/\tan I))/R_0)$$

For any site sample, constrained by the age of the rock sequence containing the site data, the ancient latitude from site to the ancient pole position on an Earth of present radius is then equal to:

P = ((R0 - Rp)ekt + Rp)) (tan-1 (2/tan I))/R0 B3

The application of this modified dipole equation to palaeomagnetic site data enables the ancient colatitude to be converted to the present geographical grid system. This then correctly locates the ancient pole position on the present-day Earth. The conventional dipole equation (equation B2) can only be used to determine the ancient latitude and ancient latitudinal zonation.

The modified dipole equation B3 can now be used to develop formulas to convert the ancient pole geographical coordinates to present geographical coordinates from ancient site mean data located on the present Earth. This enables the ancient poles to be correctly located on the present Earth, constrained by both time and ancient Earth radius.

Modified Dipole Formulae

Site mean data determined from a set of site samples are assumed by palaeomagneticians to represent a time-averaged field which compensates for any secular variation caused by non-dipole components. For a geocentric axial dipole field (figure B1) the time-averaged and structurally corrected inclination I, determined from site data establishes the ancient colatitude existing at the site when the site data were locked into the rock-record and a time-average declination D determines the direction along an ancient meridian to the ancient pole.

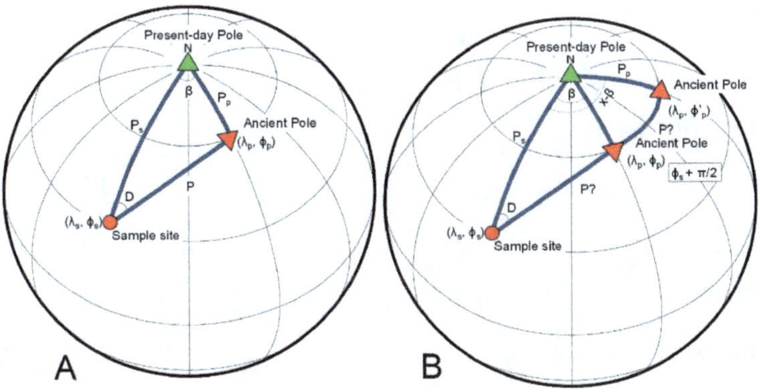

Figure B4. Determination of a magnetic pole from a magnetic field direction, using the palaeomagnetic dipole equation. Orthorhombic projection with latitude and longitude grid in 30° increments. Figure B4a: the site latitude and longitude is (λs, ϕs); the ancient magnetic pole is located at (λp, ϕp); site colatitude is P_s; colatitude of the magnetic pole is P_p; and the longitudinal difference between the magnetic pole and site is β. Figure B4b: illustrates the ambiguity in magnetic pole longitude. The pole may be at either (λp, ϕp) or (λp, ϕ'p; the longitude at ϕs + π/2 is shown by the heavy meridian line. Maxlow, 2005. After Butler, 1992.

Calculation of the ancient magnetic pole position on the present Earth surface uses spherical trigonometry, based on the dipole equation B1, to determine the distance travelled from the observing locality to the pole position (figure B4).

Details and derivation of the following conventional palaeomagnetic equations, sign conventions, and symbols for geographic locations, are adopted from Butler (1992):

1. Latitude increases from -90° at the south geographic pole to 0° at the equator and +90° at the north geographic pole.
2. Longitudes east of the Greenwich meridian are positive, while westerly longitudes are negative and,
3. (λp, ϕp) is the pole position calculated from a site-mean direction (Im, Dm) measured at site location (λs, ϕs).

The pole latitude derived from spherical trigonometry (figure B4) is:

$$\lambda p = \sin^{-1} (\sin\lambda s \cos p + \cos\lambda s \sin p \cos Dm) \quad \text{B4}$$

The longitudinal difference between pole and site, denoted by β, is positive towards the east, negative towards the west and is:

$$\beta = \sin\text{-}1 (\sin p \sin Dm/\cos\lambda p) \quad \text{B5}$$

where if:

$$\cos p \geq \sin\lambda s \sin\lambda p \quad \text{B6}$$

then the pole longitude:

$$\phi p = \phi s + \beta \quad \text{B7}$$

but if:

$$\cos p \leq \sin\lambda s \sin\lambda p \quad \text{B8}$$

then the pole longitude:

$$\phi p = \phi s + 180° - \beta \quad \text{B9}$$

For any site-mean direction (Im, Dm) the associated circular confidence limit (α95) is transformed into an ellipse of confidence about the calculated pole position with semi-axes of angular length given by (figure B5):

$$dp = \alpha 95 ((1 + 3\cos 2p)/2) = 2\ \alpha 95 (1/ (1 + 3\cos 2\ Im)) \quad \text{B10}$$

and:

$$dm = \alpha 95 (\sin p/\cos Im) \quad \text{B11}$$

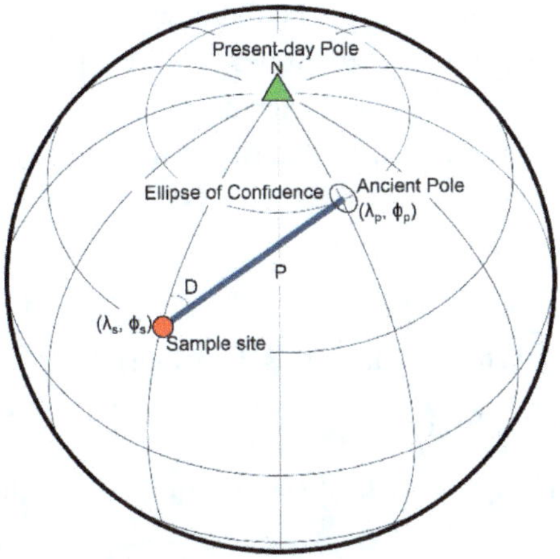

Figure B5. The ellipse of confidence about a magnetic pole position, determined using the palaeomagnetic dipole equation. For a magnetic colatitude p, dp is the semi-axis of a confidence ellipse along the great-circle path from site to pole; dm is the semi-axis perpendicular to that great-circle path. Orthorhombic projection, with latitude and longitude grid in 30° increments. (After Butler, 1992). Maxlow, 2005. After Butler, 1992

While equations B4 to B11 represent the basis for conventional Plate Tectonic determination of pole positions and derived apparent polar wander paths, they do not and cannot acknowledge any variation in angular dimension of ancient geographical coordinates as a result of a variable ancient radius or time constraint. What is assumed by these equations is that the angular dimension of the ancient geographic coordinate system, indicated by the site-data, equals the angular dimension of the geographical coordinate system represented by the present site location. By using these conventional ancient magnetic equations to determine the ancient pole position the two systems are simply added, using spherical trigonometry to give an ancient pole latitude and longitude.

To determine actual ancient pole locations on the present-day Earth each site sample must be qualified by a time and radius constraint in order to convert the ancient geographic coordinate system to the present-day system. For an Earth undergoing an exponential increase in radius from the Archaean to the present-day, derivation of ancient colatitude from site data is (equation B3):

$$p = ((R_0 - R_p)e^{kt} + R_p)) (\tan^{-1} (2/\tan I))/R_0 \qquad B3$$

By incorporating the ancient colatitude equation B3 into the pole coordinate equations B4 to B11 of Butler (1992), equations that constrain the ancient colatitude arcuate distances to the present geographical coordinates are established. These are used to determine the actual ancient pole location on the present Earth surface.

Ancient pole latitude becomes:

$$\lambda p = \sin^{-1} (\sin\lambda s \cos[((R_0\text{-}R_p)e^{kt} + R_p)) (\tan^{-1} (2/\tan I))/R_0] + \cos\lambda s \sin[((R_0\text{-}R_p)e^{kt} + R_p)) (\tan^{-1} (2/\tan I))/R_0] \cos Dm) \qquad B12$$

longitudinal difference becomes:

$$\beta = \sin^{-1} (\sin[((R_0\text{-}R_p)e^{kt} + R_p)) (\tan^{-1} (2/\tan I))/R_0] \sin Dm/\cos\lambda p) \qquad B13$$

where if:

$$\cos[((R_0\text{-}R_p)e^{kt} + R_p)) (\tan^{-1} (2/\tan I))/R_0] \geq \sin\lambda s \sin\lambda p \qquad B14$$

then the ancient pole longitude:

$$\phi p = \phi s + \beta \qquad B15$$

but if:

$$\cos[((R_0\text{-}R_p)e^{kt} + R_p)) (\tan^{-1} (2/\tan I))/R_0] \leq \sin\lambda s \sin\lambda p \qquad B16$$

then the ancient pole longitude:

$$\phi p = \phi s + 180° - \beta \qquad B17$$

and the ellipse of confidence:

$$dp = \alpha 95 ((1 + 3\cos 2[((R_0\text{-}R_p)e^{kt} + R_p)) (\tan^{-1} (2/\tan I))/R_0])/2)$$

$$= 2\, \alpha 95 (1/ (1 + 3\cos 2\, Im)) \qquad B18$$

and:

$$dm = \alpha 95 (\sin[((R_0\text{-}R_p)e^{kt} + R_p)) (\tan^{-1} (2/\tan I))/R_0]/\cos Im) \qquad B19$$

www.ingramcontent.com/pod-product-compliance
Lightning Source LLC
LaVergne TN
LVHW010533300926
849162LV00016B/1874/J
* 9 7 8 1 6 8 0 5 3 6 1 4 0 *